Vegetation Index and Dynamics

Edited by Eusebio Cano Carmona,
Ana Cano Ortiz, Riocardo Quinto Canas
and Carmelo Maria Musarella

Published in London, United Kingdom

IntechOpen

Supporting open minds since 2005

Vegetation Index and Dynamics
http://dx.doi.org/10.5772/intechopen.87465
Edited by Eusebio Cano Carmona, Ana Cano Ortiz, Riocardo Quinto Canas and Carmelo Maria Musarella

Contributors
Jon Bryan Burley, Zhi Yue, Amjad ur Rahman, Esra Gürbüz, Semih Ekercin, Shujaul Mulk Khan, Engin Eroğlu, Tuba Gül Doğan, Madan Prasad Singh, Manohara Tattekere Nanjappa, Sukumar Raman, Suresh Hebbalalu Satyanatayana, Ayyappan Narayanan, Ganesan Rengaian, Sreejith Kalpuzha Ashtamoorthy, Elena Runova, Vera Savchenkova, Ekaterina Demina-Moskovskay, Anastasia Baranenkova, Algimantas Česnulevičius, Artūras Bautrėnas, Linas Bevainis, Donatas Ovodas, Claudionor Silva, Arthur Santos, Fernando Santil, Anna Brook, Maria Polinova, Sara del Río, Ángel Penas, Raquel Alonso-Redondo, Giovanni Breogán Ferreiro Lera, Aitor Álvarez-Santacoloma, Alejandro González-Pérez, Sam Paul Mathew, Raveendranpillai Prakashkumar, Sylvester Onoriode Obigba, Maria M. Borisova-Mubarakshina, Ilya A. Naydov, Daria V. Vetoshkina, Marina A. A. Kozuleva, Daria V. Vilyanen, Boris N. N Ivanov, Natalia N. Rudenko, Rahmouni Abdelkader, Brahimi Djamel, Brahimi Abdelghani, Mesli Lotfi, Samora M. Andrew, Siwa A. Kombo, Shabani A.O. Chamshama, Clara Inés Saldamando-Benjumea, Gloria Patricia Cañas-Gutiérrez, Jorge Muñoz, Rafael Arango Isaza, Eusebio Cano Carmona, Ricardo Quinto Canas, Ana Cano Ortiz, Carmelo María Musarella

First published in London, United Kingdom, 2022 by IntechOpen
IntechOpen is the global imprint of INTECHOPEN LIMITED, registered in England and Wales, registration number: 11086078, 5 Princes Gate Court, London, SW7 2QJ, United Kingdom
Printed in Croatia

British Library Cataloguing-in-Publication Data
A catalogue record for this book is available from the British Library

Additional hard and PDF copies can be obtained from orders@intechopen.com

Vegetation Index and Dynamics
Edited by Eusebio Cano Carmona, Ana Cano Ortiz, Riocardo Quinto Canas and Carmelo Maria Musarella
p. cm.
Print ISBN 978-1-83969-385-4
Online ISBN 978-1-83969-386-1
eBook (PDF) ISBN 978-1-83969-387-8

For EU product safety concerns:
IN TECH d.o.o., Prolaz Marije Krucifikse Kozulić 3, 51000 Rijeka, Croatia, info@intechopen.com or visit our website at intechopen.com.

Meet the editors

Eusebio Cano Carmona obtained a Ph.D. in Science from the University of Granada, Spain. He is a Professor of Botany, University of Jaén, Spain. His fundamental line of research is flora and vegetation in Spain, Italy, Portugal, Palestine, the Caribbean islands, and Mexico. He has directed thirteen doctoral theses and published 250 articles, books, and book chapters. He has presented 200 papers/communications at national and international congresses. He has held a number of different academic positions, including Dean of the Faculty of Experimental Sciences at the University of Jaen, Spain, and founder and director of the International Seminar on Management and Conservation of Biodiversity. He is a member of the Spanish, Portuguese, and Italian geobotany societies. Counselor of the I.E. G., Instituto de Estudios Giennenses, Jaén.

Ricardo Jorge Quinto Canas obtained a Ph.D. in Analysis and Management of Ecosystems and is currently an invited assistant professor in the Faculty of Sciences and Technology, University of Algarve, Portugal, where he is a member of the university's Centre of Marine Sciences (CCMAR). He is also the head of the Division of Environmental Impact Assessment - Algarve Regional Coordination and Development Commission (CCDR-Algarve). His current research projects focus on botany, vegetation science (geobotany), biogeography, plant ecology, and biology conservation. Dr. Quinto-Canas has coauthored many journal publications, conference articles, and book chapters.

Ana Cano Ortiz obtained a Ph.D. in Botany from the University of Jaén, Spain. She has worked in private business, college, and high school education. She is co-director of four doctoral theses. Her main line of research is related to botanical bioindicators. She has worked in Spain, Italy, Portugal, and Central America. She has published more than 100 works in various national and international journals, as well as books and book chapters. She has also presented numerous papers and communications at national and international congresses

Carmelo Maria Musarella, Ph.D., is a biologist specializing in plant biology. He studied and worked at several European universities. He is an adjunct professor of Plant Biology at the "Mediterranea" University of Reggio Calabria, Italy. His research interests include flora, vegetation, habitats, biogeography, taxonomy, ethnobotany, endemism, and biodiversity conservation. He has published many research articles in indexed journals and books. He is a guest editor for *Plant Biosystems*. Dr. Musarella is a member of the permanent scientific committee of the International Conference on Biodiversity Conservation and Management. He has participated in several international and national congresses, seminars, and workshops and has presented several oral communications and posters.

Contents

Preface

This book contains sixteen chapters related to various methodological aspects for the investigation of vegetation, that is, the application of different methodologies to the study of plant communities and the landscape.

The book presents studies on types of indices and different ways of understanding plant dynamics. In general, the different chapters follow two methodological currents: the purely ecological and the phytosociological. In both cases, the authors rely on other branches of knowledge, such as cartography, biogeography, bioclimatology, geology, ecophysiology, and even entomology and agronomy.

Purely methodological studies are provided, while others have a more applied nature. However, in all cases, the latest scientific advances are presented, some of which are of great environmental and socioeconomic relevance.

The editors would like to thank all the contributing authors as well as Author Service Manager Ms. Jasna Bozic at IntechOpen.

Dr. Eusebio Cano Carmona and Dr. Ana Cano Ortiz
University of Jaén,
Jaén, Spain

Dr. Ricardo Quinto Canas
Faculty of Sciences and Technology,
Center of Marine Sciences (CCMAR),
University of Algarve,
Portugal

Dr. Carmelo Maria Musarella
Dipartimento di Agraria - Università "Mediterranea" di Reggio Calabria,
Reggio Calabria, Italy

Section 1

Vegetation

Chapter 1

Introductory Chapter: Methodological Aspects for the Study of Vegetation

Eusebio Cano Carmona, Ricardo Quinto Canas, Ana Cano Ortiz and Carmelo María Musarella

1. Introduction

For the study of vegetation, there are different methodologies, among which we highlight the purely ecological and phytosociological ones; the latter have acquired relevance from the second half of the twentieth century. It is in Europe where the phytosociological method has flourished; however, the ecological one has been in the United Kingdom and America. The application of the phytosociological method has led to the creation of a nomenclature code, which requires all researchers working with this methodology [1]. Thus, various geobotanical concepts of interest have emerged, such as association, alliance, order, and vegetation classes [2]. All of these have been used by different countries for the establishment of habitats in the EU.

2. Bioclimatic and biogeographic analysis

Although the knowledge of the species has been and continues to be fundamental for the further development of phytosociology [3], no less important is the strong advance, which, from the hand of Professor Rivas Martínez, has had bioclimatology, as a basis essential in the description of phytocenoses, as well as in agricultural, forestry, and livestock planning.

Bioclimatology is an ecological science, which has gained importance in recent years and which tries to highlight the relationship between living organisms (biology) and the climate (physics) on Earth. It differs from climatology in that the information, indices, and units it uses are related and delimited by species and phytocenosis (biocenosis). The development of bioclimatology as a basic discipline at the service of geobotany has been one of the most outstanding scientific aspects in recent times; the progress of this science has made it possible to better diagnose many plant communities and above all to be able to better specify the main cliserial geoseries that are observed across an altitudinal gradient.

Of the different factors that lead to the existence of certain plant ecosystems, precipitation and temperature are among the most important ones. Thus, each region or group of biogeographic regions has a peculiar altitudinal zoning of plant ecosystems; such topographic geoseries is due to the progressive decrease of the annual average temperature with the altitude (thermoclimate).

If the climate (temperature and precipitation) is correlated with the biocenotic discontinuities that appear in the mountains with altitude (altitudinal cliseries), we will see that certain rhythms or changes are fulfilled throughout the Earth as a function of temperature and precipitation (thermoclimate and ombroclimate). Consequently, based on such changes, the physical continent, which includes the bioclimatic floors, can be recognized on the one hand and the plant biological content, which includes the vegetation series, on the other. Rivas-Martínez and Loidi [4] published the bioclimatology of the Iberian Peninsula and gave a set of biogeographic indices of high interest for the study of vegetation, studies that are later perfected in various publications by these authors (positive precipitation, positive temperature annual, continental index (Ic), ombroclimatic index (Io), thermicity, and compensated thermicity index). Later, Cano et al. [5], studying various ecological and bioclimatic aspects of *Juniperus oxycedrus* L. forests, proposed an ombroedaphoxeric index (Ioex) to explain the presence of these *Juniperus* forests in bioclimatic environments not optimal for them.

The xericity of serpentines often gives rise to forests and scrublands that do not correspond to the ombrotype in the territory; plants living here develop ecophysiological and morphoanatomical adaptations to withstand the limitations. In ideal situations with good soil texture and structure and without slopes, we can assume that the water retention (WR) is maximum (100%). Otherwise, there are losses due to runoff and drainage, and the WR may therefore vary. Water is also lost through potential evapotranspiration (ETP). However, as plants have the capacity to self-regulate their losses, it can be assumed that the residual evapotranspiration e = 0.2ETP. So two parameters (i.e., e and WR) are implicated in the vegetation development, which is essentially conditioned by rainfall. Therefore, the Ombroclimatic Index (Io) does not explain the presence of plant communities that are influenced by the substrate, and we propose the new ombroedaphoxeric index (Ioex) to explain the presence of communities with *Juniperus* spp. in territories with a thermomediterranean to supramediterranean thermotype.

$$\text{Ioex} = \text{P}_\text{p}.\text{e} / \text{T}_\text{p} * WR, \tag{1}$$

where P_p = positive precipitation; T_p = positive temperature of the year; e = residual evapotranspiration, whose value is 0.2 ETP; WR = water retention in parts per unit, whose values may be 0.25, 0.50, 0.75, and 1 [5].

Parallel to the bioclimatic studies, biogeographic studies are carried out [6], which are also fundamental to the interpretation of the vegetation. In this sense, chorology or phytogeography is a branch of geography with a biological base that deals with the distribution of living organisms on Earth, a science that relates the physical to the biological. Both plant chorology or phytogeography and phytosociology are highly topical, due to the importance of plant communities in the definition and delimitation of territories; chorology and biogeography become synonymous, since they are derived from chorus (Greek) = "limited place" and logos = "science."

Biogeography is also a science derived from geobotany, which with the collaboration of other sciences, such as geography, edaphology, geology, zoology, and botany, is capable of establishing a typology or systematic of the planet's surface, through knowledge of the distribution of syntaxa, which is already the object of geobotany, botanical sociology, or phytosociology, currently known as vegetation science, whose objective is the knowledge of vascular plant communities or phytocenosis.

One of the criteria traditionally used in the recognition and delimitation of biogeographic areas with their own identity is the presence or absence of families, genera, species, and subspecies. These taxa are called endemisms, especially those whose distribution area is smaller than the biogeographic region. Endemic species and plant communities are used in the definition and delimitation of chorological units as are the provinces and sectors.

Professor Rivas-Martínez, in successive works, explains the fundamental biogeographic concepts for studying vegetation. According to Rivas-Martínez et al. [7], the main typological units in decreasing rank are kingdom, region, province, sector, district, country, landscape cell, and tesella.

The elemental unit of biogeography or "tesella" is a space or geographic surface of variable extension, homogeneous from the ecological point of view, which means that it only presents a certain type of potential natural vegetation (climatophilous, edaphoxerophilous, or edaphohygrophilous) as a mature stage of the ecosystem or biogeocenosis and, consequently, a single sequence of natural communities of substitution. The neighboring tesellas, related by an edaphic or climatic gradient, represent vegetation geoseries, which is the general expression of the zonation phenomenon. The tesella is the only biogeographic unit that can be repeated disjoint. A mosaic of tesellas or permatesellas, related in the same territory by their corresponding topographic geoserie, constitutes the landscape cell (such as river valleys, marshes, peneplains, and high mountain summit, among others). The biogeographic country is a well-delimited geographic territory, with a set of species, plant communities, and topographic geoseries.

The district must be a region characterized by the presence of peculiar associations and species that are lacking in nearby areas or districts; it is also characterized by a traditional use of the territory by man, although it does not have to be a close correlation between the natural regions, man-made, and biogeographic districts. However, experience shows that it can be useful to make both concepts coincide; in this way, the typological union between anthropogeography or human geography and biogeography (phytogeography and zoogeography) would be favored.

The sector is a large territory with a geographical entity that has its own taxa and associations, as well as its own vegetation geoseries and its own permanent and subserial communities.

The province is a large territory that, in addition to having a large number of endemisms, has climatic domains, series, geoseries, and permanent communities; a peculiar altitudinal zoning of vegetation is also characteristic of each province.

The region is a very extensive territory, which has a flora in which there are endemic species, genera, and even families. It presents particular climatic domains and territories and consequently series, geoseries (catenal set of series), and its own bioclimatic stages.

The kingdom is the supreme unit of biogeography and in it, in addition to taxonomic and ecosystem considerations, the origin of flora and fauna comes into play as well as the formation of large continents, the climate, and the paleoclimate.

Once the bioclimatology and biogeography of a territory are known, it is necessary to value a precise methodology for the study of phytocenoses. We lean toward the phytosociological method widely applied in Europe, America, and North Africa by different authors [8–10], with certain geobotanical concepts [2].

3. Conclusions

Although there are different methodologies for the study of vegetation, the phytosociological method is one of the most used worldwide, a method that relies on prior knowledge of the distribution of flora, ecological and geological aspects [11], and bioclimatic and biogeographic, and the catenal contacts between plant communities (associations), which must always be included in a hierarchical system of syntaxonomic ranges.

Author details

Eusebio Cano Carmona[1*], Ricardo Quinto Canas[2,3], Ana Cano Ortiz[1] and Carmelo María Musarella[4]

1 Department of Animal and Plant Biology and Ecology Section of Botany, University of Jaen, Jaén, Spain

2 Faculty of Sciences and Technology, University of Algarve, Faro, Portugal

3 Center of Marine Sciences (CCMAR), University of Algarve, Faro, Portugal

4 Dipartimento di AGRARIA, Università "Mediterranea" di Reggio Calabria, Reggio Calabria, Italy

*Address all correspondence to: ecano@ujaen.es

References

[1] Theurillat JP, Willner W, Fernández-González F, Bültmann H, Čarni A, Giant D, et al. International Code of Phytosociological Nomenclature. 4th ed. Applied Vegetation Science. Estonia, Brazil, Czech Republic and Australia; 2020. pp. 1-169. DOI: 10.1111/avsc.12491

[2] Cano Ortiz A, Piñar Fuentes JC, Ighbareyeh JMH, Quinto Canas R, Cano E. Didactic aspects in the teaching of geobotanical concepts. IJHSSE. 2021;**8**(4):1-6

[3] Braun BJ. Phytosociology. Bases for the Study of Plant Communities. Ed. Barcelona, Spain: H. Blume; 1979. 820 p

[4] Rivas-Martínez S, Loidi J. Bioclimatology of the Iberian Peninsula. Itinera Geobotanica. 1999a;**13**:41-47

[5] Cano E, Musarella CM, Cano-Ortiz A, Piñar Fuentes JC, Rodríguez Torres A, del Rio González S, et al. Geobotanical Study of the Microforests of Juniperus oxycedrus subsp. badia in the Central and Southern Iberian Peninsula. Sustainability. 2019;**11**(4):1111. DOI: 10.3390/su11041111

[6] Rivas-Martínez S, Loidi J. Biogeography of the Iberian Peninsula. Itinera Geobotanica. 1999b;**13**: 49-67

[7] Rivas-Martínez S, Penas Á, Díaz González TE, Cantó P, del Río S, Costa JC, et al. Biogeographic units of the Iberian Peninsula and Baelaric Islands to district level. A concise synopsis. In: Loidi J, editor. The Vegetation of the Iberian Peninsula, Plant and Vegetation. Vol. 1. Cham, Switzerland: Springer International Publishing; 2017. pp. 131-188. ISBN: 978-3-319-54784-8

[8] Cano E, Pinto Gomes CJ, Rodríguez Torres A, García Fuentes A, Torres JA, Cano Ortiz A, et al. Formations of Corynephorus caenescens in areas contaminated with heavy metals of the Iberian southwest (Spain, Portugal). In: International Congress of Phytosociology (XIX Conference of Phytosociology): Biodiversity and Territory Management. September 16-19; Tenerife. University of Laguna; 2003

[9] Cano E, Pinto Gomes CJ, Cano Ortiz A, Rodríguez Torres A, Martínez Lombardo MC. Analysis of the *Corynephorus canescens* communities in the interior of the Iberian Peninsula. In: III International Seminar on Biodiversity Management and Preservation: Current State of the Habitats Directive in the Member States: European Convergence; June 7-12; Vadillo-Castril. Spain: Jaén; 2009. pp. 55-58. ISBN: 978-84-8439-457-0

[10] Rivas-Martínez S, Fernández-González F, Loidi J, Lousa M, Penas A. Syntaxonomical checklist of vascular plant communities of Spain and Portugal to association level. Itinera Geobotanica. 2001;**14**: 5-341

[11] Molina JM. Approach to the geology of the natural park "Sierra de Andújar". Northwest of the Province of Jaén. In: Proceedings of the First Environmental Conference of the Sierra de Andújar Natural Park. Ed. Jaén Spain: Provincial Council of Jaén; 1991. pp. 15-35

Chapter 2

Using GIS and the Diversity Indices: A Combined Approach to Woody Plant Diversity in the Urban Landscape

Tuba Gül Doğan and Engin Eroğlu

Abstract

Thanks to their recreational and psychological functions as well as plant diversity, open and green spaces in a city improve the life quality of the urban inhabitants. Woody plant diversity has significant value in urban green systems. The main purpose of this study was to determine the biodiversity values and the potential of the urban green infrastructure via floristic and spatial analyses of woody plant diversity. To this aim, field studies were carried out on the open and green infrastructure in selected areas of Duzce, having different spatial characteristics. The contribution of the identified species to urban biodiversity was examined as well as the spatial characteristics of the species in terms of landscape architecture. In this study, both statistical analyses (alpha and beta diversity of the species) and GIS analyses (species density and spatial distribution) were carried out. According to the results of the research, the most common of the 173 plant species detected were *Cupressocyparis leylandii* and *Tilia tomentosa*, found in the open green areas. As a result of the study, it was found that using the floristic diversity indices and GIS jointly enabled the UFD (Urban Floristic Diversity) of the city to be defined both statistically and positionally.

Keywords: Diversity index, floristic diversity, Geographic Information System (GIS), spatial density analysis, urban plant diversity

1. Introduction

In spatial diversity for many ecosystems, plant diversity is linked to the heterogeneity of resource availability [1]. However, the diversity of plants in cities can reflect social, economic, and cultural influences as well as the traditionally described ecological theorems [2]. Although urban areas have lost biodiversity in recent years, many plants, including endangered species, can grow and develop in cities [3]. For example, the diversity of woody species in Black Sea cities of Turkey and their environs is quite considerable, with tens of thousands of plant species representing a large source of biodiversity [4]. The temperate zone located around central Turkey is an attractive field in terms of plant diversity and features a great biogeography of genes that are different from those of many countries. However, no special interest has been shown in this wealth.

Turkey, because of its geological structure, soil composition, and different features due to varying climatic conditions, is one of the few countries worldwide that include the typical plants and characteristics of three distinct flora. Turkey with its extremely rich vegetation is a center and home to many plant species and genes present in the world [5, 6]. The advantage of this situation is that many studies have been conducted examining the plant diversity of landscape architecture in Turkey. The majority of these works were carried out in natural environments and are mostly based on flora detection [7–9]. However, in urban and designed areas, researchers have not only determined the plant material, but have also studied certain plant design elements and principles [10, 11], the importance of plant material [12], plant size and forms [13], seasonal changes of plants [14], plant compositions [15], problems in planting areas [16], the current status of plants [17], and the future status of plant issues. Nevertheless, the presence of plants important for cities also contributes significantly to urban biodiversity, although studies on the density, distribution, and diversity of these plants have not been compiled. The fact should not be ignored that examining the density of plants, especially in one area, can contribute as well to many different fields such as urban permeability, air purification, and wildlife development. The open green spaces in cities represent important parts of the ecological system by hosting many animal and plant species. However, the loss of green areas as a result of urbanization threatens urban biodiversity.

Düzce is located in zone A3 according to the grid system of [18] and has quite variable geomorphological landforms, climate types, and habitats [9]. The region is rich in herbaceous and woody species and hosts 700 different plant species, 10% of which are endemic [19]. A great deal of domestic and foreign research has been conducted on the plant material in the open and green areas of the city. In these earlier studies in the field of landscaping, the research area was visited and the use of the design was interpreted by advancing in line with the identification and counting of the plants [20–22], examination of plant patterns [23], and analysis of the context of composition [24]. However, individual plants were identified on the basis of species points and their general characteristics, but their mapping with the coordinate in which they actually were located and their contributions to the urban green infrastructure in terms of floristic diversity were not examined together.

The Geographic Information System (GIS) is a computer-based system for mapping and analyzing all kinds of data that are available on the earth [25]. It is a system of equipment and methods designed to solve complex planning and management problems, including the querying, processing, analysis, modeling, visualization, and management of data located in a space. It depends on spatial data, verbal and graphical information, etc., and has a structure that stores this data in an integrated way [26]. In landscape architecture, GIS studies refer to point and spatial analysis and evaluations, and to planning and decision-making stages on a much higher scale. Based on the assumption that plant density will be a very important underpinning for many phenomena in the city, and especially for urban green infrastructure, GIS was handled in a different dimension within the scope of our research, and the data were analyzed in a GIS environment with its spatial feature.

In this study, identification and determination studies were carried out for woody plant species in open and green areas in the city center of Duzce. The spatial and functional statuses of the species were evaluated. The aim of the study was to determine the types and numbers of plants used in different places in the city center such as the parks and playgrounds, shopping centers/squares, public and residential gardens, pedestrian and vehicle roads, the cemetery, and places of worship (mosques). According to the data obtained, the contributions of plant species to urban biodiversity were evaluated in terms of their spatial characteristics and esthetic and functional aspects. To this purpose, data were collected on six different space types in the city center and then evaluated according to alpha and beta diversity indices in order to reveal the woody plant diversity (WPD) status in the research area. Here, we used the data obtained from the sample-based field study to investigate the change in plant diversity in a large area in the city center by using statistical analysis of woody plants including trees, shrubs, and hedges. Plant diversity is important in predicting the biodiversity for the entire ecosystem [2]; therefore, in this study, we focused on the spatial variation in woody plant diversity.

The diversity of plant species should be considered due to its importance in helping determine the performance and future community composition of existing species [27, 28]. Studies have been conducted on the relationships between species diversity and habitat characteristics, and these have generally focused on the relationships between alpha diversity and habitat characteristics [29–31]. Within the scope of our research, the diversity status of the plant species detected in the open green areas of the Düzce city center was evaluated at alpha and beta levels.

The parameters of species diversity, urban green infrastructure relations, and positional characteristics of the plants were all considered together in this study, which included:

I. Determining the UFD (Urban Floristic Diversity), not only statistically, using floristic diversity indices, but also by using GIS as an important tool to positively establish the UFD,

II. Determining the presence of natural and exotic species at the levels of diversity and spatial characteristics that characterize the urban landscape in its environment,

III. Determining the contribution of green infrastructure potential to the city in terms of the spatial and characteristic relationships among the positional plant traits and discussing the floristic diversity results.

By exhibiting a dynamic structure and creating livable urban spaces, trees and shrubs contribute to urban esthetics and ecology and are the symbols of the modern city [31]. With this study, different plant species in various areas in the provincial city center of Düzce were examined, quantitatively studied, and their effects expressly seen. The study examined a total of 35 samplings of open and green areas, both public and private, located in the city. These included Monument Park, Avni Akyol Park, Celalaeddin Özdal Park, Düzce High School Park, İnönü Park, Urban Park, Konak Park, Küçüksu Park, the city cemetery, and areas around Özdilek Shopping Center, Krempark Shopping Center, Karaca Mosque, Cedidiye Mosque, Gürcü Osman Mosque, Hamidiye Mosque, Çağsu Hospital, Atatürk Hospital, Courthouse, Düzce Municipality, Düzce Cultural Center, Provincial Directorate of Youth Services and Sports, Provincial Public Library, Forestry Directorate, Meteorology Station, Düzce Governorship, Bus Terminal, Kuyumcuzade Boulevard, Düzce-Akçakoca Highway, D100 Highway and Kervan crossroads, İstanbul Street, Bolu Street, Haydar Gördebil Boulevard, Nezih Tütüncüoğlu Boulevard, Atatürk Boulevard, and Rasim Betir Boulevard. Species identified in these areas were esthetically and functionally evaluated via various analyses in order to determine their density and distribution.

2. Materials and methods

The main subject materials were the plants located in the open green areas in the Duzce central district. Plants were identified and plant compositions were evaluated. The plants in the sample areas were used to reveal the floristic diversity of the Duzce city center.

Within the scope of the research, the plants in the open green areas of Düzce were determined and analyzed. A total of 35 public spaces and residential areas (65 sample areas) were included in the study. Field studies lasted about 7 months and were carried out between March–September 2017. It was carried out on an area of approximately 40 km^2.

2.1 Research area

The research area encompassed the city of Düzce in the central district of Düzce Province, which is located between 40° 37′ and 41° 07′ north latitude and 30° 49′ to 31° 50′ east longitude [32] (**Figure 1**). In general, lime-free brown forest soils, yellow/red podsolics, and light podsol soils are widespread in the majority of the Düzce basin and the soil depth is medium to deep [9, 33].

Düzce Province is located in the Western Black Sea Region on the low wet and humid coast. Total rainfall ensures that the green cover remains constant except for the rocky areas [32]. The average temperature in Düzce is 13.41°C and the average annual rainfall is 840 mm [34]. The vegetation period begins in April and lasts until the end of October [9]. According to the land-use classification of the Düzce region, 24,369 ha are Class I land (9.36%), 8,148 ha Class II (14%), 6,546 ha Class III (2.52%), and 17,548 ha Class IV (77%) [35, 36]. Land classes I-IV in the region are generally agricultural land [37]. Düzce is rich in herbaceous and woody species and hosts 700 different plant species, 10% of which are endemic [19].

It has been repeatedly accepted that the number of plant species in the cities is higher than the number in the surrounding area [38–42]. Within the feature of the system known as the 'green infrastructure', which is expected to encircle the city like an ecological network, are different species with different characteristics in open green areas [43]. One of the most important factors contributing to the green

Figure 1.
Research area: (1) parks and recreation areas; (2) public open and green spaces; (3) shopping centers/squares; (4) cemetery and mosques; (5) residential sampling points.

infrastructure of a city is a high level of urban biodiversity and its density and distribution [44, 45].

This research was concerned with providing floristic information from various areas of Düzce in order to determine the important contributions to the urban landscape.

2.2 Determination of sample areas and floristic analysis

The research was carried out in public and private open and green spaces in Düzce city center. The boundary of the research area was determined by the east - west D100 highway. The border line in the north was the Özdilek Shopping Center and in the south, the Düzce Bus Terminal and the city cemetery were the last landmark points included in the research area. Our site selection was influenced by features such as intensive use areas and landscaping, or areas in the city partially demonstrating this potential. To this aim, defined spaces in the city were the subject of the research. The 35 sample areas and the identified residential areas represented the urban open green areas existing throughout the urban landscape (**Table 1**).

The research was conducted in two stages: field work (on-site observation) and office studies (analyses) (**Table 2**). Identification of species and number of woody plant species, usage areas, usage purposes, usage intensities, and seasonal change potentials of the species were determined in the research area.

Since science is a set of documents with generalizability, the extent to which the research is studied affects the reality of the research to the same extent [46]. In some cases it is possible to reach all the examples in the research area, but

Parks and Recreation Areas	Monument Park, Celalettin Özdal Park, Düzce High School Park, İnönü Park, Avni Akyol Park, Urban Park, Konak Park, Küçüksu Park
Shopping Centers/ Squares	Özdilek Shopping Center, Krempark Shopping Center
Public Open and Green spaces	Courthouse, Atatürk Hospital, Municipality of Düzce, Düzce Cultural Center, Duzce Youth Services and Sports Directorate, Çağsu Hospital, Provincial Public Library, Forest Management Directorate, Meteorology Station, Düzce Governorship, Bus Terminal
Cemetery and Mosques	Cedidiye Mosque, Gürcü Osman Mosque, Hamidiye Mosque, Karaca Mosque, The city cemetery
Roads	Kuyumcuzade Boulevard, Nezih Tütüncüoğlu Boulevard, Düzce – Akçakoca highway, Atatürk Boulevard, D100 highway and Kervan Crossroads, İstanbul Street, Bolu Street, Rasim Betir Boulevard, Haydar Gördebil Boulevard
Residential Areas	A total of 65 residential area samples

Table 1.
Sample areas subject to research.

Field observation	a. Determining woody plant diversity (WPD) b. Determining positional status of species c. Determining spatial characteristics of species d. Conducting density analysis of species
Office studies	a. Evaluation of spatial and functional characteristics of species b. Determination of spatial distribution of species c. Determination of species diversity d. Evaluation of results

Table 2.
Research plan.

sometimes this is impossible or too much time is required to examine all the values in the research universe. In fact, when a certain amount of information is sufficient, dealing with masses of information is meaningless [47]. In the research area, sampling was done in all open and green areas having different spatial characteristics, except for the residential areas, in which 7902 residences were subjected to stratified sampling. For example, in accordance with the 'random sampling' method, in the neighborhoods where the population and the number of dwellings were high, over-sampling was performed, and in the neighborhoods where they were lower, less sampling was performed [48].

By dividing the total number of dwellings of each neighborhood by the total number of dwellings in the research area, the weight of the dwellings of a neighborhood was obtained in relation to the total research area. After this process, by multiplying the resulting value by 77, the number of housing samples that should be taken from each neighborhood was calculated [49], with $p = 1\text{-}\alpha\ 77$ as the sampling size and a confidence interval of 90–95%. As a result of the operations performed, a total of 65 residential sampling points were determined. These included those in the neighborhoods of Aziziye (13), Azmimilli (3), Burhaniye (3), Camikebir (2), Cedidiye (4), Cumhuriyet (2), Fevziçakmak (6), Hamidiye (2), Kültür (9), Kiremitocağı (4), Nusrettin (5), Uzunmustafa (6), Çay (2), and Şerefiye (4).

Within the scope of the study, species detection was carried out for plants belonging to each sample area in the research area. In the field studies, a field introduction table was prepared in which general evaluations of each area and each species were discussed. Features such as the height, phenological characterization, and seasonal status of each plant were recorded in the field introduction table. In order to perform the spatial analysis of the identified woody species in GIS, data regarding point and positional status were needed. For this purpose, a hand-held Global Positioning System (GPS) with high resolution (1–5 cm) sensitivity was used. The locations of trees and shrub groups, as the material of the study, were determined by the hand-held GPS device and marked on a previously printed satellite image of each area, along with their numbers.

2.3 Determination of woody plant diversity (WPD)

In order to determine the diversity status of the species, the data obtained from the field studies were converted into a Microsoft Excel file. After the detection and identification studies of the plant species, statistical data were digitized and used to characterize the structural characteristics of each plant [50]. The basis of this analysis was the existence of species and their numbers in the sample areas [51]. In other words, "Is that plant present in the sample area or not, and if so, how many?" Afterwards, classifications were made according to diversity indices in order to evaluate species diversity [52].

According to [50], the Shannon, Margalef, and Berger-Parker indices are generally used for the evaluation of species diversity. In order to determine the distribution of species in the area, analyses were made using numerical classifications. Community Analysis Package (CAP) 1.4.1 software was used to investigate the relationships among species [50, 51]. Whittaker [53], explained that species diversity can be evaluated from three perspectives as alpha (α), beta (β) and gamma (γ) diversity. Alpha and gamma diversity are defined as inventory diversity and share the same characteristics. The only difference between them is that of the scale; alpha represents a sample area or habitat, while gamma diversity represents the ecosystem. Beta diversity is directly linked to alpha and gamma diversity. Beta diversity indices are used to compare the variation between two different areas (sample area, plant association, ecosystem, etc.) [53]. In this study, alpha (α) and beta (β) diversity indices were used. There are many indices used to determine alpha species diversity. Although the direct determination of the number of species may be an index value, the Shannon Wiener, Simpson's D, Margelef D, Berger-Parker Dominance, McIntosh D, Brilouin D, Fisher's Alpha, and Q Statistics are different indices commonly used in determining alpha diversity [22, 52, 54–56]. As suggested in many studies, in this study, the Shannon-Wiener, Simpson's D, Margelef D, and Berger-Parker Dominance indices were used, respectively.

2.3.1 Shannon-wiener function (H)

The Shannon-Wiener Function (H) was obtained using Eq. (1).

$$H = -\sum\{p_i \log(p_i)\} \tag{1}$$

According to the formula, the proportional value of the species is expressed by p. The natural logarithm (ln) of the proportion of species is taken and this value is multiplied by the number of species. The Shannon-Wienner (H) gives the negative multiplication value of the sum of the products and the number of "ln" values of all species [57].

2.3.2 Simpson's index (D)

This diversity index was proposed by Simpson (1949) [58] and is used to determine the likelihood of a second sample from a population being the same as the original [52]. The index is expressed by the formula D = 1/C. In Eq. (2) the greater the value obtained, the greater the proportionality.

$$C = \sum_{i}^{s} p_i^2, p_i^2 = \frac{Ni\ (Ni - 1)}{Nt\ (Nt - 1)} \tag{2}$$

Ni is the number of species corresponding to the number i and Nt is the total number of individuals in the sample.

2.3.3 Margalef D

The Margalef D Index is calculated with the formula in Eq. (3).

$$D = (S - 1)/(\text{In N}). \tag{3}$$

Here, S is the number of species and N is the total number of individuals in the sample [52].

2.3.4 Evenness (Berger-Parker dominance) index

The Evenness/Berger-Parker Dominance Index is calculated with Eq. (4).

$$d = \frac{Nmax}{Nt}. \tag{4}$$

Here, Nt represents the total species ratio and Nmax is the most dominant species in this ratio. The evenness index, like the others, is used to calculate diversity results [54–57].

In the calculations based on alpha diversity indices, open green areas with the same spatial characteristics were considered as a single group and the arithmetic means of the index values were taken and a general evaluation made about the species diversity of the places such as parks, public gardens and roadsides. At this stage, the 'arithmetic mean' which is a statistical evaluation method, was used.

2.4 Density and spatial distribution analysis of plant species

Woody plant data for points obtained from all sample areas were used to evaluate the spatial relationships via GIS. The study carried out for this purpose consisted of the following steps:

A database was created in ArcGIS 10.4 to evaluate the information obtained from the sample areas via GIS. First, point data (coordinates) of the plants obtained from the field studies and recorded in the field introduction table were then transferred to the ArcGIS environment. Subsequently, entries in each field and the positional data of each plant species were realized in the database. The headings of the plant data in the database were as follows:

Species name: The plant name analysis, the number of plants, and the area covered were determined and the density map was prepared, taking into consideration the community.

Plant family: Plants were analyzed by considering density according to families.

Phenological properties: Visualization and analysis of the plants were conducted in ArcGIS and they were classified according to their phenological status using the table. We coded each of the identified phenological features as: Species in the foreground with leaf beauty = 1, Species in the foreground with beauty of flowers = 2, leaves and flowers = 3, leaves and fruits = 4, leaves and stems = 5, flowers and stems = 6, leaves, flowers and fruits = 7.

Seasonal situation: According to the seasonal status of the plants, the GIS database was coded as evergreen (1) or deciduous (2) and displayed on ArcGIS.

2.5 Spatial density analysis

Density analysis was performed on the database. After entering the required information in point layers, point density was selected under Arctoolbox 'Spatial Analyst Tools-Density'. In the screen that opens, the point layer we wanted to analyze in the input features section was selected and the field in the population field section was selected. In the output raster section, the result of the density and distribution analysis was completed by selecting the map name and the desired location to be recorded. As a result of these processes, density maps were created for each sample area.

2.6 Spatial interpolation method

In order to see the distribution of the species identified in the sample areas, we conducted spatial distribution analyses in the GIS environment. The 'Analysis Tools - Proximity - Create Thiessen Polygons' command located under the Toolbox was applied to the database. The spatial interpolation method was used to determine the spatial distribution maps of the individual research areas according to the species, family, seasonal conditions and phenological properties of the plants and then the results were evaluated.

3. Results

In evaluating the floral design of the research area, the sample areas identified in the city center were examined separately and as a whole for the entire research area and then were analyzed and the results of the analysis interpreted. In the research area, 173 plant species were found (**Table 3**) from the different sample areas (**Figure 2**). Although most of the identified species were tall trees, 38 shrub species

Species	Botanical family	Number of species
Abelia grandiflora	Caprifoliaceae	22
Abies nordmanniana subsp. nordmanniana	Pinaceae	45
Abies nordmanniana subsp. bornmuelleriana	Pinaceae	3
Acer campestre	Aceraceae	2
Acer negundo	Aceraceae	361
Acer negundo 'Flamingo'	Aceraceae	8
Acer palmatum 'Atropurpurea'	Aceraceae	7
Acer platanoides	Aceraceae	1
Acer platanoides 'Crimson King'	Aceraceae	122
Acer pseudoplatanus	Aceraceae	27
Aesculus hippocastanum	Hippocastanaceae	32
Ailanthus altissima	Simoribaceae	33
Albizia julibrissin	Leguminosae	2
Berberis thunbergii 'Atropurpurea'	Berberidaceae	71
Berberis thunbergii 'Atropurpurea Nana'	Berberidaceae	200
Betula pendula	Betulaceae	11
Buddleia davidii	Buddlejaceae	3
Buxus sempervirens	Buxaceae	173
Buxus sempervirens 'Nana'	Buxaceae	68
Calocedrus decurrens	Cupressaceae	9
Calocedrus decurrens 'Aurea'	Cupressaceae	14
Callistemon citrinus	Myrtaceae	2
Carpinus betulus	Betulaceae	10
Catalpa bignonioides	Bignonaceae	29
Cedrus atlantica	Pinaceae	59
Cedrus atlantica 'Glauca'	Pinaceae	41
Cedrus atlantica 'Glauca Pendula'	Pinaceae	2
Cedrus atlantica 'Glauca Pyramidalis'	Pinaceae	65
Cedrus deodora	Pinaceae	21
Cedrus deodora 'Aurea'	Pinaceae	11
Cedrus deodora 'Pendula'	Pinaceae	2
Cedrus libani	Pinaceae	45
Cercis siliquastrum	Leguminosae	143
Chaenomeles japonica	Rosaceae	7

Species	Botanical family	Number of species
Chamaecyparis lawsoniana	Cupressaceae	130
Chamaecyparis lawsoniana 'Columnaris Glauca'	Cupressaceae	1
Chamaecyparis lawsoniana 'Ellwoodii'	Cupressaceae	25
Chamaecyparis lawsoniana 'Fleckellwood'	Cupressaceae	5
Chamaecyparis nootkatensis 'Pendula'	Cupressaceae	5
Chamaecyparis pisifera 'Boulevard'	Cupressaceae	1
Cornus alba	Cornaceae	24
Corylus avellana	Betulaceae	1
Cotinus coggygria	Anacardiaceae	2
Cotinus coggygria 'Royal Purple'	Anacardiaceae	21
Cotoneaster dammeri	Rosaceae	17
Cotoneaster franchetti	Rosaceae	46
Cotoneaster lacteus	Rosaceae	3
Cotoneaster microphylla	Rosaceae	8
Crataegus oxyacantha	Rosaceae	2
Cryptomeria japonica 'Elegans'	Taxodiaceae	2
Cupressocyparis leylandii	Cupressaceae	2066
Cupressocyparis leylandii 'Aurea'	Cupressaceae	15
Cupressocyparis leylandii 'Fastigiata'	Cupressaceae	37
Cupressocyparis leylandii 'Pyramidalis'	Cupressaceae	10
Cupressus arizonica	Cupressaceae	136
Cupressus arizonica 'Glauca'	Cupressaceae	44
Cupressus macrocarpa 'Gold Crest'	Cupressaceae	87
Cupressus sempervirens	Cupressaceae	55
Cupressus sempervirens 'Pyramidalis'	Cupressaceae	1
Cycas revoluta	Cycadaceae	5
Cydonia oblonga	Rosaceae	1
Diospyros kaki	Ebenaceae	2
Elaeagnus angustifolia	Elaeagnaceae	17
Eriobotrya japonica	Rosaceae	3
Euonymus japonica	Celastraceae	62
Euonymus japonica 'Aurea'	Celastraceae	164
Euonymus japonica 'Aurea Nana'	Celastraceae	27
Ficus carica	Moraceae	18
Forsythia × *intermedia*	Oleaceae	31
Fraxinus angustifolia	Oleaceae	73
Fraxinus excelsior	Oleaceae	32
Ginkgo biloba	Ginkgoaceae	4
Hedera helix	Araliaceae	31
Hedera helix 'Aurea'	Araliaceae	1

Species	Botanical family	Number of species
Hibiscus syriacus	Malvaceae	148
Hydrangea macrophylla	Saxifragaceae	64
Ilex aquifolium 'Golden Queen'	Aquifoliaceae	1
Juglans regia	Juglandaceae	17
Juniperus chinensis 'Pfitzeriana Glauca'	Cupressaceae	1
Juniperus chinensis 'Stricta'	Cupressaceae	2
Juniperus communis 'Compressa'	Cupressaceae	1
Juniperus horizontalis	Cupressaceae	329
Kerria japonica	Rosaceae	1
Koelreuteria paniculata	Sapindaceae	1
Lagerstroemia indica	Lythraceae	72
Laurocerasus officinalis	Rosaceae	79
Laurus nobilis	Lauraceae	15
Ligustrum japonicum	Oleaceae	45
Ligustrum vulgare	Oleaceae	423
Ligustrum vulgare 'Aurea'	Oleaceae	1
Liquidambar styraciflua	Altingiaceae	6
Lonicera caprifolium	Caprifoliaceae	7
Lonicera japonica	Caprifoliaceae	2
Lonicera japonica 'Chinensis'	Caprifoliaceae	6
Magnolia grandiflora	Magnoliaceae	24
Mahonia aquifolium	Berberidaceae	4
Malus domestica	Rosaceae	2
Morus alba	Moraceae	29
Morus nigra 'Pendula'	Moraceae	13
Musa × paradisiaca	Musaceae	7
Nerium oleander	Apocynaceae	11
Olea europaea	Oleaceae	16
Philadelphus coronarius	Hydrangeaceae	1
Phoenix canariensis	Arecaceae	1
Photinia x fraseri 'Red Robin'	Rosaceae	200
Picea abies	Pinaceae	8
Picea glauca	Pinaceae	3
Picea glauca 'Conica'	Pinaceae	23
Picea orientalis	Pinaceae	147
Picea pungens	Pinaceae	13
Picea pungens 'Glauca'	Pinaceae	5
Picea pungens 'Globosa Nana'	Pinaceae	9
Picea pungens 'Hoopsii'	Pinaceae	11
Pinus brutia	Pinaceae	8

Species	Botanical family	Number of species
Pinus griffithii	Pinaceae	2
Pinus nigra	Pinaceae	69
Pinus pinea	Pinaceae	37
Pinus sylvestris	Pinaceae	140
Pittosporum tobira	Pittosporaceae	3
Pittosporum tobira 'Nana'	Pittosporaceae	11
Platanus occidentalis	Platanaceae	1
Platanus orientalis	Platanaceae	174
Populus alba	Salicaceae	1
Populus nigra	Salicaceae	1
Prunus avium	Rosaceae	1
Prunus cerasifera	Rosaceae	2
Prunus cerasifera 'Pissardii Nigra'	Rosaceae	195
Prunus cerasus	Rosaceae	18
Prunus domestica	Rosaceae	1
Prunus laurocerasus 'Otto Luyken'	Rosaceae	15
Prunus persica	Rosaceae	8
Prunus serrulata 'Kanzan'	Rosaceae	3
Pseudotsuga menziesii var. glauca	Pinaceae	1
Pseudotsuga menziesii var. viridis	Pinaceae	2
Pyracantha coccinea	Rosaceae	14
Pyracantha coccinea 'Nana'	Rosaceae	42
Pyrus communis	Rosaceae	5
Quercus ilex	Fagaceae	5
Quercus robur	Fagaceae	4
Robinia pseudoacacia	Leguminosae	115
Robinia pseudoacacia 'Umbraculifera'	Leguminosae	274
Rosa sp.	Rosaceae	418
Rosmarinus officinalis	Lamiaceae	30
Salix babylonica	Salicaceae	4
Salix caprea 'Pendula'	Salicaceae	1
Sequoia sempervirens	Taxodiaceae	6
Sophora japonica	Leguminosae	3
Syringa vulgaris	Oleaceae	8
Taxus baccata	Taxaceae	41
Taxus baccata 'Fastigiata'	Taxaceae	2
Taxus baccata 'Pyramidalis'	Taxaceae	1
Thuja occidentalis	Cupressaceae	1
Thuja occidetalis 'Aurea'	Cupressaceae	1
Thuja occidentalis 'Fastigiata'	Cupressaceae	2

Species	Botanical family	Number of species
Thuja occidentalis 'Globosa Nana'	Cupressaceae	14
Thuja occidentalis 'Golden Globe'	Cupressaceae	73
Thuja occidentalis 'Pyramidalis'	Cupressaceae	2
Thuja occidentalis 'Sunkist'	Cupressaceae	5
Thuja orientalis	Cupressaceae	132
Thuja orientalis 'Aurea'	Cupressaceae	21
Thuja orientalis 'Nana'	Cupressaceae	61
Thuja orientalis 'Compacta Aurea Nana'	Cupressaceae	36
Thuja orientalis 'Compacta Nana'	Cupressaceae	1
Thuja orientalis 'Pyramidalis'	Cupressaceae	53
Thuja orientalis 'Pyramidalis Aurea'	Cupressaceae	190
Thuja plicata	Cupressaceae	25
Tilia cordata	Tiliaceae	1
Tilia platyphyllos	Tiliaceae	3
Tilia tomentosa	Tiliaceae	324
Viburnum opulus	Caprifoliaceae	6
Viburnum tinus	Caprifoliaceae	10
Wisteria sinensis	Fabaceae	1
Yucca flamentosa	Agavaceae	26

Table 3.
Abundance values of the species recorded in sample areas for Duzce city.

were also determined. *Cupressocyparis leylandii* M.L., *Tilia tomentosa, Acer negundo* L., Rosa sp., and *Juniperus horizontalis* were the most common species in the research area.

The results of the research are presented in two parts, as a statistical analysis and as a spatial analysis performed in GIS.

3.1 Diversity and distribution of species (diversity analysis)

In order to determine the WPD in the open and green areas of Düzce city center, the distribution and diversity statuses of the species in the research area were obtained as a result of the statistical evaluations for alpha diversity (**Table 4**). **Table 4** shows that:

- According to the Shannon Index, species diversity was observed the most in the area of the Governorship and in residential gardens, and the least along the Düzce - Akçakoca highway.

- According to Simpson's Index, species diversity was observed the most in residential gardens, around the Governorship and in Monument Park, and the least along the Düzce - Akçakoca highway.

- According to the Margalef Index, species diversity was observed the most in residential gardens and around the Governorship, and least along the Düzce - Akçakoca highway.

Figure 2.
(A) Avni Akyol Park; (B) Düzce municipality; (C) D100 highway and Kervan crossroads; (D) İnönü park; (E) İstanbul street; (F) Düzce - Akçakoca inter-city highway.

- According to the Evenness index, species diversity was observed the most in residential gardens, Monument Park, and İnönü Park, and the least along the Düzce - Akçakoca highway.

The arithmetic means of the alpha diversity index values were taken by considering the same spatial open green areas as a single group. A general evaluation was made about species diversity in the context of space (**Figure 3**). According to the Shannon Index, the highest species diversity was observed in public areas and the least in residential areas. According to Simpson's Index, the most species diversity was again seen in public areas and the least in shopping centers. According to the Margalef Index, the most species diversity was observed in parks and the least species diversity in shopping centers. According to the Evenness Index, the diversity of species was highest in public areas and lowest in residential gardens.

Samples	Sample Areas	Shannon-Wiener	Simpson's	Margalef D	Evenness
1	1.1.	2.99	20	6.34	0.58
	1.2.	2.56	13	4.68	0.50
	1.3.	2.48	12	4.43	0.48
	1.4.	2.89	18	5.88	0.56
	1.5.	4.03	56	13.66	0.78
	1.6.	1.79	6	2.79	0.35
	1.7.	2.89	18	5.88	0.56
	1.8.	2.94	19	6.11	0.57
	1.9.	2.94	19	6.11	0.57
	1.10.	2.30	10	3.91	0.45
	1.11.	2.40	11	4.17	0.47
2	2.1.	2.83	17	5.65	0.55
	2.2.	3.85	47	11.95	0.75
	2.3.	3.66	39	10.37	0.71
	2.4.	1.79	6	2.79	0.35
	2.5.	2.77	16	5.41	0.54
	2.6.	2.94	19	6.11	0.57
	2.7.	3.09	22	6.79	0.60
3	3.1.	2.48	12	4.43	0.48
	3.2.	3.26	26	7.67	0.63
4	4.1.	2.83	17	5.65	0.55
	4.2.	2.56	13	4.68	0.50
	4.3.	1.79	6	2.79	0.35
	4.4.	1.79	6	2.79	0.35
	4.5.	3.50	33	9.15	0.68
	4.6.	0.69	2	1.44	0.13
	4.7.	2.48	12	4.43	0.48
5	5.1.	3.74	42	10.97	0.73
	5.2.	2.20	9	3.64	0.43
	5.3.	2.56	13	4.68	0.50
	5.4.	2.08	8	3.37	0.40
	5.5.	1.79	6	2.79	0.35
6		4.14	63	14.97	0.80

Table 4.
Diversity of plant species: 1: Public open and green spaces (11 sample areas); 2: Roads (7 sample areas); 3: Squares (2 sample areas); 4: Parks (7 sample areas); 5: Cemetery and mosques (5 sample areas), 6: Residential area gardens.

The distribution and diversity status of the species in the research area according to the results of the statistical evaluations made for the beta diversity are expressed graphically in **Figure 4**. According to this graph, the greatest changes in the field were Cody's βc values.

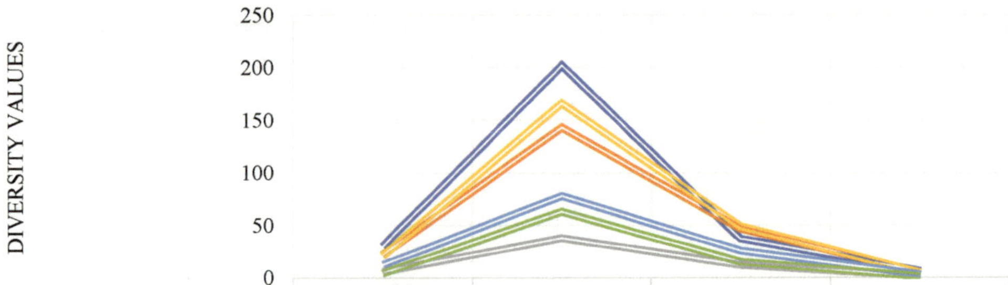

	SHANNON-WİENER	SİMPSON	MARGALEF D	EVENNESS
Public Areas	30.2328	202	36.9698	5.87329
Roads	22.178	143	46.628	4.308
Squares	5.743	38	12.099	1.115
Parks	20.946	166	49.075	4.069
Cemetery and Mosques	12.371	78	25.445	2.403
Residential Gardens	4.1431	63	14.965	0.80488

Figure 3.
Diversity values of species according to space types (the arithmetic means of the alpha diversity index values were taken by considering the same spatial open green areas as a single group).

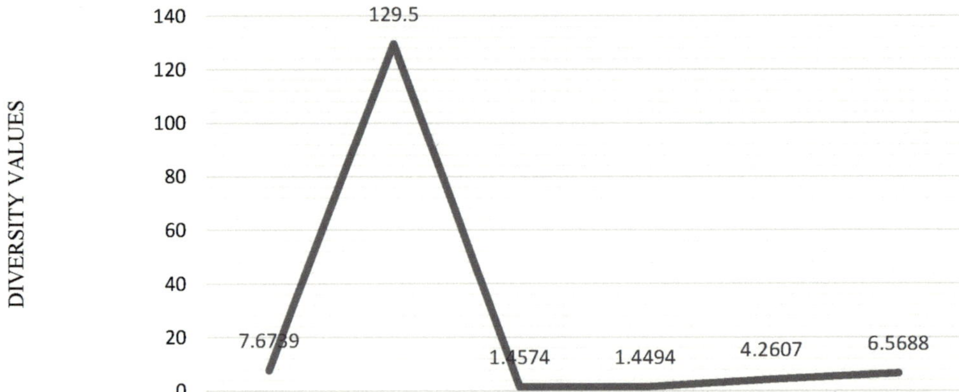

	Whittaker Bw	Cody Bc	Routledge Br	Routledge Bi	Routledge Be	Wilson & Shmida Bt
Beta Diversity	7.6739	129.5	1.4574	1.4494	4.2607	6.5688

Figure 4.
Beta index and diversity distribution graph.

3.2 Results of data evaluation on GIS environment

According to the results of the analysis, the density of the point data obtained from the sample areas was determined according to the species, families, seasonal conditions and phenological properties of the species;

The most common species in the sample areas was *Cupressocyparis leylandii* (Hybrid Cypress), followed by *Rosa* sp. and *Tilia tomentosa* (Silver Linden) (**Figure 5A**). Species represented by a single individual included *Acer platanoides* L., *Chamaecyparis lawsoniana* 'Columnaris Glauca', *Chamaecyparis pisifera* 'Boulevard', *Corylus avellana, Cupressus sempervirens* 'Pyramidalis', *Cydonia oblonga, Ilex aquifolium* 'Golden Queen', *Juniperus chinensis* 'Pfitzeriana Glauca', *Juniperus communis* 'Compressa', *Kerria japonica, Koelreuteria paniculata, Ligustrum vulgare* 'Aurea', *Philadelphus coronarius, Phoenix canariensis, Platanus occidentalis, Populus alba, Populus nigra, Prunus avium, Prunus domestica, Pseudotsuga menziesii* var.

Glauca, *Salix caprea* 'Pendula', *Taxus baccata* 'Pyramidalis', *Thuja occidentalis, Thuja occidetalis 'Aurea', Thuja orientalis 'Compacta Nana', Tilia cordata, Wisteria sinensis.*

In Avni Akyol Park and İnönü Park and around the D100 highway-Kervan crossroads species density was high. On the other hand, in the residential areas and along the Düzce - Akçakoca highway and Bolu Street, the density of species was low.

When the distribution of the identified species was considered according to families, Cupressaceae was the most common. İnönü Park and the D100 highway-Kervan crossroads were observed as the areas with the highest family density depending on the species density (**Figure 5B**). When the phenological properties of the species were considered, the plants with highest density in the area were those having 'beauty of leaves' (**Figure 5C**). When the seasonal status of the species was taken into consideration, it was seen that the density of evergreen species was higher than that of deciduous species (**Figure 5D**).

The spatial distribution of the species was similar to the results of the density analysis. The results of the spatial distribution of plants on a species basis were significant. In the research area, *Cupressocyparis leylandii* was the most widespread species, spreading over 2.8 km^2, followed by *Photinia x fraserii* 'Red Robin', and *Tilia tomentosa*, *Acer platanoides* L., *Cotinus coggygria.*, *Cupressus sempervirens* L. 'Pyramidalis', *Ilex aquifolium* L. 'Golden Queen', *Taxus baccata* 'Pyramidalis', and *Thuja occidentalis* 'Aurea' were the least distributed species in the area.

When the spatial distribution analysis results of floristic diversity according to families were evaluated, the Rosaceae family, dominant over an area of approximately 5.9 km^2 and represented by 24 species in the sample areas, was the most highly distributed family, followed by the Cupressaceae family, with 16 species represented. The Aquifoliaceae, Arecaceae, and Sapindaceae families were the least encountered families (**Figure 6**).

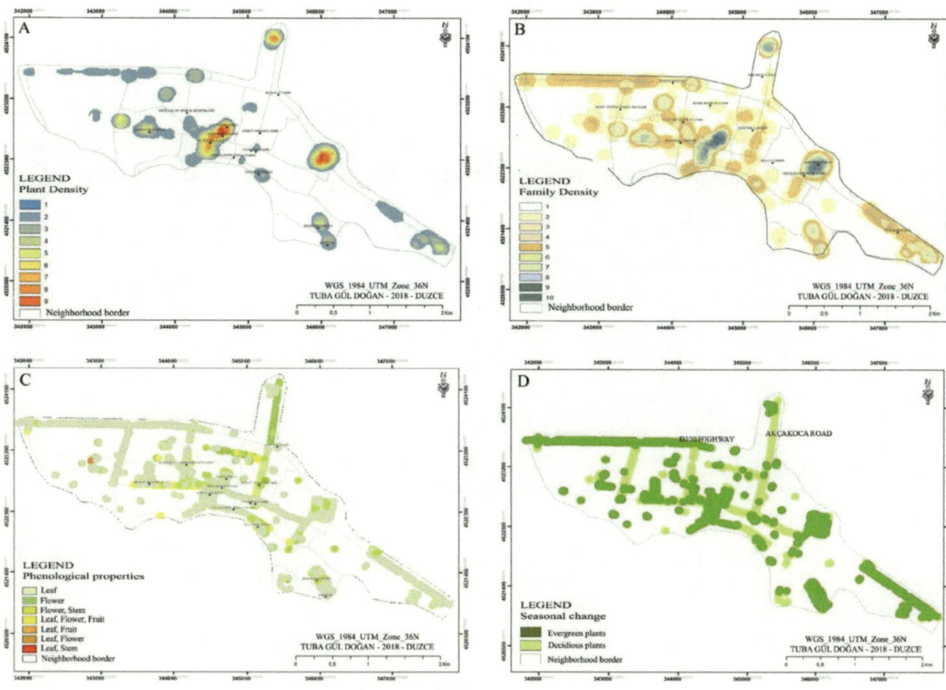

Figure 5.
Density analysis of floristic diversity: (A) according to species; (B) according to families; (C) according to phenological status; (D) according to seasonal conditions.

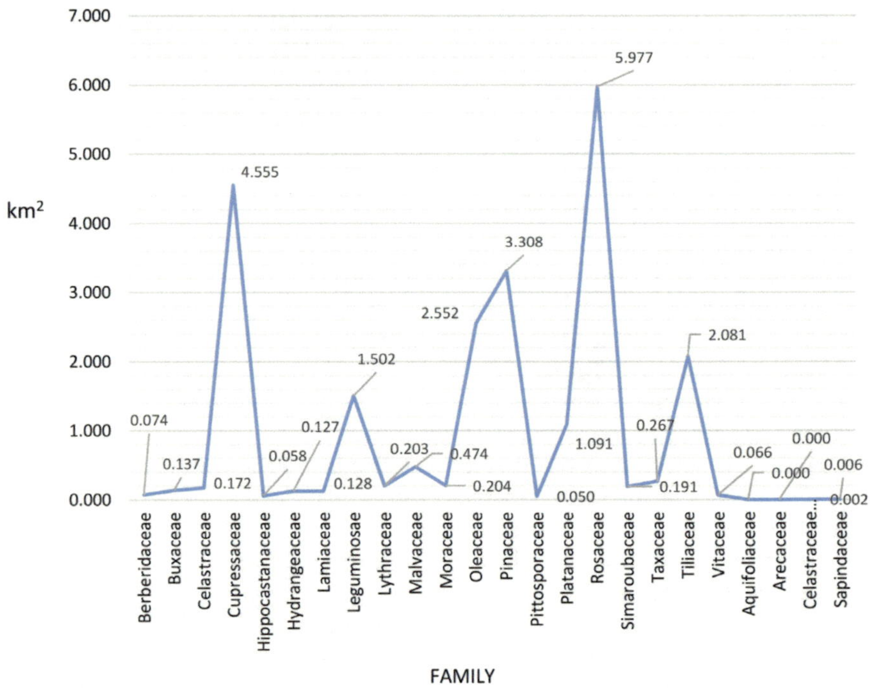

Figure 6.
Spatial distribution of woody plant diversity (WPD) according to families.

In the spatial distribution analysis of WPD, the plants with the most important leaf beauty had the widest distribution, with around 75% of the species consisting of *Euonymus japonica* 'Aurea', *Photinia x fraseri* 'Red Robin', and *Prunus laurocerasus* 'Otto Luyken'. Species having leaf and stem beauty (*Lagerstroemia indica* L., *Olea europaea* L.) were represented by 0.2% and had the least distribution (**Figure 7**).

Just as in the results of the density analysis, in the spatial distribution analysis of the plants according to phenological properties, those with leaf beauty were in the foreground and had the widest distribution. Evergreen species exhibited wide distribution compared to deciduous species (**Figure 8**). It is thought that the use of

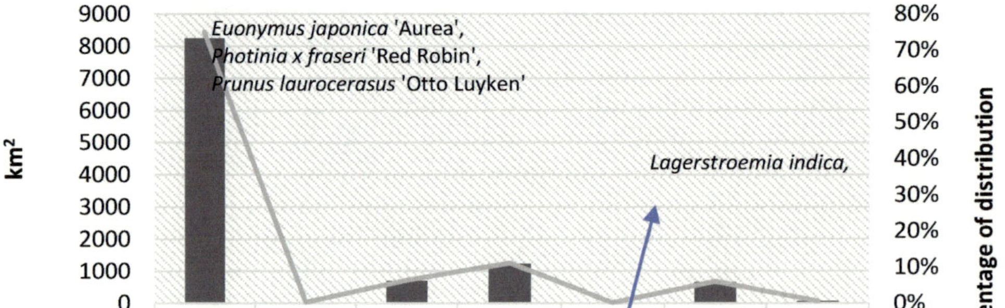

	1: Leaf	2: Flower	3: Leaf, Fruit	4: Leaf, Flower	5: Leaf, Stem	6: Leaf, Flower, Fruit	7: Flower, Stem
Series1	8253	31	702	1219	24	659	72
Series2	75%	0.30%	6.40%	11.10%	0.20%	6.01%	0.65%

Phenological Status

Figure 7.
Distribution of floristic diversity according to phenological status.

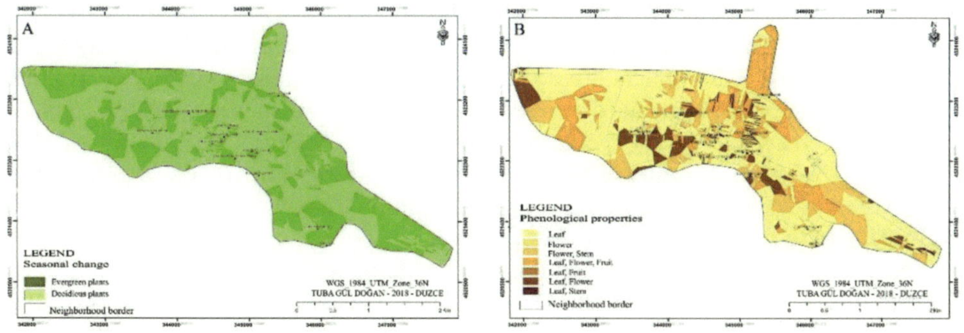

Figure 8.
Spatial distribution analysis of WPD: (A) according to seasonal conditions; (B) according to phenological status.

Cupressocyparis leylandii along the road in the sample area of the D100 highway resulted in the evergreen species being in the foreground of the density analysis. However, when the same results were obtained in the spatial distribution analysis, it is possible to conclude that evergreen species were scattered throughout the research area.

As a result of the analyses carried out in order to reveal density variation of the species identified in the sample areas according to the seasonal conditions, it was found that deciduous species were used extensively in Monument Park and İnönü Park, and along İstanbul Street and the Düzce – Akçakoca highway, while the evergreen species were concentrated around the D100 highway-Kervan crossroads and in the city cemetery. The Kervan crossroads, the Governorship and the Provincial Directorate of Forestry were seen as areas where evergreen and deciduous species were combined and dense.

In the park and recreation areas, the density of species was mostly observed in İnönü Park, while the least density was observed in Celaleddin Özdal Park, where six different species were identified. According to the density analysis of WPD in public spaces, it was found that the density was highest in the vicinity of the Governorship. As a result of the analysis applied to determine the density of floristic diversity along the roads, the Kervan crosssroads on the D100 highway was seen as the area where the density increased the most. As a result of the density analysis, which was in parallel with the diversity analysis, the Düzce - Akçakoca highway was determined as the road with the least density. In the category of cemetery and worship areas, the density of the city cemetery was seen to be increasing.

4. Conclusions

The green texture created by the rich plant diversity in cities also provides a living space for different species [59]. According to Nero, urban biodiversity is crucial to building resistant and sustainable cities [60]. In recent years, biodiversity has gained importance in cities. Consequently, our study focusing on WPD was carried out in the central district city of Düzce in areas with different spatial usage potential. The results on existing flora were revealed to be parallel with previous studies on this urban landscape [7, 14]. The number of plant species in the selected sampling spaces in the research area was quite high in terms of species richness. The wealth of species can be mentioned especially in the open green areas of the Düzce city center.

Open green areas in a city representing the diversity of species include parks, public gardens, urban forests, and cemeteries [61]. In addition, highway plantings, which are defined as artificial corridors [62], are the passageways connecting green areas in cities [63], and species diversity is high in these areas. This was not the case in our research area. A single species (*Cupressocyparis leylandii*) was encountered along the D100 highway, one of the main urban corridors in Düzce. Moreover, the D100 highway was included in the list of exemplary areas, verified by diversity analyses, where there was little diversity.

Species studies were also carried out on the plants located on the banks of Asar Creek within the boundaries of the research area. The diversity of vegetation along a watercourse normally reflects the diversity in the physical environmental conditions [64, 65]. However, the diversity of vegetation located on the banks of the Asar Creek was different from that of the overall research area. The main reason for this is because the destruction of natural habitats is low and the creek exhibits a relationship with the immediate environment, although excessive degradation has occurred in the areas landscaped with unnatural species. Urban areas are places where structural and physical changes are intense.

According to Eroğlu et al., at Düzce intersections in general, the species were randomly positioned in the space and although they were correct in some points in terms of species selection, they were positioned in a scattered way and too many color types were used, contrary to the regulation principles [7]. The findings of this study conducted in 2005 are similar to the results of our research. At the Düzce intersections discussed within the scope of the study, attention was drawn to wrong type selection and designs that were not suitable for the space. According to Acar et al., the diversity of plants in urban landscape areas plays an important role in the protection of urban nature and in the determination of planning and policies [50]. Accordingly, their study in Trabzon shows parallels with our research. The results of the diversity analysis obtained from the research area of Düzce show that diversity was high on the Shannon-Wiener, Evenness, and Margalef indices, but not on Simpson's Index. In the scope of our research, vegetative arrangements were found on private properties, but the diversity in public housing was higher. In other housing types, natural landscape effects were observed.

In order to determine the WPD in open and green spaces in the city center, sample areas were examined and the results were evaluated. According to this, *Tilia tomentosa* was the most frequently used tree in Düzce parks, and thus determined the general characteristics of these parks. Species diversity was quite high in the parks, which were mostly composed of tall trees; however, planting criteria were generally not followed and the plants did not display their natural forms. The situation was no different with road, median and junction arrangements. The planting at central intersections was considered a negative factor due to the random planting of species below eye level. At the Kervan crossroads, there was a dense variety of species which were generally positioned according to the correct design criteria. This could be considered as a very accurate design if it were to be used as a recreation area. However, because of the characteristics of the space, the design was found to be unsuitable for the intersection landscape. Since the *Cercis siliquastrum* L. planted along the Düzce - Akçakoca highway is not suitable for roadside use, it was evaluated as the wrong species selection for this area. There are many plant types in public spaces, ranging from tall trees to shrubs, but poor and inadequate designs were often encountered in the garden outside the Governorship. The dense, hard ground and random location of species around the Municipality Building were considered as negative features. These factors add a negative esthetic value to the image of the city and the municipality.

Hybrid Cypress (*Cupressocyparis leylandii*) was the most common species in the research area, while Silver Linden (*Tilia tomentosa*) determined the general characteristic of Düzce urban parks. *Cupressocyparis leylandii* was identified as the dominant species in the research area and showed the highest distribution. Within the boundary of the research area there is a high use of *Cupressocyparis leylandii* for 8 km between the Ankara and the Istanbul sections of the D100 highway.

It is possible to discuss the diversity in the open and green areas of central Düzce in terms of the spatial distribution and diversity of species. The Governorship and residential areas were the richest in terms of species diversity and the Düzce - Akçakoca highway was observed as the area with the least diversity. When the sample areas of similar characteristics were evaluated as a group, public areas and parks were rich in species diversity, whereas residential gardens and shopping centers/squares were the areas where diversity was low, according to different diversity indices. İnönü Park, which is one of the most important and intensely used recreational areas of the city, was observed as having a rich variety of species.

As a result of the study, it was found that using the floristic diversity indices and GIS jointly enabled the UFD of the city to be defined both statistically and positionally.

It is possible to make current and past evaluations of a place using a database known as the Urban Information System (UIS), as an important new concept that can contribute to the creation of more livable and accessible cities. Once the infrastructure is prepared, plant species analysis, space analysis, layer analysis, and spatial distribution analysis can be carried out in an area to determine which layers (trees, shrubs, ground cover, etc.) are dominant. In addition, with the use of the ArcGIS program, the location of plants can be determined and current photographs of the plants uploaded to that point along with information about their appearance, all of which are easily accessible. In addition, the newly proposed UIS can contribute to the evaluation and interpretation of the concept of 'Green Infrastructure', which has played such an important role in the urban model in recent years. It will also be able to propose methods and locations for implementation of applications having the potential to respond to the green infrastructure of a city. It is seen as playing an important role in shaping the infrastructure of a system that, working from this standpoint, could be established for stakeholder institutions, organizations, and municipalities that will need such information within the UIS.

Acknowledgements

The author would like to thank Prof. Dr. Cengiz Acar for analyzed the data on diversity and Dr. Ahmet Salih Değermenci for all his support for this study. This research was self-funded.

Conflict of interest

The authors declare that the research was conducted in the absence of any commercial or financial relationships that could be construed as a potential conflict of interest.

Author details

Tuba Gül Doğan and Engin Eroğlu*
Department of Landscape Architecture, Düzce University, Faculty of Forestry, Düzce, Turkey

*Address all correspondence to: engineroglu@duzce.edu.tr

References

[1] Chesson, P. General theory of competitive coexistence in spatially-varying environments. Theoretical Population Biology. 2000. 58:211–237. DOI: 10.1006/tpbi.2000.1486

[2] Hope, D., Gries, C., Zhu, W., Fagan, W. F., Redman, C. L., Grimm, N. B., Nelson, A. L., Martin, C. and Kinzig, A. Socioeconomics drive urban plant diversity, PNAS (Proceedings of the National Academy of Sciences of the United States of America. 2003. 100(15): 8788–8792. DOI: 10.1073/pnas.1537557100

[3] Heywood, V. H. The nature and composition of urban plant diversity in the Mediterranean. Fl. Medit. 2017. 27: 195–220. DOI: 10.7320/FlMedit27.195

[4] Avcı, M. Türkiye'nin Bitki Çeşitliliği Ve Coğrafi Açıdan Değerlendirmesi. In: Akkemik, Ü. editor. Türkiye'nin Doğal Egzotik Ağaç ve Çalıları I. 1nd ed. Ankara: Publication of the General Directorate of Forestry. 2014. p.28–53.

[5] Anşin, R. Eastern Black Sea Region Flora and Floristic Contents of Prime Vegetation Types. [thesis]. Trabzon: Karadeniz Technical University; 1980.

[6] Terzioğlu, S. Flora and Vegetation of Uzungöl (Trabzon-Çaykara) and its Environment. [thesis]. Trabzon: Black sea Technical University; 1998.

[7] Eroğlu, E., Akıncı Kesim G., Müderrisoğlu H. Determination of plants in open and green areas in Düzce and evaluation of these plants according to some planting design principles. Journal of Agricultural Sciences, 2005. 11(3):270–277.

[8] Eroğlu, E. and Acar, C. Alpine plants of some highlands near Trabzon and their usage potentials regarding landscape architecture. Duzce University Journal of Forestry. 2009. 5 (1):42–59.

[9] Aksoy, N., Özkan, N. G., Aslan, S. and Koçer, N. Düzce ili Bitki Biyolojik Çeşitliliği, Endemik, Nadir Bitki Taksonları ve Koruma Statüleri. Ertuğrul A. editor. Düzce'de Tarih ve Kültür. Bursa: *History and Culture in Duzce*, Duzce Municipality Cultural Publications. 2014. p: 361–375.

[10] Vila-Ruiz, C. P., Ackerman, E. M., Bartolomei, R. S., Montiel, D. G., Lastra, L., Figuerola, C. E. and Caban, J. F. Plant species richness and abundance in residential yards across a tropical watershed: implications for urban sustainability. Ecology and Society, 2014. 19(3): 22. DOI: 10.5751/ES-06164-190322

[11] Paganova, V. and Jurekova Z. Woody plants in landscape planning and landscape design, *Landscape Planning*, 1nd ed. InTech: Slovak. 2015. 200–216 p.

[12] Ceylan, E. Analyzıng the effects of rapıd urbanızatıon on olıve trees wıth the help of remote sensıng and geographıcal ınformatıon systems: çeşme regıon. [thesis]. İzmir: Soil Science and Plant Nutrition Department, Ege University; 2013.

[13] Karaşah, B., Var, M. Discussion of planting design in urban pattern in terms of scale-form in Trabzon and its county. Bartın Faculty of Forestry Journal. 2012. 14:1–11.

[14] Eroğlu, E. and Demir, Z. A New Perspective to Open and Green Spaces in Urban; The Project Study of Ordu Urban Forest and Botanical Park. Biological Diversity and Conservation, 2016. 143–153.

[15] Emecen, Y. Identification and evaluation of changes in the landscape in the region of Sariyer by using landscape metrics. [thesis]. İstanbul: Department of Landscape Architecture, Istanbul University; 2015.

[16] Dana, E., Vivas, S. and Mota, J. Urban vegetation of Almeria city a contribution to urban ecology in Spain. Landscape and Urban Planning. 2002. 59: 203–216. DOI: 10.1016/S0169-2046 (02)00039-7.

[17] Eroğlu, E., Acar, C. and Aktaş, E. Kent içerisindeki açık ve yeşil alanlara yeni bir soluk; ordu şehir ormanı ve botanik parkı peyzaj proje çalışması. Artium, 2016. 4(2):30–42

[18] Davis, P.H. Flora of Turkey and the East Aegean lslands. Edinburgh: Edinburgh Univ. Press; 1965–1985.

[19] Aksoy, N., Özkan, N. G., Aslan, S. and Koçer, N. The Endemic Plants of Duzce and Their Conservation Status. In: XII. Optima Meeting, Proceedings; 22–26 March, Antalya. Turkey: 2010. p. 148.

[20] Aronson, M. F. J., Handel, S. N., LaPuma, I. P. and Clemants, S. E. Urbanization promotes non-native woody species and diverse plant assemblages in the New York metropolitan region. Urban Ecosystem. 2015. 18:31–45. DOI: 10.1007/s11252-014-0382-z. 2015.

[21] Yang, J., Yan, P., He, R. and Song, X. Exploring land-use legacy effects on taxonomic and functional diversity of woody plants in a rapidly urbanizing landscape, Landscape and Urban Planning. 2017. 162: 92–103. DOI: 10.1016/j.landurbplan.2017.02.003

[22] Kahveci, H. and Acar, C. Distribution and Floristic Composition of Coastal Vegetation in Northeastern Turkey, American Journal of Engineering Research (AJER). 2018. 7 (12): 205–220

[23] Swan, C., Johnson A. and Nowak, D. Differential organization of taxonomic and functional diversity in an urban woody plant metacommunity. Applied Vegetation Science. 2016. DOI: 10.1111/avsc.12266.

[24] Van Heezik, Y., Barrat, I. P., Dickinson, J. M., Freeman, C., Porter, S., Johnstone P. D. and Wing, J. Biodiversity of Coleoptera and other invertebrates in urban gardens: a case study in a New Zealand city. Insect Conservation and Diversity. 2015. 8 (5): 428–437. DOI:10.1111/icad.12120.

[25] Geographical Information System. [Internet], 2005. Available from: http://cbs.ormansu.gov.tr /cob2011/wp-content/ uploads/2011/05/cbs1.pdf. [Accessed: 2012–06]

[26] Orhan, M. D. Landscape elements database design within the scope of GIS based implementations in Turkey: Ankara Castle surroundings example. [thesis]. Ankara: Department of Landscape Architecture, Ankara University; 2007.

[27] Xu, Y., Chen, Y., Li, W., Fu, A., Ma, X., Gui, D., Chen, Y. Distribution pattern of plant species diversity in the mountainous Region of Ili River Valley, Xinjiang, Environmental Monitoring and Assessment. 2011. 177(1–4): 681–94.

[28] Wang, X. G., Hao, Z. Q., Ye, J., Zhang, J., Li, B.H. & Yao, X.L. Spatial pattern of diversity in an old- growth temperate forest in Northeastern China. Acta Oecologica. 2008. 33:345–354.

[29] Özkan, K. Relationships Between Tree and Scrub Species Diversity and Physiographic Site Factors in Çariksaraylar Site Section Groups, Beyşehir Watershed. Anadolu University Journal of Science and Technology. 2006. 7(1):157–166.

[30] Özkan, K., Gülsoy, S. Ecological Land Classification and Mapping Based on Vegetation Environment Hierarchical Analysis - A Case Study of Buldan Forest district (Turkey). Polish Journal of Ecology, 2010. 58(1): 55–67.

[31] Öner, N., Bilgili, B.C., Çorbacı, Ö.L. Evaluating the Potential Site Locations

of Natural Pine Species for Landscape Design Using GIS in Turkey, Fresenius Environmental Bulletin. 2015. 24(8): 2477–2483.

[32] Duzce Province Environmental Status Report. Duzce: 2009.

[33] Görcelioğlu, E., Günay, T., Karagül, R., Aksoy N. & Başaran, M.A. 19–21 may 1998 Western Black Sea flood causes, precautions to be taken and suggestions (Scientific Committee Report). 2nd ed. Ankara: TMMOB The Chamber of Forest Engineers Publication; 1999.

[34] Serbes, A. B., Kaplan, A. The Survey of Pollen and Spore Dispersal in the Atmosphere of Duzce City. Karaelmas Science and Engineering Journal. 2014. 4 (2):46–58.

[35] Yaltırık F., İşgüzar H., Küçükkoca A.H. Düzce İlçesi ve Orman İşletmesi. İstanbul: Ülkü Printing; 1953.

[36] Duzce Agricultural Drought Action Plan. T. C. Ministry of Food, Agriculture and Livestock, Duzce, Turkey: 2013.

[37] Özmen, S., Yıldırım, M., Şahin, B. Assessment of Water and Soil Resources in Duzce Area in Terms of Agricultural Use, Journal of Adnan Menderes University Agricultural Faculty. 2015. 12 (2):9–13.

[38] Walters, S. M. The next twenty years. In: Perring F. (ed.),*The Flora of Changing Britain*. Hampton. 1970. 136–141.

[39] Haeupler, H. Statistische Auswertung von Punkrasterkarten der Gefasspflanzenflora Süd-Niedersachsens. Scr. Geobot. 1974. 8(1):141.

[40] Wittig, R. and Durwen, K.J. Das ökologische Zeigerwertspektrum der spontanen Flora von Grossstädten im Vergleich zum Spektrum ihres Umlandes. *Natur. U. Landsch*. 1981. 56: 12–16.

[41] Rapoport, E.H., Diaz-Betancourt, M.E. & Lopez-Moreno, I.R.L. Aspectos de la ecologia urbana en la ciudad de Mexico. Mexico: Editorial Limusa; 1a edition. 1983.

[42] Pysek, P. Factors affecting the diversity of flora and vegetation in central European settlements, Plant Ecology. 1993. 106:89–100.

[43] Prance, G.T., Dixon, G.R., Aldous, D.E., Biodiversity and Green Open Space. Horticulture: Plants for People and Places. 2014. 2: 787–816.

[44] McDonnell, M.J., and Hahs, A.K. The future of urban biodiversity research: Moving beyond the 'low-hanging fruit'. Urban Ecosystems. 2013. 16(3):397–409.

[45] Williams, N.S.G., Schwartz, M.W., Ves, P.A., McCharty, M.A. Hahs, A.K., Clemants, S.E., Corlett, R.T., Duncan, R. P., Norton, B.A., Thompson, K., McDonnell1, M.J. A conceptual framework for predicting the effects of urban environments on floras, Journal of Ecology. 2009. 97:4–9.

[46] Karasar, N. Scientific Research Method. Ankara: Nobel Publishing; 2005.

[47] Gökçe, B. Research in Social Sciences. Ankara: War Publications; 1988.

[48] Ercan, M. Statistics in Scientific Research. Ministry of Forest Poplar and Rapidly Developing Species Forest Trees Research Institute. 2nd ed. Izmit, Turkey; 1997.

[49] Kalıpsız, A. Statistical Methods. Istanbul:Istanbul University Faculty of Forestry Publications;1981. 294–558 p.

[50] Acar, C., Eroğlu, E. and Sarı, D. Planting design strategies in urban landscape planning and design studies. III. National Landscape Architecture

Congress; 2007; Antalya. Turkey. p. 87–94.

[51] Eroğlu, E., Kaya, S. and Özçelik, Z. Investigation of plant diversity in an urban area having historical features aspect of floristic and aesthetic quality. Journal of Forestry. 2016. 12(2):163–177.

[52] Gülsoy, S. ve K. Özkan. Importance of biodiversity from the ecological standpoint and some diversity indexes. Suleyman Demirel University Faculty of Forestry Journal. 2008. A(1):168–178.

[53] Whittaker, R.H. Evolution and measurement of species diversity. Taxon. 1972. 21:213–251.

[54] Carboni, M., Carranza, M. L. and Acosta, A. Assessing conservation status on coastal dune: a multiscale approach. Landscape and Urban Planning. 2009. 91:17–25.

[55] Margalef, R. Perspectives in Ecological Theory. Chicago: University of Chicago Press. 1968.

[56] Luca, E. D., Novelli, C., Barbato, F., Menegoni, P., Iannetta Mand Nascetti, G. Coastal dune systems and disturbance factors: Monitoring and analysis in central Italy. Environmental Monitoring and Assessment. 2011. 183:437–450.

[57] Shannon, C. E. A mathematical theory of communication. The Bell System Technical Journal. 1948. 27: 379–423.

[58] Simpson, E.H., Measurement of diversity. Nature. 1949. 163: 688, DOI: 10.1038/163688a0.

[59] Garbutt, R., Sparks, T. Changes in the botanical diversity of a species rich ancient hedgerow between two surveys. Biological Conservation, 2002. 106: 273–278.

[60] Nero, B. F. Woody species and trait diversity functional relations of green spaces in Kumasi, Ghana. Urban Ecosystems. 2019. 22:593–607.

[61] Shashua-Bar, L., Hoffman, M. E. Vegetation as a climatic component in the design of an urban street: An empirical model for predicting the cooling effect of urban green areas with trees. Energy and Buildings. 2000. 31 (3):221–235.

[62] Uslu, A., Shakouri, N. Supporting facilities of Green Infrastructure and Biodiversity in Urban Landscape, Turkish Journal Of Scientific Reviews, 2013. 6 (1):46–50.

[63] Hamada, S. and Ohta, T. Seasonal variations in the cooling effect of urban green areas on surrounding urban areas. Urban Forestry & Urban Greening. 2010. 9(1):15–24.

[64] Gardiner, J.L. River Projects and Conservation: A Manual for Holistic Appraisal. John Wiley and Sons Inc., UK. 1994.

[65] Eroğlu, E., Ak, M. K., Aksoy, N., Keten, A., Keskin, Ş., Atmaca, E., Yıldız, O. and Kaya, S. Visual and Ecological Description of Landscape Character of Urban Water Corridors; Duzce Asar Stream, TUBITAK Tovag 3001 Project Result Report. 2018.

Chapter 3

Classical and Modern Remote Mapping Methods for Vegetation Cover

Algimantas Česnulevičius, Artūras Bautrėnas, Linas Bevainis and Donatas Ovodas

Abstract

Plant classification is quite complex and multilevel. All living organisms are divided into domains, kingdoms, types, classes, ranks, families, tribes, and species. This classification complexity is also reflected in the classification of biogeographic maps, which is much simpler. Based on floristic dependence, vegetation is grouped by connecting it into spatial (territorial) complexes. This paper presents the interfaces of mapping methods with taxonomic vegetation types at different hierarchical levels. At the same time, examples of vegetation mapping techniques from national and thematic atlases of different countries are presented in this article. UAV aerial photographs are widely used for local mapping of vegetation areas. The authors of this article propose a new methodology that can be used to assess the ecological condition of young trees and the volume of mature forest wood. The methodology is based on the separation of tree crown areas in UAV aerial photographs and photo color analysis. For automated area calculation of young trees, a PixRGB software has been developed to determine the area of pixels of the same color in aerial photographs. The software is based on the comparison of young tree crown area calculations in AutoCAD software and area measurements of individual color spectrum pixels. In the initial stage, aerial photographs are transformed to the exact size of the photographed area. Transformations were performed with an error of less than 2–3 cm. The transformation of the spectrum of aerial photographs allowed to concentrate the color of the image of young trees in a relatively narrow color range. Studies performed in 2019–2020 to assess the ecological condition of trees and the amount of wood using UAV INSPIRE 1 and PixRGB color analysis software showed the effectiveness of the applied methodology.

Keywords: methods of vegetation mapping, UAV, aerial imagery, automated color analysis

1. Introduction

Plant classification is quite complex and multilevel. All living organisms are divided into domains, kingdoms, types, classes, ranks, families, tribes, and species. This classification complexity is also reflected in the classification of biogeographic maps, which is much simpler. Phenomena describing vegetation and fungus are mapped in geobotanical maps. The distribution of natural vegetation and its change is usually mapped in these maps. Vegetation types and species composition are determined by ecological conditions that vary in different parts of the Earth and determine the productivity of the entire biological ecosystem. Biological productivity is determined by a complex of environmental conditions sets. On the other hand, the biological abundance of plants directly affects the environmental conditions. The basic elements of the areas covered by vegetation (vegetation cover) are the respective plant species, which differ in their appearance, stems, leaves, size. These indicators determine the appearance and structure of the entire local vegetation cover. Based on floristic dependence, vegetation is grouped by connecting it into spatial (territorial) complexes.

The first biogeographical maps focused on forest mapping. Later, the content of these maps was expanded: mapped shrubs, wetlands, meadows, detached trees. Contemporary biogeographical maps acquired modern content in the 19th century, when classification of vegetation indicators began to be applied to their compilation. The first floristic biogeographic mapping methodology was applied by the Danish botanist Joachim Frederik Schouw (1789–1852), who developed the basics of phytogeography. He compiled the phytogeographical World atlas (Pflanzengeographischer Atlas zur Erläuterung von Schouws Grundzügen einer allgemeinen Pflanzengeographie) with 22 global maps. This atlas was published in 1823 in Berlin.

Between 1837 and 1848, six geobotanical maps were printed in the Berhaus Atlas. In a later edition from 1886 to 1892, geobotanical maps covered the entire fifth volume of the atlas. Five biogeographic maps were published by Alvin J. Johnson in Family Atlas (1843). The Atlas of Gymnasiums and Real Schools by Richard Andree and Friedrich Wilhem Putzger (Germany, 1879) contains biogeographical and World vegetation zone maps. Richard Lüddecke and Herman Haak in Deutscher Schulatlas Mittelstufe (1895) published a floristic map of Europe.

The vegetation cover of a specific area is assessed and mapped in two aspects:

1. Floristic, when the vegetation cover is associated with the species composition of the plants.

2. Phytocenotic, when assessing the totality of plant species of a certain habitat with its characteristic structure and environment.

The plant species is considered to be the basic vegetation classification element - phytotaxone. From the phytocenotic point of view, the heterogeneity of the vegetation cover is assessed. It is highly dependent on the area of vegetation cover: for the larger the area characterized the greater heterogeneity of the vegetation cover. On the other hand, for each study area, are distinguished of diagnostic plant species, basis of which the vegetation areas are assigned to the respective formations. The groups of plant species that dominate the vegetation habitats reflect the relevant biocenotic homogeneity or heterolyticity of the site and are taxonomic indicators of vegetation cover. Cenotic vegetation derivatives are at several hierarchical levels: species, varieties, facie's, plant complexes, plant formation, plant range, plant class, and floristic region (**Figure 1**).

At the same time, are applied the different classifications of vegetation cover, which reflects the indicators of plant density and their height. Based on this classification, the territories divided on different types of vegetation land cover: forests, shrubs, semi-shrubs, meadows.

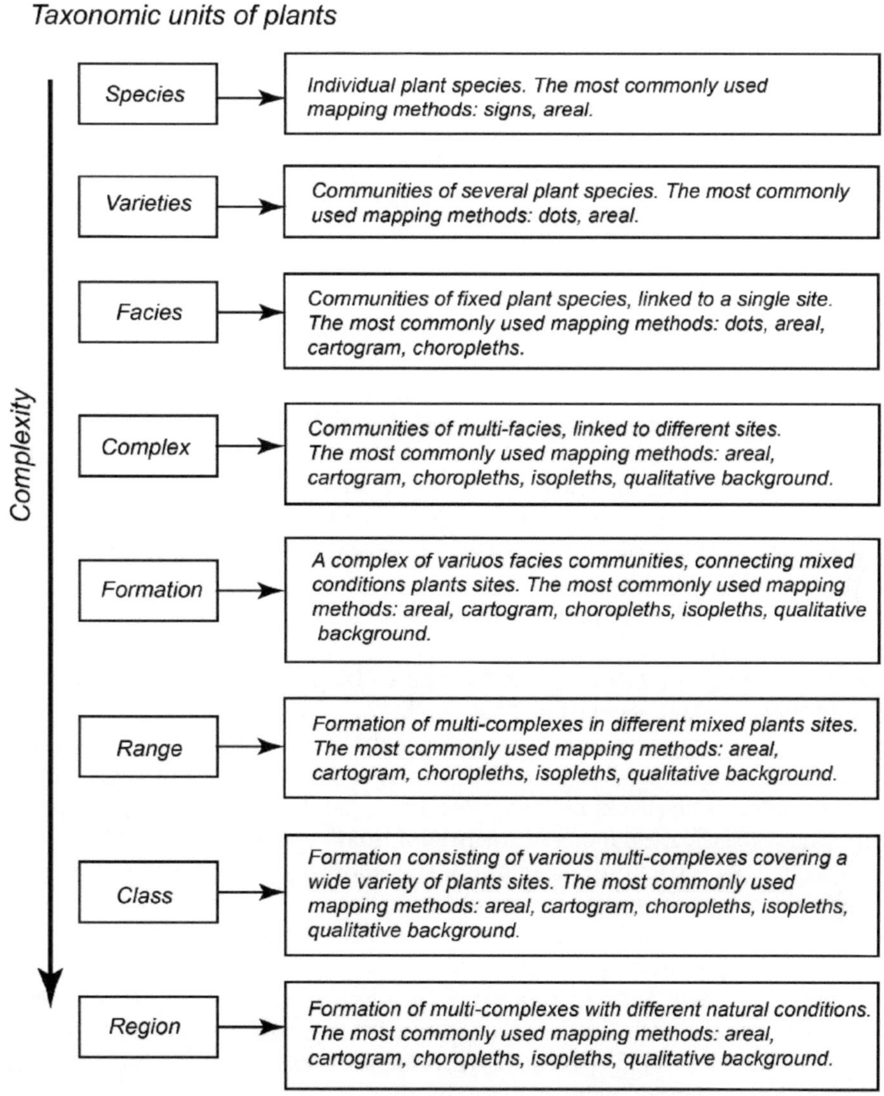

Figure 1.
Phytocenotic types of vegetation.

2. Methodology

Thematic maps of geobotanical, forestry, relevant sections of national atlases and separately published maps were analyzed for the evaluation of traditional vegetation mapping methods. The main source of information was thematic atlases, in which vegetation cover was mapped in many aspects: dispersal of vegetation types and complexes, dispersal of plant association, dispersal of individual plant species, vegetation resources and dispersal of vegetation resources, dispersal of rare and protected vegetation species, economic use of vegetation types and complexes.

The analysis was performed using a summary of quantitative indicators. Were calculated different mapping methods of maps and summarize the collected statistics data, which allowed to evaluated the most widely used traditional mapping methods. Publications of other countries researchers have also been used for the analysis [1–4].

The analysis of the application of modern research methods was performed by summarizing numerous publications on the using of multispectral aero- and space images in the separation of vegetation type aerials or individual plant species. The possibilities of applying remote sensing methods are widely studied in numerous publications on forest management, assessing the amount of wood in forests, the ecological condition of trees. The authors of the article discuss in more detail one of the methodological variants that they applied in the assessment of the ecological condition of trees and wood volume.

3. Vegetation mapping techniques and image generalization

The sign method is used to map the location of special (rare, unique) plants. In the case where the relevant plant species do not form a continuous cover and the plants are scattered over a wide area, the habitat method is used to map the distribution of the species. The boundaries of the habitats are drawn in lines of different colors and structures, and inside them are drawn signs, reminiscent of a plant species, or the names of plant species are written. Using cartodiagrams, cartograms and habitat methods are mapped the number of plants, the number of their habitats sites, the number of plant species or the covered area (**Figures** 2 and 3).

Areas of natural vegetation where dominated by plants of one species shall be assigned to the relevant vegetation type. In phytocenological maps, homogeneous territorial formations are mapped using the qualitative background method. This method of mapping is supplemented by symbols (images, letters or geometric signs), numbers, raster backgrounds (dots, lines, small figures of simple shape) or inscriptions (classification name of the predominant species or vegetation type). Using the qualitative background method, the colors of areas of different vegetation types are selected on the basis of isomorphism: "cold" colors (blue, green) are used to mark areas of wet-loving plants, "warm" colors (yellow, pink) are used to mark areas of dry-loving plants.

In cases where vegetation communities form mixed structures (types), strip coloring or raster and color techniques are applied together. Rare plant habitat sites, small areas of concentrated species distribution and taxonomically important plant sites generally are mapped using the sign method.

The spatial structure of the vegetation is mapped using identical methods, and the methods of cartograms and cartographies are used to quantify the spatial structure, giving the absolute values of rare plants in the standard area (cartodiagrams) or the percentage structure of plant species (choroplets, cartograms). Various methods of vegetation mapping have been used in the Atlas of Forests of the USSR [11]. A detailed analysis of vegetation mapping methods for Italian territory has been performed by Franco Pedroti [12].

An important aspect is the generalization of vegetation maps. It is performed by grouping and connecting small territorial areas into higher-ranking larger formations. Particular attention is appoint for latitudinal and vertical differentiation and spatial spread of vegetation type. For compiling of vegetation maps are distinguished geographical variations of plant species, plant species adapted to grow at different heights or characterized by differences in seed and fruit abundance. Grouping of taxonomic areas is performed by combining homogeneous vegetation types into more complex spatial formation. This also determines the mapping methods: the areas of vegetation types mapped by choropleths, isopleths and habitat methods on large-scale maps and by non-scale signs on small-scale maps. Landscape complexes or natural regions are mapped on the smallest scale

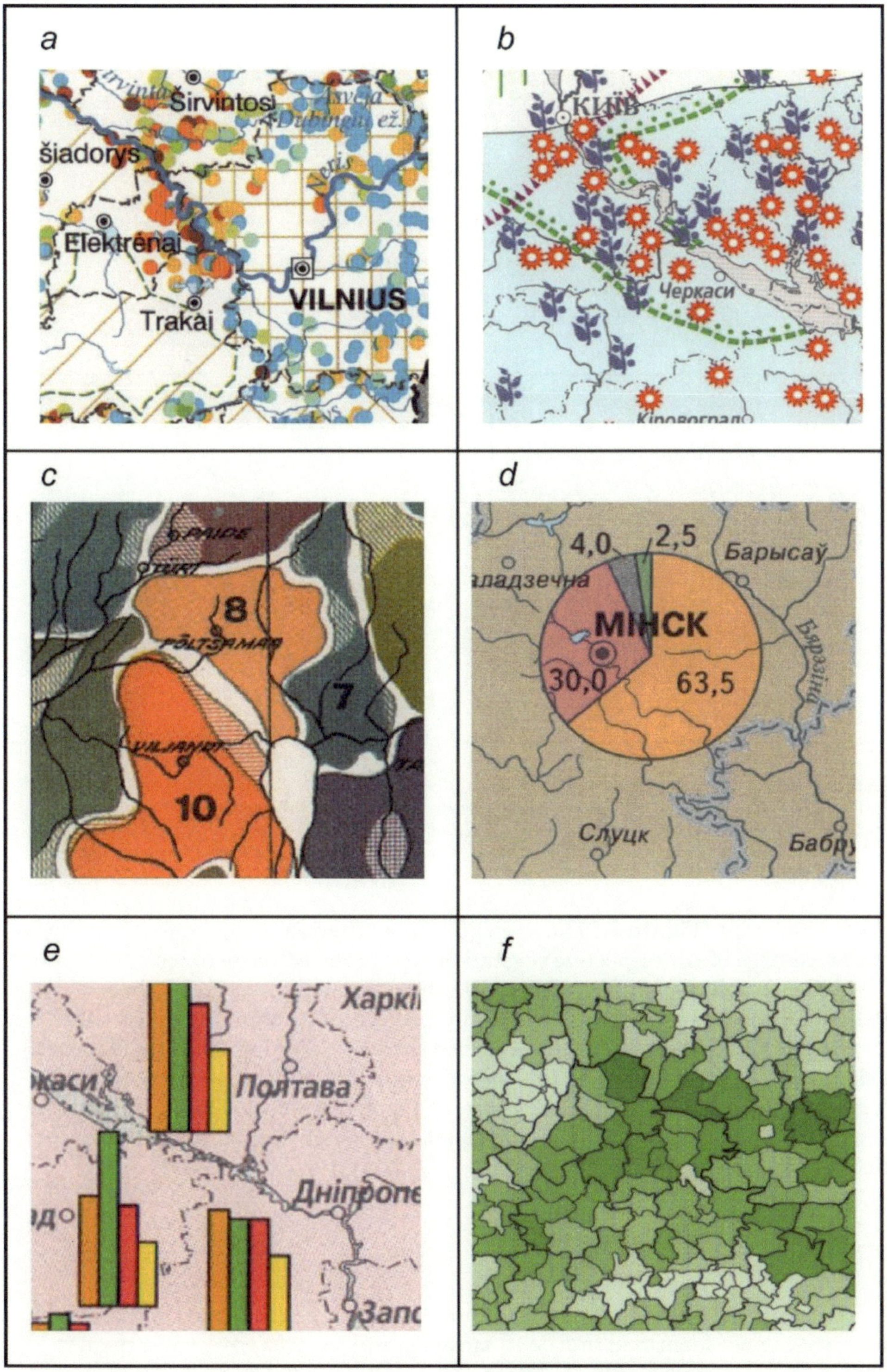

Figure 2.
Examples of plant mapping methods: a – Dots (fragment of natural meadows and pastures map from National Atlas of Lithuania. 2017), b – Dots and areas (fragment of Areals of herbal plants map from National Atlas of Ukraine, 2008), c – Qualitative background (fragment of regions of vegetation map from [5]), d – Cartodiagram (fragment of afforestation map from [6]), e – Cartodiagram (fragment of forests age map from National Atlas of Ukraine, 2008), f – Cartogram (isopleths) (fragment of forestation map from [7]).

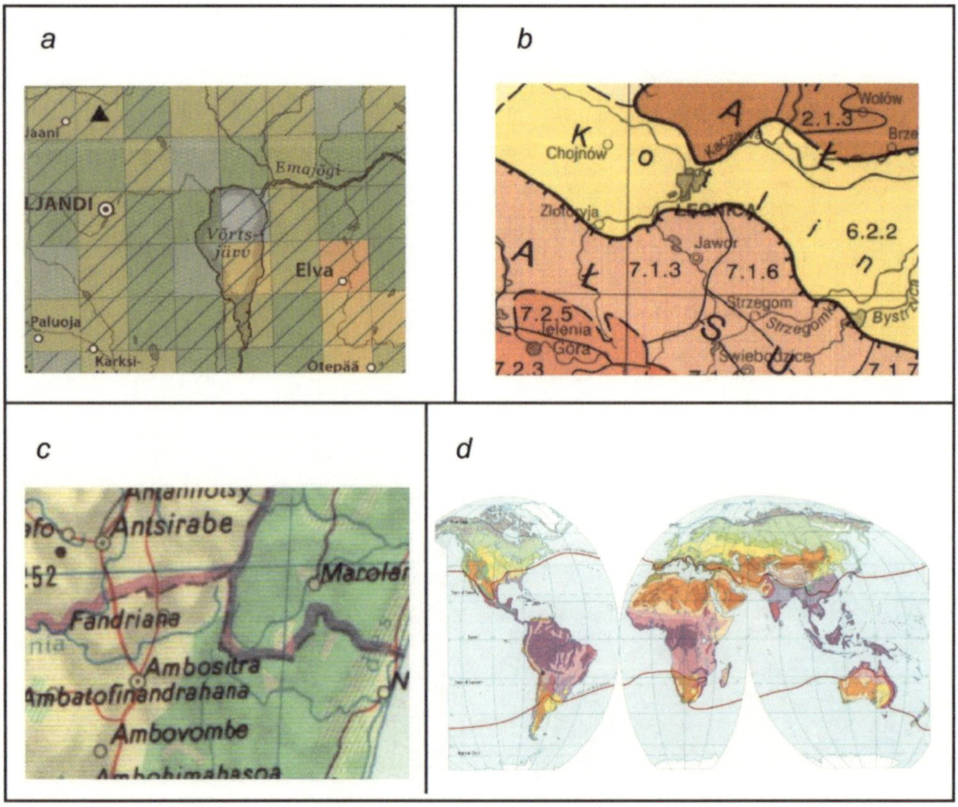

Figure 3.
Examples of plant mapping methods: a – Formal cartogram (isopleth) (fragment of Flora map from [5]), b – Qualitative background (fragment of Geobotanical division of Silesia map from [8]), c – Areas (fragment of Madagascar general map from Haack [9]), d – Qualitative background and areas (fragment natural vegetation map from [10]).

vegetation maps (**Figure 1**). The main grouping criterion in the vegetation maps is the percentage of the respective vegetation type in the whole mapped area.

In cartographic generalization process small vegetation types and their individual habitats sites are combined into complex formations or completely eliminated. The area of the complex carved surface is dominated by small mosaic habitats of vegetation types. In this case, the generalization of vegetation types is carried out on the basis of the diversity of relief complexes. This means that vegetation types are combined into complexes based on orographic criteria rather than floristic ones.

4. Sources of information

Classical floristic maps are compiled on the basis of field research, where typical plant species are identified in 10 × 10 km areas and the area covered by them is drawn. The boundaries of the habitats sites are determined on the basis of the distribution of the respective soils and the geomorphological features of the relief. A unified transverse cylindrical Mercator projection is often used to construct global floristic maps.

Floristic information is collected through a terrestrial plant distribution inventory and the use of multispectral aero- and space images. Its essence: a well-studied area in terms of vegetation species is photographed, followed by a floristic interpretation of the studied areas based on external identification features (color, color saturation, contour image). Interpretation of multispectral aero- and spatial images allows the identification of vegetation types areas, but the identification of individual plant species is complicated and unreliable. It is particularly difficult to distinguish individual tree species in multi-species tropical and equatorial forests.

At the beginning of the 21st century, unmanned aerial vehicles (UAV) were widely used to assess the prevalence of relatively small plant species areas. They can be used to obtain chromatographic and multispectral large-scale images in which species identification is simpler and more reliable. Aerial photographs of UAV are most widely used in forest management, assessing the amount of wood in forests, the ecological condition of trees.

5. Methodological features of UAV aerial photography

The application of UAV images to vegetation mapping is carried out in several directions:

1. Assessment of plant species composition.

2. Assessment of the ecological status of plants.

3. To assess the amount of wood in growing forests.

1. Mapping of plant species composition (individual tree species) may be performed only in large (M 1: 10 000) or very large (M 1: 5 000) scales. Alternatively, when the scales of the maps are smaller, the mapped area is divided into phytocenotic combinations areas. The boundaries of these combinations are mapped on a map. In this case, the map resembles a "frock - coat", where each patch is a territorial combination of phytocenosis. Maps with a scale of 1:25,000 and smaller can be used to phytocenoses contours mapping, the target sizes of which are measured in hundreds of meters or kilometers. Where phytocenosis contours are smaller (several tens of meters), it can be mapped only at the scale of topographic plans: 1: 1,000–1: 5,000.

2. An accurate indicator of wood volume calculation is necessary for the assessment of the ecological condition of trees [13]. It is related to the diameter at breast height (DBH) parameter of the tree trunk and the total tree height (H) [14–17]. Many studies have shown that biomass models with DBH and H parameters are more reliable in determining tree biomass and assessing its viability [18–21].

3. Most common ground-based method is clinometric measurements that does not require cut off of tree. This method has one drawback: it is necessary to see simultaneously the base of the trunk and its top during the measurement. In boreal forests such a requirement is difficult to realize, and in equatorial forests it is practically unenforceable. In addition, the accuracy of over-ground trees height measurements deteriorates with increasing tree height [22].

Tree height is one of the most important indicators for evaluation of forest wood resource. Along with other key characteristics (diameter at chest height (DBH), tree species), tree height allows evaluate other important indicators: tree age, wood amount, biomass. In this case, the accuracy of the measurements becomes a very important factor. For the assessment of wood amount, the accuracy index becomes as important or even more important than the determination of tree species [23–25].

Measurement accuracy is enhanced by laser scanning (LiDAR) based on active sensing technology. This method has several advantages [26]:

1. A direct 3D image is obtained where measurements can be made with centimeter accuracy.

2. Determine the real surface on which the forest grows and thus adjust the height of the trees.

3. Application of automated calculation methods to facilitate large area surveys.

On the other hand, laser scanning equipment is quite heavy, so the possibilities of such a method for UAV are limited.

6. Methodology for applying UAV aerial photographs to assess the ecological condition of trees and wood amount

Various methodologies are currently used to assess the ecological condition of trees and wood amount. Their application possibilities are determined by the available equipment and its price: UAV and its carrying capacity, camera capabilities, software used for aerial images analysis. The methodology for assessing the ecological condition and trees wood amount developed at the Institute of Geosciences of Vilnius University. It is based on the use of widely available and relatively inexpensive equipment.

Aerial images obtained with the UAV DJI Inspire 1, with an RGB Zenmus X3 camera, are used to calculate the wood volume. The technical parameters of equipment are given in **Table 1**.

UAV orthophoto and perspective images were used to calculate wood amount. In the overlapping images, the crowns of individual trees were separated and the height of the trees was determined using photogrammetric methods. For more accurate calculation of the scale of photo images, Ground control points (GCP) were used. The position of GCPs were additionally measured using a GPS device Trimble R4. Based on the coordinates of the GCPs, the whole 3 D tree crown model was refined. Pix4D Mapper photogrammetric software was used to calculate the height and volume of tree trunks [27].

The crown of a particular tree was treated as a regular cone that could be divided into plan and profile segments (**Figure 4**). Knowing the area S_1, it is possible to calculate the diameter of the base of the cone of a tree crown (formulas (1) and (2)):

$$R = \sqrt{\frac{S_1}{\pi}} \tag{1}$$

$$D = 2R \tag{2}$$

where S_1 is the base area of the tree crown (m2), R is the radius of the cone (m), D is the diameter of the cone (m).

When the area of the crown cone S_2 is known, the height of the cone can be calculated (formula (3)):

$$L = \frac{S_2}{R} \tag{3}$$

where L is the height of the crown cone (m), S_2 is the area of the crown cone (m^2), R is the calculated radius of the base of the cone (m).

The obtained indicators allow estimate the volume of the tree crown (the total amount of green tree mass) (formula (4)):

$$V = \frac{1}{3}\pi R^2 L \tag{4}$$

Hovering accuracy (GPS mode)	Vertical: 0.5 m; Horizontal: 2.5 m
Max angular velocity	Pitch: 300°/s; Yaw: 150°/s
Max tilt angle	35°
Max ascent and descent speed	5 m/s; 4 m/s
Max speed	22 m/s
Max wind speed resistance	10 m/s
Max service ceiling above take-off point	120 m
Type and model	X3; FC350
Total and effective pixels	12.76 M; 12.4 M
Max capacity	64 GB
Maximal image size	4000x3000
ISO Range	Photo - 100-1600; Video - 100-3200.
The electronic shutter speed	8 s - 1/8000 s
A field of view (FOV)	94°
Supported file formats	Photo: JPEG, DNG; Video: MP4/MOV (MPEG-4 AVC/H.264)

Table 1.
The specifications of UAV DJI Inspire1 and RGB Zenmus X3 camera.

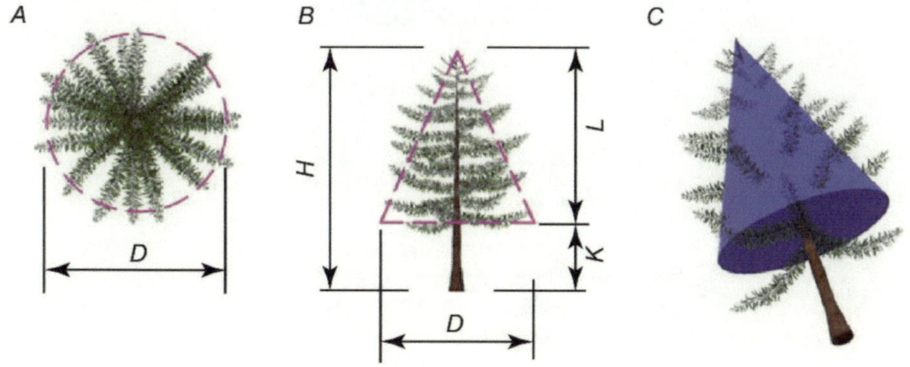

Figure 4.
Quantitative indicators of the tree crown: A - top view, B - side view, C - crown cone 3D.

where V is the volume of the crown cone (m^3), R is the calculated radius of the base of the cone (m), L is the calculated height of the crown cone (m).

The photogrammetric methods allow determined tree height and calculate the height of the tree trunk (K) up to the base of the tree crown.

$$K = H - L \quad (5)$$

where K is the height of the trunk to the base of the crown (m), H is the height of the tree measured photogrammetric methods (m), L is the calculated height of the crown cone (m).

The calculation of tree trunk volume was performed analogously to the calculation of tree crown volume (formulas (1)–(4)). In this case, a very important condition is to measure as accurately as possible to the ground.

Crown areas S_1 and S_2 are calculated using the specialized software Pixel Color Analysis (PixRGB) (**Figures 5** and **6**), which also allows to calculate the trunk thickness. The total crown volume is calculated together with the part of the trunk in the crown, so for more accurate estimation, the trunk volume is eliminated from the calculated cone volume.

To reduce the number of analyzed pixels, only parts of the analyzed tree and GCPs are distinguished in the original photo images (**Figure 7**). The original scaling and resolution of the images are not changed during the calculations. This means that the pixels in all the selected parts will be the same size and will correspond to the scaled images adjusted with the help of GCPs. The software (PixRGB) allows automatically determine the size of one pixel area and calculate the areas of the extracted parts.

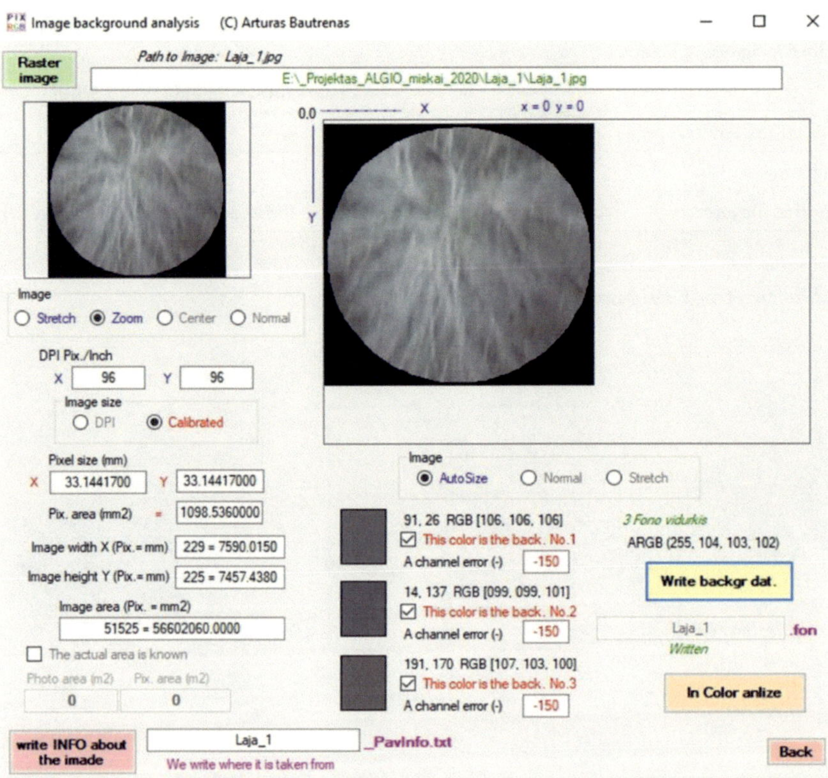

Figure 5.
The window of tree crown element analysis (software PixRGB).

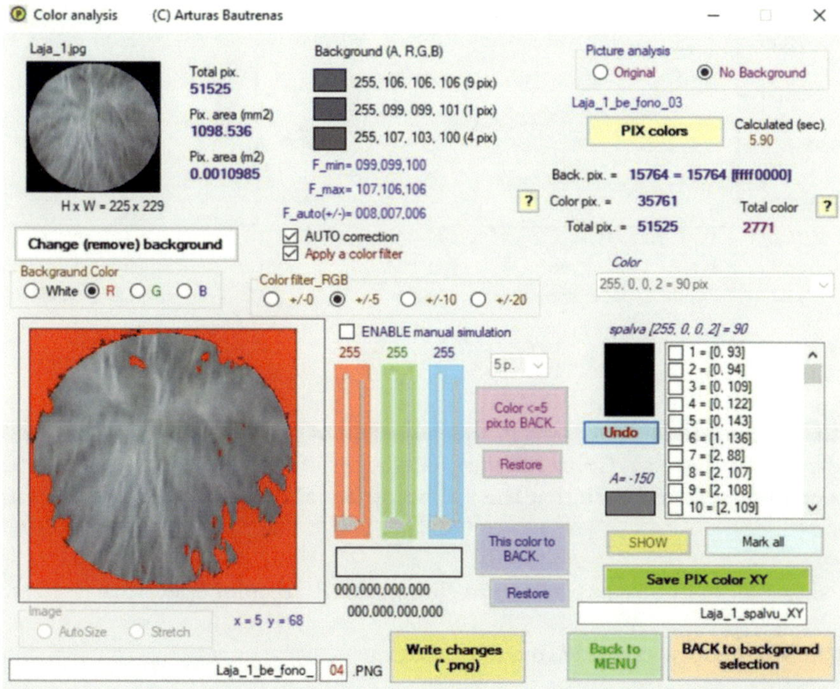

Figure 6.
The window of color analysis (software PixRGB).

Figure 7.
Parts or photo-images: A - side view, B - top view, C – control, D – subtracted side view, where pixels in red marked rejected as background, D – top view, where pixels in red marked rejected as background.

Several levels of iteration were applied to color analysis. Repeated iteration showed that a satisfactory accuracy result was achieved after the third level of iteration (**Figure 8**).

The extraction of background pixels and their elimination provides information about the area covered by the crown of a particular tree. This indicator is related to the ecological condition of the tree: the denser the crown shown the better the condition of the tree. Such a methodology allows with sufficient accuracy assess the condition of forest seedlings (very young trees), which is important from the forest management point of view.

Elimination of background pixels also allows highlight the trees crown color and connected it with the trees species composition. Studies have shown that the crown colors of deciduous and coniferous trees in the green spectrum range vary depending on the species composition.

Spectral transformation of the photo-images was performed with Adobe Photoshop software by Mode, Adjustments, Hue/Saturation, Color Balance, Brightness/Contrast, Selective Color, and Channel Mixer tools. The Mode tool changed the color separation model. The most suitable for UAV images analysis

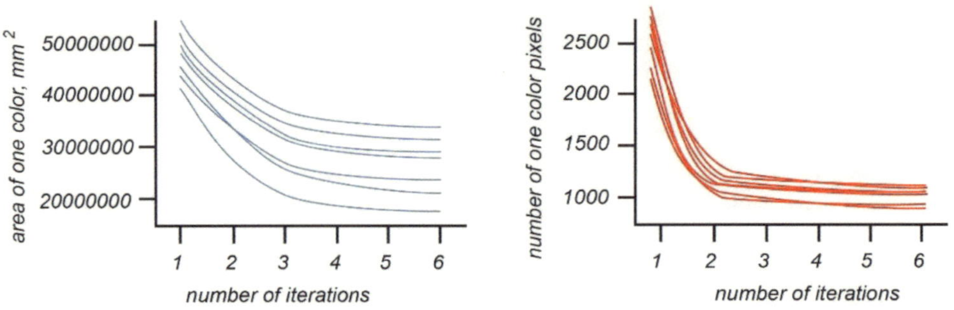

Figure 8.
Effect of repeatedly iterations.

is the RGB color model. The tools in the Adjustment group allowed to change the Color Intensity (Levels), Color Balance, Brightness/Contrast, Hue/Saturation.

Experiments have shown that the following aerial image transformations are necessary:

1. Using Adobe Photoshop software changes to RGB color spectrum indicators.

2. The Mode tool sets the RGB color model.

3. Using Adjustments toolbox will made following changes:

    - The Input Levels tool in the (Levels group) changed the whole RGB color spectrum. The Output Levels tool narrows the RGB colors spectrum.

    - The Hue/Saturation tool sets the final RGB color model.

In the transformed aerial photographs, calculations of the area covered by young tree crowns were performed using AutoCAD software and created investigated forest plot database.

For automated calculation of young trees areas, a PixRGB software has been identified the area of the same color pixels in aerial images. The software is based on the comparison of young tree crown area calculations in AutoCAD software and one color pixel area measurements. In the initial stage, aerial images are transformed to the exact size of the photographed area. Transformations were performed with an error of less than 2–3 cm.

The transformation of the areal image color spectrum allowed to concentrate the image color of young trees in a relatively narrow color range. In addition, an error of +/– 5 color coordinates was allowed, the magnitude of which was determined after testing several analysis variants. After the analysis, pixels of different shades were distinguished, eliminating pixels of a specific color, the location of which obviously does not correspond to the analyzed areas. The remaining color pixels formed the total area covered by all young tree crowns.

Comparing the calculations of the area covered by young tree crowns performed with AutoCAD software and applying the automatic analysis of pixel areas of the defined color, it was found that the error rate does not exceed 3%.

Investigations, which performed in 2019–2020 to assess the ecological condition of trees and the amount of wood using UAV INSPIRE 1 and the PixRGB software showed the effectiveness of the applied methodology. The species composition of the trees was assessed with the help of aerial images taken during the summer, and the amount of wood was assessed with the help of images taken during the autumn.

7. Discussion

The geobotanical map reflects the heterogeneity space of vegetation cover and is the basis for geobotanical zoning. Hierarchical vegetation cover classification is most commonly used when developing vegetation cover maps. The main element of the content of geobotanical maps is a set of autotrophic plants, which is typical of the habitats of the area. Vegetation mapping in classical ways is based on a chain of corresponding actions: the area is segmented into phytocenoses and their combinations; their boundaries are fixed in the area; based on the position of the topographic objects, the boundaries are transferred to the working map; creation of the final map version. The maps created in this way are reminiscent of a “patch wrap” in which elemental phytocenotic territorial structures are distinguished. The detail content of a map, which created in this way will largely depend on the scale of the map. Detailed elemental phytocenoses mapping is possible only on a very large scale (1: 5,000).

When creating smaller-scale (1: 10,000 or 1: 50,000) maps, the primary image must be generalized by combining elemental phytocenoses into microcomplexes and mesocomplexes. Their boundaries are usually fixed on the basis of the boundaries of the topographic objects of the area; terrain lines, hydrographic network lines, roads, settlement boundaries and the like. In small-scale maps, the degree of content generalization is even higher, so that vegetation meso-complexes are joined into macro-complexes or mega-complexes. A very wide range of plant communities (phytocenoses and their meso-combinations) are found in each of these contours. The legend described only the most common plant communities for that contour. It should be noted that such a map is only approximate characterized to the vegetation area, and the individuality of small areas in such maps completely disappears.

In this case, various combinations of mapping techniques are used to highlight the unique individual characteristics of the vegetation. Such methods are widely used in vegetation maps of small-scale thematic atlases and national atlases ([8, 28], Atlas Republiky [6, 29–31]). In many cases, synthetic derivatives are also used to map the productivity of natural vegetation, wood growth in forests, and so on.

Remote vegetation mapping poses other problems: satellite image quality, cloud areas covering the surface image, accurate identification of vegetation types and species, UAV aerial imagery for small areas only, accuracy and reliability of extrapolation of data collected in Ref. areas. These problems are presented in numerous publications on land cover structure and vegetation area analysis [32], forest inventory and timbering volume estimation [33], increment of adjustment of wood volume [34], accuracy and reliability of laser scanning models for forest wooden evaluation [35], complex application of UAV aerial images and ground-based measurement methods [26, 36], assessment of underwood influence on UAV aerial images [37], for the application of high-resolution space images to the mapping of grassland plant communities [38].

All the mentioned research directions are focused on the accuracy of measurements and data, lower time costs. The methodology of tree ecological condition and wood volume assessment proposed by the authors is based on the use of medium-class UAVs and the application of widely available equipment for image analysis. In addition, the presented methodology is suitable for estimating the amount of wood growth at all seasons of the year. Images taken in spring and autumn have the highest accuracy in identifying tree species.

Of course, the application of remote sensing methods is very effective, but the reliability of the identification of individual areas of the space images that are used for the analysis of land use/land cover data is a problem. Faster use of remote sensing data and methods helps to make the process of spatial analysis faster and more powerful, but the increased complexity of the methods also creates the potential for higher errors.

8. Conclusions

Mapping of individual plant groups and vegetation cover has old traditions. Forest distribution was mapped in the Finnish National Atlas. This atlas is the first national atlas in the World, published in 1899 (Atlas [39]). Vegetation maps published in the some national atlases of the countries in the first half of the 20th century: Czechoslovakia (Atlas Republiky [29]), France (Atlas [40]), Germany [41], Great Britain [42], Romania [43]; Argentina [44], Mexico [45], Brazil [46] and others). These maps provide floristic and phytocenotic information.

Subsequent national and regional atlases and specialized thematic atlases have mapped the distribution of individual vegetation species, usually rare and protected. Traditional habitat and sign methods were used to map them.

In the second half of the 20th century, monochromatic (black - white) and panchromatic aerial images were used for floristic mapping, and from the 1980s, were also used space images. Landsat, SPOT, Terra (EOS AM-1), Aqua (PM-1), IKONOS, QuikBird satellite sensors were introduced for the mapping of vegetation cover developed in the eighties, with the help of which high-resolution images were created. Recently, images that can determine the normalized difference vegetation index (NDVI) have been widely used for phytocenotic mapping of vegetation. This is a convenient way to map the phytocenotic state of vegetation, but has its limitations. Restrictions relate to classifications of land cover, heterogeneity of floristic or whole vegetation classifications [47–52].

UAVs are used for floristic and phytocenotic mapping of small vegetation areas. The possibilities of their use were expanded by panchromatic images, which allow to identify individual plant species (mostly trees), to evaluate the phases of plant phenological process and to evaluate the amount of wood in forests by photogrammetric methods.

Author details

Algimantas Česnulevičius*, Artūras Bautrėnas, Linas Bevainis and Donatas Ovodas
Vilnius University, Faculty of Chemistry and Geosciences, Institute of Geosciences, Lithuania

*Address all correspondence to: algimantas.cesnulevicius@gf.vu.lt

References

[1] Lentz S., Moser J. 2013. The national Atlas of Germany – knowledge transfer via maps. *Kartographische Nachrichten*. 63. 148 – 153. https://link.springer.com/article/10.1007/BF03546127

[2] Omerling J. 1995. New forms, contents, and structures for european national atlases. Cartographic Perspectives. 20. 12-20. https://doi.org/10.14714/CP20.890

[3] Ostrowski W., Ostrowski J. 2014. A comparison of the form and thematic structure of selected European national atlases. Polish Cartographical Review. 45 (1). 15-33. https://journals.indexcopernicus.com/search/article?articleId=667899.

[4] Polakova N.O. 2010. Model of the semiotic-thematic structure of the National Atlas of Ukraine. Vistnyk Heodezji ta Kartohrafii. 6. 19-22.

[5] The National Atlas of Estonia. 2019. Tartu: Tartu University. 424 p. https://pood.regio.ee/en/national-atlas-estonia.

[6] National Atlas of Belarus. 2002. Minsk: Belarussian state commity of Land Resurses, Geodesy and Cartography. 292 p. https://www.worldcat.org/title/natsyianalny-atlas-belarusi/oclc/53461332.

[7] Statistical Atlas of Poland. 2018. Warsaw: GUS. 216 p. https://stat.gov.pl/en/topics/other-studies/other-aggregated-studies/statistical-atlas-of-poland,17,1.html.

[8] Atlas of Lower Silesia and Opole. 2008. Second edition. Wrocław: Wrocław University and Polish Academy of Sciences. 131 p. https://katalog.ue.wroc.pl/162000387855/dokumenty-kartograficzne/atlas-slaska-dolnego-i-opolskiego.

[9] Haack Weltatlas. 1982. Gotha: VEB Herman Haack Geographisch – Kartographische Anstalt. 408 p. https://www.zvab.com/servlet/BookDetailsPL?bi=15527821807&tab=1&searchurl=hl%3Don%26kn%3Dhaack%2Bweltatlas%26sortby%3D20.

[10] Philip's Great World Atlas. 2001. London: George Philip Limited. 184 p. https://www.thebooktique.org/product/5331/Philips-Great-World-Atlas-Books.

[11] Atlas of Forests of the USSR. 1973. Moscow: State Board of Geodesy and Cartography. 222 p. https://daac.ornl.gov/cgi-bin/dsviewer.pl?ds_id=692.

[12] Pedroti F. Plant and Vegetation Mapping. 2013. Springer: Heidelberg – New York – Dordrecht – London. 294 p. https://doi.101007/978-3-642-30235-0.

[13] Konópka, B.; Pajtík, J.; Noguchi, K.; Lukac, M. Replacing Norway spruce with European beech: A comparison of biomass and net primary production patterns in young stands. For. Ecol. Manag. 2013. 302. 185-192. https://doi:10.3390/f11020202.

[14] Sileshi, W.G. A critical review of forest biomass estimation models, common mistakes and corrective measures. For. Ecol. Manag. 2014. 329. 237-254. https://dx.doi.org/10.1016/j.foreco.2014.06.026.

[15] Temesgen, H.; Affleck, D.; Poudel, K.; Gray, A.; Sessions, J. A review of the challenges and opportunities in estimating above ground forest biomass using tree-level models. Scand. J. For. Res. 2015. 30. 326-335. https://doi.10.1080/02827581.2015.1012114.

[16] Weiskittel, A.R.; MacFarlane, D.W.; Radtke, P.J.; Affleck, D.L.R.; Temesgen, H.; Woodall, C.W.; Westfall, J.A.; Coulston, J.W. A Call to Improve Methods for Estimating Tree Biomass

for Regional and National Assessments. J. For. 2015. 113. 414-424. https://doi.org/10.5849/jof14-091.

[17] Yuen, J.Q.; Fung, T.; Ziegler, A.D. Review of allometric equations for major land covers in SE Asia: Uncertainty and implications for above- and below-ground carbon estimates. For. Ecol. Manag. 2016, 360, 323-340. https://doi.org/101016/j.foreco.2015.09.016.

[18] Bi, H.; Murphy, S.; Volkova, L.; Weston, C.; Fairman, T.; Li, Y.; Law, R.; Norris, J.; Lei, X.; Caccamo, G. Additive biomass equations based on complete weighing of sample trees for open eucalypt forest species in south-eastern Australia. For. Ecol. Manag. 2015, 349, 106-121. http://dx.doi.org/10.1016/j.foreco.2015.03.007.

[19] Dong, L.; Zhang, L.; Li, F. Additive biomass equations based on different dendrometric variables for two dominant species (*Larix gmelini* Rupr. and *Betula platyphylla* Suk.) in natural forests in the Eastern Daxing'an Mountains, Northeast China. Forests 2018, 9, 261-284. https://doi:10.3390/f9050261.

[20] Kralicek, K.; Huy, B.; Poudel, K.P.; Temesgen, H.; Salas, C. Simultaneous estimation of above- and belowground biomass in tropical forests of Vietnam. For. Ecol. Manag. 2017. 390. 147-156. https://doi.org/10.1016/j.foreco.2017.01.030.

[21] Meng, S.; Liu, Q.; Zhou, G.; Jia, Q.; Zhuang, H.; Zhou, H. Aboveground tree additive biomass equations for two dominant deciduous tree species in Daxing'anling, northernmost China. J. For. Res. 2017. 22. 233-240. https://doi.org/10.1080/13416979.2017.1333277.

[22] da Silva, G.F., Curto, R.D.A., Soares, C.P.B., Piassi, L. de C., Evaluation of height measurement methods in natural forests. Rev. Arvore. 2012. 36, 341-348. https://doi.org/10.1590/S0100-67622012000200015.

[23] Kearsley, E., de Haulleville, T., Hufkens, K., Kidimbu, A., Toirambe, B., Baert, G., Huygens, D., Kebede, Y., Defourny, P., Bogaert, J., Beeckman, H., Steppe, K., Boeckx, P., Verbeeck, H. Conventional tree height–diameter relationships significantly overestimate aboveground carbon stocks in the Central Congo Basin. Nat. Commun. 2013. 4. 2269. https://doi.org/10.1038/ncomms3269.

[24] Luoma, V., Saarinen, N., Wulder, M.A., White, J.C., Vastaranta, M., Holopainen, M., Hyyppa, J., 2017. Assessing precision in conventional field measurements of individual tree attributes. Forests. 2017. 8. 38. https://doi.org/10.3390/f8020038.

[25] Tompalski, P., Coops, N.C., White, J.C., Wulder, M.A.Simulating the impacts of error in species and height upon tree volume derived from airborne laser scanning data. For. Ecol. Manag. 2014. 327. 167-177. https://doi.org/10.1016/j.foreco.2014.05. 011.

[26] Wang Y., Lehtomäki M., Liang X., Pyöorälä J., Kukko A., Jaakkola A., Liu J., Feng Z., Chen R., Hyyppä J. Is field-measured tree height as reliable as believed – A comparison study of tree height estimates from field measurement, airborne laser scanning and terrestrial laser scanning in a boreal forest. ISPRS Journal of Photo grammetry and Remote Sensing. 2019. 147. 132-145. https://doi.org/10.1016/j.isprsjprs.2018.11.008.

[27] Pix4D mapper 2.1 manual. 2019. https://support.pix4d.com/hc/en-us/articles/204272989-Offline-Getting-Started-and-Manual-pdf

[28] Atlas of the Republic of Poland. 1993 – 1997. Warszawa. 624 p.

https://www.igipz.pan.pl/zsigik-projects-atlas-republic-poland.html.

[29] Atlas Republiky Ćeskoslovenske. 1935. Praha: Ćeske Akademie Ved a Timeni, 110 p. https://www.atlaseum.com/atlases/atlas-republiky-ceskoslovenske-1935.

[30] National Atlas of Lithuania. 2014. Vol. 1. Vilnius: National Land Service and Vilnius University. 142 p. https://www.geoportal.lt/geoportal/web/en/the-national-atlas-of-the-republic-of-lithuania#savedSearchId={7D5AA860-4893-4D9B-965D-D030A0EE0277}&collapsed=true.

[31] National Atlas of Ukraine. 2007. Kiev: State Scientific – Production Enterprise„Cartograhy". 450 p. http://wdc.org.ua/en/services/national-atlas-ukraine.

[32] Rwanga S.S., Ndambuki J.M. 2017. Accuracy Assessment of Land Use/Land Cover Classification Using Remote Sensing and GIS. **International Journal of Geosciences.** 8 (4). 611-622. https://doi.org.10.4236/ijg.2017.84033.

[33] Hou, Z.; Xu, Q.; Vauhkonen, J.; Maltamo, M.; Tokola, T. 2016. Species-specific combination and calibration between area-based and tree-based diameter distributions using airborne laser scanning. Canadian Journal of Forest Research. *46*, 753-765, https://doi.org.10.1139/cjfr-2016-0032.

[34] Jurjević L., Liang X., Gašparović M., Balenović I. 2020. Is field-measured tree height as reliable as believed – Part II, A comparison study of tree height estimates from conventional field measurement and low-cost close-range remote sensing in a deciduous forest. ISPRS Journal of Photogrammetry and Remote Sensing 169. (2020) 227-241. https://doi.org/10.1016/j.isprsjprs.2020.09.014

[35] Kotivuori E., Korhonen L., Packalen P. 2016. Nationwide airborne laser scanning based models for volume, biomass and dominant height in Finland. Silva Fennica vol. 50 no. 4 article id 1567. 1 - 28 p. http://dx.doi.org/10.14214/sf.1567.

[36] Yua X., Hyyppäa J., Vastaranta M., Holopainen M., Viitala R. 2011. Predicting individual tree attributes from airborne laser point clouds based on the random forests technique. ISPRS Journal of Photogrammetry and Remote Sensing 66 (2011) 28-37. http://doi:10.1016/j.isprsjprs.2010.08.003

[37] Yao W., Krull J., Krzystek P., Heurich M. 2014. Sensitivity Analysis of 3D Individual Tree Detection from LiDAR Point Clouds of Temperate Forests. *Forests* 2014, *5*, 1122-1142; http://doi:10.3390/f5061122.

[38] Jędrych M., Zagajewski B., Marcinkowska-Ochtyra A. 2017. Application of Sentinel-2 and EnMAP new satellite data to the mapping of alpine vegetation of the Karkonosze Mountains. Polish Cartographical Review. Vol. 49, 2017, no. 3, pp. 107-119. https://doi.org/10.1515/pcr-2017-0011.

[39] Atlas de Finlande. 1899. Helsingfors: Société de Géographie de Finlande. 19 p. https://www.doria.fi/bitstream/handle/10024/101486/Atlas_de_Finlande_texte_1.pdf?sequence=1&isAllowed=y.

[40] Atlas de France. 1931. Paris: Comite National de Geographie. 82 p. https://www.abebooks.fr/edition-originale/ATLAS-FRANCE-COMITE-NATIONAL-GEOGRAPHIE-Editions/22635982179/bd.

[41] Atlas der Deutschen Volkskunde. 1929. Marburg: N.G.Elwet. 254 p. https://ixtheo.de/Record/107242990X.

[42] The National Atlas. 1942. London: Ordnance Survey. 258 p. https://books.google.lt/books/about/Great_Britain.

html?id=rT_8wgEACAAJ&hl=en&output=html_text&redir_esc=y.

[43] Atlas Fisiografic si Statistic al Romaniei. 1930. Bucureşti: Institutul Geological Romaniei. https://rcin.org.pl/Content/40655/WA51_50350_r1964-z1_Dokumentacja-Geogr.pdf.

[44] Atlas geografico de la Republica Argentina. 1940. Buenos Aires: Ediciones Peuser, 21 p. https://articulo.mercadolibre.com.ar/MLA-810873974-atlas-de-la-republica-argentina-_JM#position=7&type=item&tracking_id=37507c6c-1217-434a-a680-a5f80c899da8.

[45] Atlas geográfico de los Estados Unidos Mexicanos. 1942. Mexico: Dirección De Geografía, Meteorología e Hidrología. https://catalog.hathitrust.org/Record/001272382.

[46] Atlas Estatistico do Brasil. 1941. Rio de Janeiro: Carlos Augusto Ribeiro Campos. 112 p. http://memoria.org.br/pub/meb000000419/pequenoatlascafe5/pequenoatlascafe5.pdf.

[47] Dansereau P. Vegetation study and recording. Proc Ninth Pacif. Sci Congr. 1962. 4, 64-75. https://catalog.hathitrust.org/Record/007185524.

[48] Degraaf R.M, Chadwick N.L. Habitat classification: a comparison using avian species and guilds. Environ Manage. 1984. 8, 511-8. file:///F:/1%20uzduotis/Degraaf-Chadwick1984_Article_HabitatClassificationAComparis.pdf.

[49] Harms K.E., Condit R., Hubbell S.P., Foster R.B. Habitat associations of trees and shrubs in a 50-ha neotropical forest plot. J Ecol. 2001. 89. 947-59. https://doi.org/10.1111/j.1365-2745.2001.00615.x.

[50] Salovaara K.J, Thessler S. Riffat N. M., Tuomisto H. Classification of Amazonian primary rain forest vegetation using Landsat ETM+ satellite imagery. Remote Sens Environ. 2005. 97. 39-51. https://doi:10.1016/j.rse.2005.04.013.

[51] Thenkabail P.S, Hall J., Lin T., Ashton M.S, Haris D., Enclona E.A. Detecting floristic structure and pattern across topographic and moisture gradients in a mixed species Central African forest using IKONOS and Landsat-7 ETM+ images. Int J Appl Earth Obs Geoinf. 2003. 4. 255-70. https://doi./10.1016/S0303-2434(03)0000-0.

[52] Xie Y., Sha Z., Yu M. Remote sensing imagery in vegetation mapping: a review. Journal of Plant Ecology. 2008. 1 (1). 9 – 23. https://doi/10.1093/jpe/rtm00.

Chapter 4

Assessment of the State of Forest Plant Communities of Scots Pine (*Pinus sylvestris* L.) in the Conditions of Urban Ecosystems

Elena Runova, Vera Savchenkova,
Ekaterina Demina-Moskovskaya and Anastasia Baranenkova

Abstract

Siberian cities are characterized by one feature: many of them have preserved natural woodlands during construction, which on the one hand give a completely unusual, unique appearance to cities, on the other hand, trees suffer from recreational load, high levels of pollution and other anthropogenic factors. To assess the condition of pine stands, 3 test areas (0.5 ha, 0.1 ha and 1.9 ha) were laid. All considered plantings of natural origin are areas of woodland that were preserved during the construction of the city and are subject to recreational and industrial pollution. The test sites belong to areas with a high anthropogenic load, as they are located along highways and in close proximity to residential and public buildings and are part of parks with a high recreational load. The average age of trees is 70–80 years. The sanitary condition of the massif and its landscape characteristics are also determined. The critical condition of the massif is established, requiring sanitary logging and other forestry measures that could reduce recreational and anthropogenic loads.

Keywords: Scots pine (*Pinus sylvestris* L.), dendrometric characteristics, sanitary condition, damage to the trunk and crown, care

1. Introduction

Place of research: Irkutsk region, Bratsk city, residential district Energetik. On the territory of Bratsk and the Bratsk region, southern taiga and taiga natural complexes of Central Siberia prevail [1, 2]. In the vegetation of the city territory, forests of natural origin and urban plantations stand out. The dominant breed in forests of natural origin is Scots pine (*Pinus sylvestris* L.) - 57% of the total composition of the woodland, hanging birch (*Betula pendula* Roth.) And fluffy birch (*Betula pubescens* Ehrh.) Make up 17%, Siberian larch (*Larix sibirica* Ledeb.) - 6%, aspen (*Populus tremula* L.) - 16%, in much smaller quantities there are ordinary spruce (*Picea abies* L.), Siberian spruce (*Picea obovata* Ledeb.), goat willow (*Salix caprea* L.), shrub alder (*Duschekia fruticosa* Rupr.), common grouse (*Sorbus aucuparia* L.) and Siberian grouse (*Sorbus sibirica* Hedl.).

Inner-city vegetation is artificially created communities that are not self-regulating systems, they need constant care, which in most cases they do not receive. The predominant breed in urban planting is balsamic poplar (*Populus balsamifera* L.), In much smaller quantities are represented hanging birch (*Betula pendula* L.) and fluffy birch (*Betula pubescens* Ehrh.) - 11%, tree-shaped caragana (*Caragana arborescens* Lam.) - 6%, squat elm (*Ulmus pumila* L.) - 4%, Siberian mountain ash (*Sorbus sibirica* Hedl.) - 4%, Siberian larch (*Larix sibirica* Ledeb.) - 3%, apple berry (*Malus baccata* L.) - 3% of the total. The remaining representatives of trees and shrubs make up 2% or less of the total.

The city of Bratsk is located on the banks of the Bratsk and Ust-Ilimsk water reservoirs formed on the Angara River during the construction of hydroelectric power plants and is an agglomeration of dispersed residential areas separated by significantly vast forests and water spaces. Residential areas, various in size and degree of improvement are former villages that arose near industrial enterprises under construction. The area of the city is 428 km^2.

Bratsk is located in the north-west of the Irkutsk region in the central part of the Angara ridge. The city arose in 1955, in connection with the construction of the Bratsk hydroelectric station, north of the old village of Bratsk (Bratsk, Bratskoye), founded as an ostrog in 1631.

Climatic conditions are not favorable for diverse vegetation in urban plantations and forest areas. The climate is sharply continental with long harsh winters (down to - 35-45 ° C) and short hot summers (up to +25–30 ° C). During the year and day, the temperature here can vary within large limits. The cold period lasts an average of six months (from the second decade of October to the third decade of April). The average long-term frost-free period in the central part of the city is 94 days. The first frosts are recorded on September 8, the last - on June 5. About 370 mm of precipitation falls per year. The characteristics of the soils of the region include their fineness due to the large dissection of the relief and the diversity of the lithological composition of the rocks, the low temperature regime due to deep seasonal freezing and slow thawing, insufficient moisture due to the small amount of precipitation and spring waters that roll down the still thawing soils and dirt. Soils are subject to wind and water erosion, which reduces the content of humus and reduces fertility. Soils lack organic and mineral fertilizers and need new agricultural techniques [2].

2. Research methodology and methods

To conduct studies on the state of forest woody vegetation, 3 test areas (0.5 ha, 0.1 ha and 1.9 ha) were laid with a number of trees from 150 to 664 plants. All considered plantations of natural origin are areas of the forest preserved during the construction of the city, subject to recreational effects and industrial pollution. Test areas belong to territories with a high anthropogenic load, as they are located along roads and in close proximity to residential and public buildings and are part of parks with a high recreational load. The average age of trees is 70–80 years.

The purpose of this research work is to study the conditions for the growth and development of Scots pine masses (*Pinus sylvestris* L.). in the center of the village of Energetik in the city of Bratsk with high attendance. The massif was preserved during the construction of the village of Energetik, the age of the trees is approximately the same - 70-80 years. Studies were carried out according to generally accepted methods adopted in forest dendrometry [3, 4] (**Figure 1**).

At the test areas, a continuous inventory of trees was carried out with measuring diameters at the height of the chest with a measuring fork, the height of each tree using a altimeter - an eclimeter, the condition of the crown, the state of the roots

Figure 1.
Test area №1.

and trunk, the curvature of the trunk was determined visually. Kraft classes and a sanitary assessment class were also defined. The class of sanitary assessment is as follows: 1 - healthy (without signs of weakening); 2 - weakened; 3 - severely weakened; 4 - drying; 5 - fresh dry; 5 (a) - fresh wind; 5 (b) - fresh brown; 6 - old dry; 6 (a) - old wind; 6 (b) - old brown; 7 - emergency trees.

Next, the area of the entire project area is defined using the compass and measuring ribbon. For instrumental evaluation, 10 model trees were selected for each trial area. To assess the internal state of the wood, the following were used: Arbotom ® pulsed tomograph and Resistograph ® micro-drilling device of the German company Rinntech. For instrumental evaluation, 10 model trees were selected for each trial area. The principle of operation of Arbotom ® is based on determining the velocity of the pulse through the wood between sensors. Vibration sensors are arranged in series to measure trunk circumference. Measurements were taken at chest height. After the sensors are connected to the power supply unit and the manufacturer's specialized software, a number of shocks are applied to each sensor, after which information about the pulse speed between them is recorded by the computer. The standard deviation limit was set at 10%. The software presents the results in the form of matrix values, linear graphs and plane graphs (tomograms) (**Figure 2**).

The principle of operation of the Resistograph ® device is based on the determination of wood resistance to drilling. The Resistograph ® design uses an ultra-thin drill (1 mm) to reduce damage to the study object. To level the error taking into account the point profile, 2 perpendicular measurements were made on each model tree, the results of which were averaged [5–9].

Processing of the obtained data with isolation of destruction zones is carried out by visual evaluation of the obtained graphs taking into account the averages for the tree in question and for the total sample, as well as the distribution pattern of the

Figure 2.
Arbotom® wood condition measurement.

Figure 3.
Test area №3.

Test Area Number	Breed	Percentage of total trees,%
1	Scots pine (*Pinus sylvestris* L.)	91
	Siberian larch (*Larix sibirica* Ldb.)	5
	Birch fluffy (*Betula pubescens* Ehrh.)	4
2	Scots pine (*Pinus sylvestris* L.)	100
3	Scots pine (*Pinus sylvestris* L.)	100

Table 1.
Breed composition of wood stands on test areas.

relative density of wood. Direct correlation of instrument readings with wood density is not possible due to the absence of fixed graded scales.

An agglomerative hierarchical clustering method was chosen for statistical processing of results. Given the nature of the sampling, the Euclidean distance was adopted as the distance between objects:

$$\rho(x, x') = \sqrt{\sum_i^n (x_i - x_i')^2}, \tag{1}$$

where i are signs; n - number of characteristics.

Clustering was carried out using a complete communication algorithm (far neighbor method), according to which the degree of proximity is estimated by the degree of proximity between distant objects of the clusters (**Figure 3**).

Data is entered in Excel and processed using statistics methods [4]. All test areas are pure or almost pure single-tier pine trees (**Table 1**). In this regard, further evaluation was carried out only on Scots pine (*Pinus sylvestris* L.).

As can be seen from **Table 1**, common pine, as the most flexible in the biological and ecological aspect, tree species predominate in the composition of woodlands.

3. Research result

The results of the studies and treatment of the experimental materials obtained are as follows: Average biometric parameters of wood stands on trial areas are given in **Table 2.**

As can be seen from **Table 2**, the trees in all test areas are severely weakened, there are stunted and drying trees and even old dry. Since the trees are single-age and single-tier, the classification by the degree of growth and dominance by Kraft classes was used. The distribution by Kraft class is shown in **Table 2**. Weakened trees according to the Kraft classification are (IV, Va, Vb) in trial areas of more than 20 percent, which indicates the depressed state of the tree stand, which is also confirmed by sanitary assessment points, which average 2.8 points in trial area No. 1; 3.2 points on trial area No. 2 and 2.5 points on trial area No. 3.

Dynamics of average height of trees by thickness stages is shown in **Figures 4–6.**

As can be seen from **Table 2** and **Figures 4–6**, all woodlands have a clearly underdeveloped height and at the age of 70–80 years belong to the woodlands of 4–5 bonitet classes (on the scale of Professor Orlov), which indicates low productivity of woodlands and very unfavorable growing conditions. Productivity (Orlov bonitet classes are determined by the ratio of height and diameter at a certain age). The low bonitet class indicates poor growing conditions (soil or climatic). If we compare the average height of trees of the same age of class III bonitet (through which most forests of the Angara region grow), then the studied trees lack by 3–9

Test Area Number	Average age, years	Average diameter, cm	Average height, m	Vearage height of crown base, m	Root condition	Sanitary grade,%	Kraft grade, %
1	70–80	30.21 ± 1.2	17.2 ± 0.9	7.2 ± 0.3	Not visible	1–14 2–11 3–55 4–17 5–3	I – 18 II – 12 III – 42 IV – 26 V – 2
2	70–80	26.24 ± 0.8	16.4 ± 0.7	7.0 ± 0.2	Not visible	1–3 2–17 3–58 4–7 5–14 6–1	I – 15 II – 28 III – 36 IV – 10 Va – 10 V6 – 1
3	70–80	21.6 ± 1.1	10.1 ± 0.6	5.3 ± 0.5	Visible – 87 Not visible – 575	1–0 2–82 3–1 4–10 5–0 6–7	I – 12 II – 8 III – 39 IV – 21 Va – 19 V6 – 1

Table 2.
Biometric indicators of the studied Scots pine tree (Pinus sylvestris *L.).*

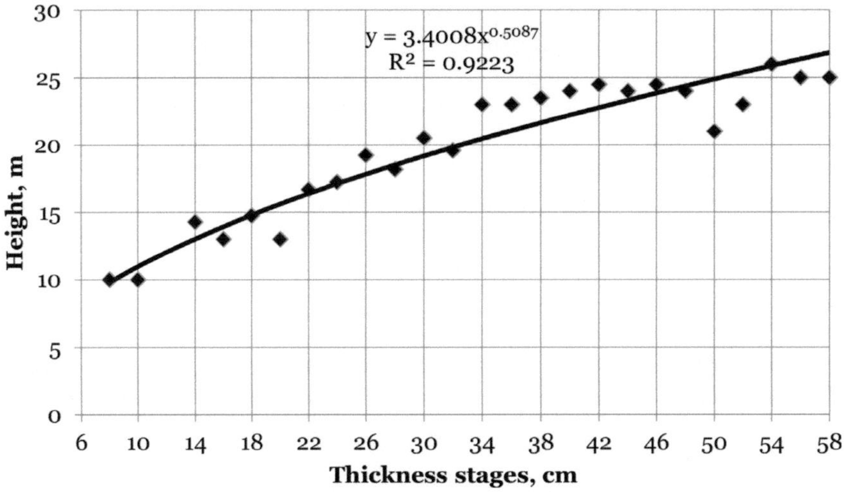

Figure 4.
Height distribution by thickness stages for site No. 1. The figure shows the relationship between the thickness of trees determined at a height of 1.3 meters (abscissa axis) and the height of trees in meters (axis of ordinates) on the first test area.

meters in height, which is very significant, especially on test area No. 3, where the lowest height and the highest recreational load.

When analyzing the condition of tree crowns, it is worth noting that deviations in crown shapes from the norm (flag-shaped, compressed or cut crown) are found in 50% of trees on site No. 1, 35% on site No. 2 and 53% on site No. 3. The percentage of crown condition is shown in **Figures 7–9**.

During measurements, the presence of trunk defects was noted in 95% of cases at site No. 1; 99% - at site No. 2; 96% - at site No. 3. At the same time, it should be noted that in most cases there were 2 or more types of external wood defects on the trunk (**Figures 9** and **10**; **Table 3**).

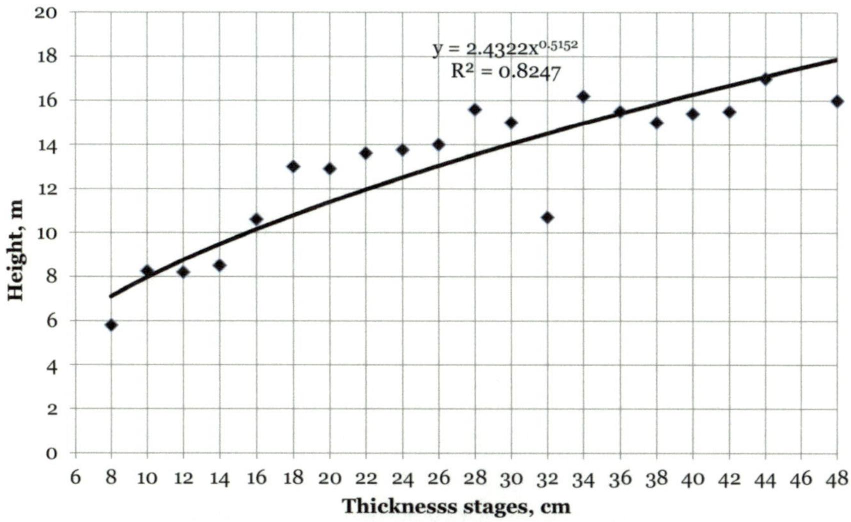

Figure 5.
Height distribution by thickness stages for site No. 2. The figure shows the relationship between the thickness of trees determined at a height of 1.3 meters (abscissa axis) and the height of trees in meters (axis of ordinates) on the second test area.

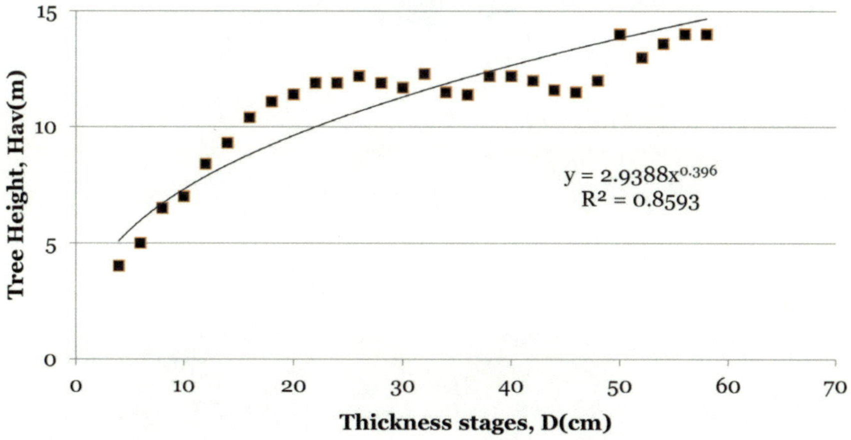

Figure 6.
The distribution of heights by thickness stages for site No. 3. The figures show the correlation between the thickness of trees determined at a height of 1.3 meters (abscissa axis) and the height of trees in meters (ordinate axis) on the third test area.

All model trees studied are ripe and restless. Most trees have a slope or curvature of the trunk and crown asymmetry. Visible wood defects are often observed - dryness, mechanical damage. The trunk height is from 12 to 19 meters, the trunk diameter is from 30 to 70 cm. **Table 4** shows the taxation indicators of the three trees most characteristic of the study object.

The nature of the tomograms obtained indicates a heterogeneous distribution of the wood density of the studied model trees. The speed varies from 912 m/s to 2018 m/s. The maximum frequency of occurrence falls in the range of 1003–1349 m/s. The degraded wood content ranges from 12 to 79% (average content is 30%). According to Resistograph ®, the average drilling resistance of the sample model trees was 121. The degraded wood content ranges from 26% to 85% (average content is 50%). In the vast majority of cases (96% of the sample), resistogram

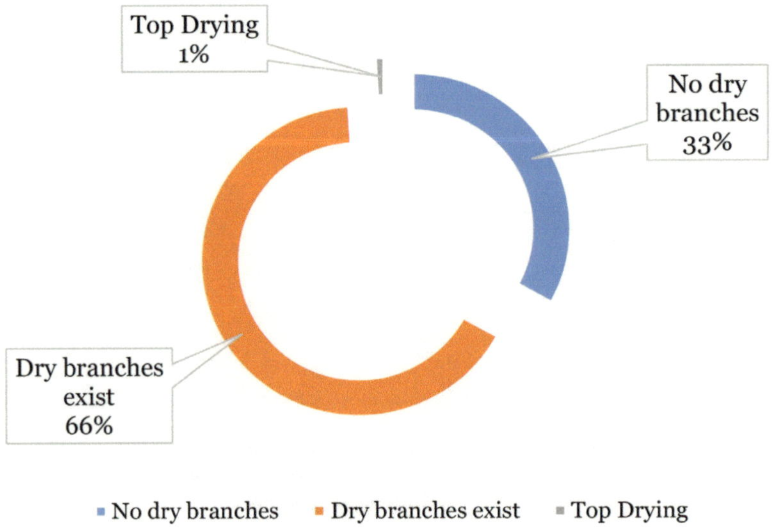

Figure 7.
Presence of signs of drying of the crown of trees of site No. 1, % of the sample.

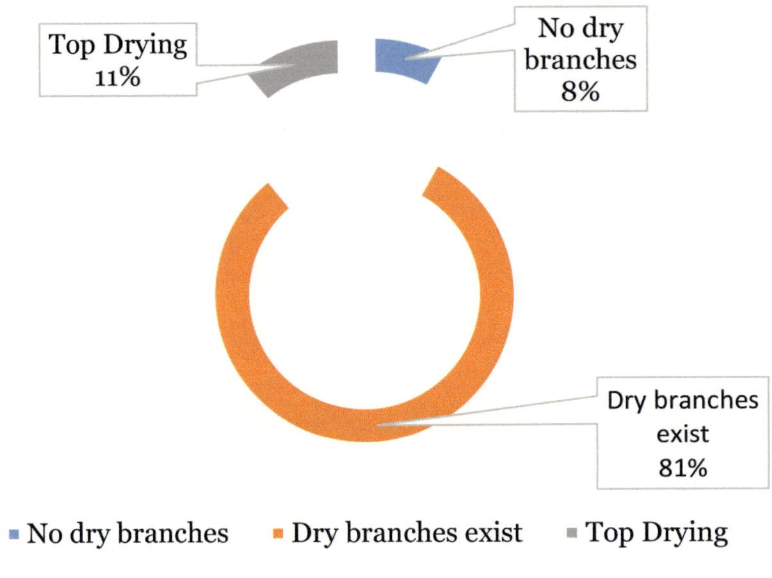

Figure 8.
Presence of signs of drying of the crown of trees of site No. 2, % of the sample.

readings reflect a significantly higher percentage of trunk destruction. This is a consequence of incomplete accounting of the area of peripheral areas of sickness during profile analysis by drilling.

Particularly strong oppression of trees can be traced on test area No. 3. Therefore, its characteristics should be discussed separately.

As can be seen from **Table 5**, the trees on this test area develop accordingly to the dwarf type, the bonitet class (woodland productivity) is only V, that is, the lowest at a given diameter. The average class of sanitary assessment is 3.6, which indicates almost the decay of the tree. **Table 5** shows the main forms of tree crowns and their number in pieces and percentages. It should be noted that on the 1.9 hectare site there are quite a large number of trees, namely 662 plants, in some areas the fullness of the tree stand significantly exceeds the maximum equal to 1.0.

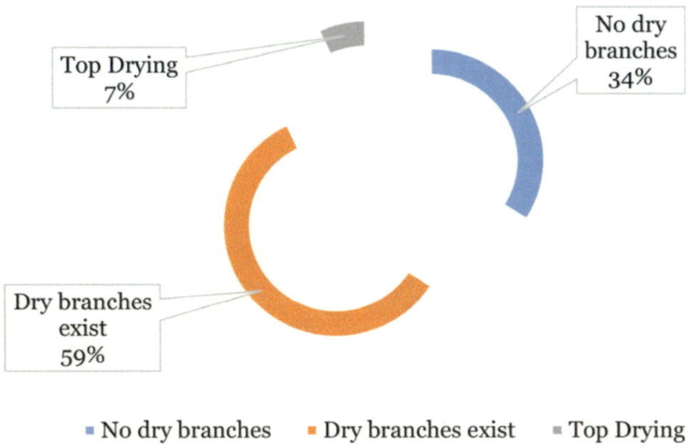

Figure 9.
Presence of signs of drying of the crown of trees of site No. 3, % of the sample.

Figure 10.
Mechanical damage to tree trunks.

Test Area Number	Number of kinds of trunk defects	Share in sample,%
1	No defects	5
	1 kind of defect	14
	2 or more kinds of defects	81
2	No defects	1
	1 kind of defect	26
	2 or more kinds of defects	73
3	No defects	4
	1 kind of defect	30
	2 or more kinds of defects	66

Table 3.
Number of trunk defects in test areas.

№ of the tree	6	7	8
D_{main} cm	38	44	52
$D_{1,3}$ cm	32	38	44
H_{tr}, m	12.5	18.8	19.0
Tree age	68	72	78
Visible defects of the trunk	Drywall, trunk slope	Drywall, trunk slope	Drywall, trunk slope
Average pulse speed, m/s	936	966	1011
Average drilling resistance, relative units.	137	134	142
Disturbed wood content according to Arbotom ®,%	78	52	15
Disturbed wood content according to Resistograph ®,%	50	57	69

Table 4.
Taxation characteristics of model trees.

Type	Quantity	%
Flag-shaped	326	49.10
Oval	311	46.84
Round shaped	3	0.45
No crown or dried up	24	3.61
Total	664	100

Table 5.
Distribution of the number of trees by crown forms in test area No. 3.

As the table shows, almost all crowns, namely 95.94% of the total number of trees, have either a flag-shaped crown or a strongly compressed oval crown. **Figure 11** shows in more detail the distribution of crowns in form.

Table 6 and its results indicate extreme soil compaction as a result of long-term recreational load, there is no living soil cover characteristic of the forest, so it is impossible to determine the type of forest and type of forest conditions. More than 13% of trees have a highly bare root system, which has mechanical damage and does

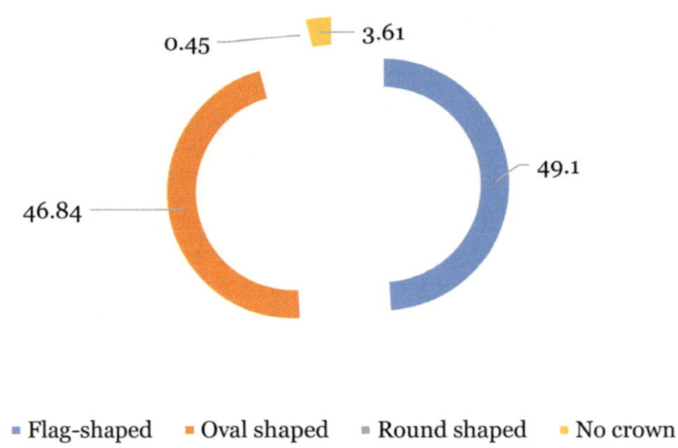

Figure 11.
Crown form distribution chart, %.

Type	Quantity	%
Bare rooted	88	13,11
Not visible	576	86,59

Table 6.
Conditions of tree roots on test area No. 3.

Type	Quantity	%
partial drywall, rot	268	40,36
partial drywall	91	13,70
partial drywall, mechanical defects	101	15,21
partial drywall, mechanical defects, rot	15	2,26
partial drywall, rot, cancerous object	1	0,15
healthy	28	4,22
rot	85	12,80
rot, cancerous object	2	0,30
mechanical defects	28	4,226
mechanical defects, cancerous object	1	0,15
mechanical defects, rot	43	6,48
cancerous object	1	0,15
Total	664	100

Table 7.
Identification of the quality status of the tree trunk in test area No. 3.

not allow plants to develop fully. The state of trees in terms of the quality of trunk wood is especially manifested, absolutely all trees have numerous mechanical and mushroom damages (**Table 7**).

The trunks have a drywall to a greater extent, which was formed from mechanical action on the trunks, drywall with open damaged wood led to the presence of mushroom lesions and rotting, up to the hollow. **Figure 12** shows the distribution of the number of trees by type of stem injury.

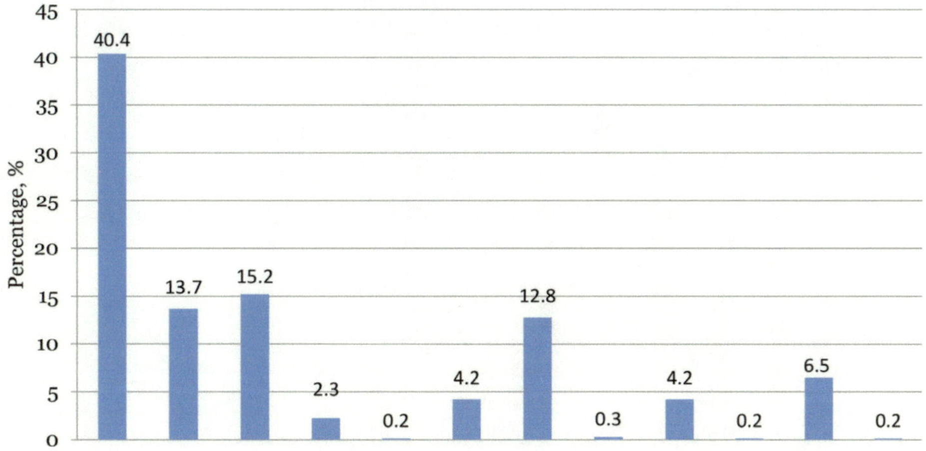

Figures 12.
Diagram of stem damage distribution by number of trees, %.

In addition to damage to the trunk wood, the trees under study had a different degree of crown damage, which was expressed by the presence of a large number of dry branches (**Table 8**).

Figure 13 shows the percentage distribution of trees by crown condition. On the axis of ordinates are represented percentages, on the axis of abscissa - categories of trees as per crown.

Type	Quantity	%
No dry branches	223	33.58
Dry branches exist	394	59.34
Half dried tree, top drying	1	0.15
Dry tree	41	6.17
Top drying	3	0.45
Half dried tree	2	0.30
Total	664	100

Table 8.
Presence of dry branches of investigated trees on test area No. 3.

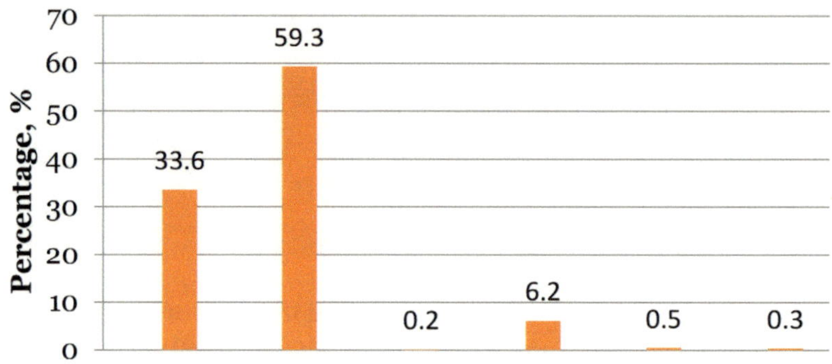

Figure 13.
Distribution of the number of trees by the degree of damage to the crown (presence of dry branches), % on test area No. 3.

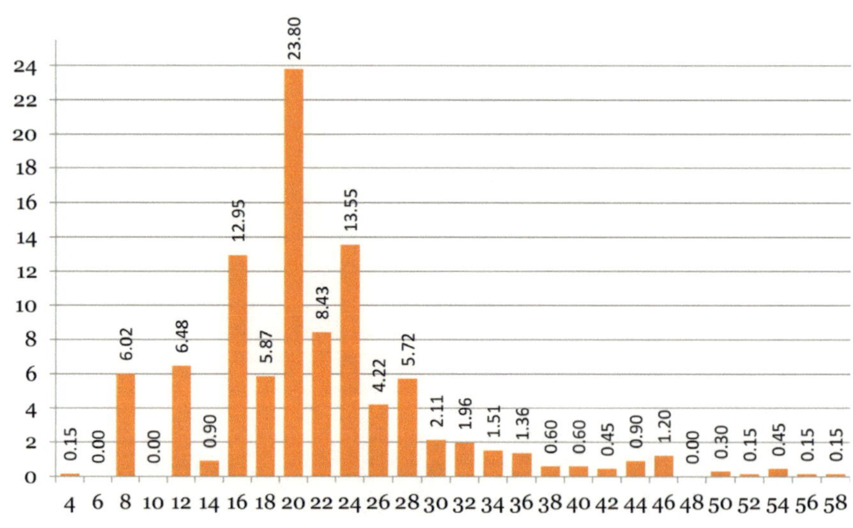

Figure 14.
Distribution of number of trees in% by diameters in test area No. 3.

Another evidence of the extreme oppression of trees in the pine massif is the nature of the distribution of trees along the thickness stages. In normal, healthy wood, this distribution is close to the normal distribution curve. Let us look at **Figure 14**, which shows a diagram of the distribution of trees by diameter at a height of 1.3 meters.

The **Figure 12** shows the percentage of trees by diameter on test area No. 3. Percentages are represented along the ordinate axis, along the abscissa axis - the thickness of trees at a height of 1.3 meters. It can be seen what a large range of diameters in the woodland is from 4 centimeters to 58 at a uniform age, which indicates strong intraspecific competition of trees.

4. Discussion of the results

Based on the conducted studies, it can be concluded that the results of studies of the Arbotom ® and Resistograph ® devices of the German company Rinntech [7–17] are quite often found in scientific publications. However, in most studies, the assessment of the state of the stem wood using appropriate instruments is performed separately [5, 7, 8, 11, 14]. The data are comparable with the data of well-known scientists in terms of quantitative indicators [7, 8, 10, 12–14], which confirms the reliability of our studies. We tried to compare the data of the readings of the two devices and compare them in our work. Cluster analysis methods were also used to simultaneously compare the readings in order to more accurately quantify the condition of trees and predict the appearance of emergency trees that are dangerous to human life in the city under wind loads. Most often, urban plantings were described using visual or dendrometric characteristics [16–18], but only visual assessment did not determine how long a particular tree or plant species could exist in an urban environment without loss of viability and signs of accidents. That is, the visual assessment method cannot determine the degree of damage to the tree by internal rot and the degree of development of rot, up to the formation of a hollow [16–18]. In this work, visual and measurement methods for assessing the state of forest areas included in the urban environment were also carried out, which is unique, since it is typical only for the northern regions of Russia, where cities are relatively poorly built (50–70 years). A fairly high correlation was found between dendrometric parameters, the state of the roots and tops of trees, and the presence of internal trunk defects. The reduction in the life expectancy of trees in urban conditions under recreational loads and high levels of atmospheric pollution has been proven.

5. Chapter conclusions

Based on the studies carried out, the following conclusions can be drawn:

1. Natural forests preserved 60–65 years ago during the development of the city of Bratsk certainly perform esthetic and sanitary protective functions. But at the same time they themselves are subjected to strong anthropogenic effects. Typically, there are no forest plants of living soil cover, in some cases the trees are dead cover due to the high degree of trampling. Plants grow in 4–5 bonitet class, have a height of almost twice as high as trees of the same age of 1 bonitet class.

2. Trees have a large percentage of shape of the trunk defects, primarily drywalls and prophecy, which is associated with mechanical effects on tree trunks.

3. Tree crowns are cut, often have a flag-shaped crown, a large number of dry branches.

4. Studies have proved the presence of internal defects in all studied model trees. It is possible to conclude the general oppression of woody vegetation of the test areas under consideration.

5. Among the studied model trees, trial area No. 2 is noticeably distinguished in terms of wood hardness. However, in the general picture of the distribution of internal defects, significant selectivity between sites is not observed. Thus, one can conclude that the conditions for the growth of the woodland are relatively equal and the green spaces of various areas of the urban ecosystem of the city are evenly oppressed.

6. Under conditions of industrial pollution and increased recreational loads, processes of earlier aging of trees occur up to their natural death. In the forest environment, the life expectancy of common pine is from 350 to 600 years in Russia [10], and in the urban environment without proper care and with a high level of anthropogenic load, already at the age of 70–80 years there are pronounced signs of tree aging, which are manifested in the presence of dry branches, dry trees, the presence of internal rots, stem pests.

7. When compiling an assessment of the state of the plantation as a whole, it is advisable to recommend comparing the data of the two devices both according to the parameters of the expert assessment (proportion of disturbed wood) and according to the parameters of automated measurements (instrument data) to compile the most complete picture of the state of dendrocenosis.

8. The work performed is of great practical importance, as it allows to identify emergency trees that can be exposed to wind and windbreak under heavy wind loads, and to carry out timely replacement with younger and healthier trees.

9. All these signs indicate the need for additional studies of the internal condition of the wood of model trees of the sites considered by instrumental methods. In view of the high value of urban forests, non-destructive testing methods are recommended. Woodlands require a whole range of forestry measures to preserve these unique objects of the urban environment - such as sanitary cutting of dead trees, cutting off dry branches, introducing a fertile layer of land, and treating mechanical damage to the trunk.

10. The developed proposals should be used in the care of green spaces in urban urban ecosystems, especially in the care of areas of natural forests that are located inside urban development. Regular monitoring of the condition of such plantings is planned to be carried out not only for Pinus sylvestris L., but also for other tree and shrub species growing under anthropogenic loads.

6. Grainude

I express my gratitude to my colleagues in the work who took part in the collection of experimental material I.A. Garus, A.N. Mukhacheva.

Author details

Elena Runova[1]*, Vera Savchenkova[2], Ekaterina Demina-Moskovskaya[2]
and Anastasia Baranenkova[2]

1 Bratsk State University, Bratsk, Russian Federation

2 Mytischi Branch of Bauman Moscow State Technical University, Moscow, Russian Federation

*Address all correspondence to: runova0710@mail.ru

References

[1] Runova E. M., Anoshkina L. V., Gavrilin I. I. State of woody vegetation in urban ecosystems on the example of Bratsk: monograph.- BrSU publishing house, 2017. – 80 p.

[2] Geographical encyclopedia of the Irkutsk region: commonly. Essay/ed. L. M. Korytny; Irkut Government. Region, Institute of Geography named after VB Sochava SB RAS, Irkut. State un-t. – Irkutsk: Institute of Geography. V.B. Sochavy SB RAS, 2017.–336 p.

[3] OST 56–69-83. Trial areas, forest inventory. Bookmarking methods. Enter. For the first time 1984.01.01. M. Publishing house of standards, 1984.61 p.

[4] Runova E.M., Zhang S.A., Puzanova O.A., Savchenkova V.A. Dendrometry: Textbook for the ACT.- St. Petersburg: Lan Publishing House, 2020. -160 p. ISBN 978–5–8114-5916-2.

[5] Johnstone D., Moore G., Tausz M., Nicolas M. The measurement of wood decay in landscape trees//Arboriculture & Urban Forestry 36(3). 2010. P. 121–127.

[6] Chubinsky A.N., Tambi A.A., Fedyaev A.A., Fedyaeva N. Yu., Kulkov A.M. Directions of using physical methods for controlling the structure and properties of wood// Systems. Methods. Technology. 2015.2 (26). S. 152–158.

[7] Tyukavina O.N. The speed of passage of a sound pulse in pine wood//Bulletin of the Northern (Arctic) Federal University. Series: Natural Sciences. – Arkhangelsk, 2014 .–S. 78–85.

[8] Lavrov M.F. Determination of quality indicators of wood by drilling method/M.F. Lavrov, D.K. Chakhov, I. A. Doctors//Bulletin of the Moscow State Forest University. – Forest Bulletin. – 2014. No. 5. – S. 196–201.

[9] Runova E.M., Garus I.A., Mukhacheva A.N. The use of instrumental methods in assessing the state of the trunks Pinus sylvestris L. Forest engineering journal, No. 3, 2020, p. 72–85. Bibliography: p. 83–84 (15 names). - DOI: 10.34220/issn.2222-7962/2020.3/8.

[10] Rinn F. Eine neue Bohrmethode zur Holzuntersuchung//Holz-Zentralblatt. 1989. №15 (34). s. 529–530.

[11] Literature review of acoustic and ultrasonic tomography in standing trees/Arciniegas A., Prieto F., Brancheriau L., Lasaygues P. //Trees. 2014. № 28(6). p. 1559–1567.

[12] Rinn F. Holzanatomische Grundlagen der Schall-Tomographie an Bäumen//Neue Landschaft. 2004. № 7/ 04. s. 44–47.

[13] Rinn F. Statische Hinweise im Schall-Tomogramm von Bäumen//Stadt und Grün. 2004. № 7/2004. s. 41–45.

[14] Melnichuk I. A., Yasin N., Cherdantseva O. A. Diagnostics of the internal state of *TILIA CORDATA* MILL trees. with the use of the complex of acoustic ultrasound tomography equipment "ARBOTOM °" / /Bulletin of the Russian University of Friendship of Peoples. Series: Agronomy and animal husbandry. 2012.№ S5. P. 25–32.

[15] Palchikov S. B., Antsiferov A. B. Assessment of the state of trees affected by xylotrophic fungi using Resistograph and Arbotom devices. assessment of the condition of trees affected by xylotrophic fungi using Resistograph and Arbotom devices. Eurasian Union of Scientists (EUU) # 4 (25), 2016. Biological Sciences series, pp. 121–125

[16] Kolomyts E. G., Rosenberg G. S., Glebova O. V. The natural complex of the big city: landscape-ecological analysis. - Moscow: Nauka, 2000 – - 286 p.

[17] Mozolevskaya E. G. Methodological recommendations for assessing the viability of trees and the rules for their selection and assignment to felling and transplanting. Moscow: MGUL, 2003. 28s.

[18] Petrovskaya P. A., Stolyarova A. G. Basic principles of urbanized territory improvement / / Vestnik Rossiyskogo universiteta druzhby narodov. Series: Agronomy and animal husbandry. 2013. № 5. S. 86–92.

Chapter 5

Landscape Genetics and Phytogeography of Criollo Avocadoes *Persea americana* from Northeast Colombia

Clara Inés Saldamando-Benjumea, Gloria Patricia Cañas-Gutiérrez, Jorge Muñoz and Rafael Arango Isaza

Abstract

Avocado (*Persea americana*) Mill represents one of the most consumed fruits around the world. This species has been differentiated into three main races Guatemalan, Mexican and West Indian according to several molecular markers. However, the interaction between genotypic and phenotypic traits of this crop is still unknown. For this reason, a landscape genetics analysis was made in 90 criollo trees from Northeast Colombia (Antioquia) with 14 microsatellites, sequencing of 3 nuclear loci, endo-1-4-D-glucanase (Cell), Chalcone synthase (CHS) and serine-threonine-kinase (STK) and 28 morphological traits. High genetic diversity was found suggesting a hybrid origin of criollo trees. Morphological variation showed intermixed racial features. FST = 0.03, p =0.001 (measured with microsatellites) suggested low genetic differentiation. According to STRUCTURE, K = 2 for both microsatellites and concatenated nuclear sequences. Criollo trees were assigned together with the Guatemalan and Mexican races. Pearson correlation was significant between expected heterozygocities and elevation. Mantel test was low (r2 = 0.0097, p = 0.015) but significant demonstrating isolation by distance. Grafting is suitable between criollo trees and Hass variety is possible since both avocados are produced within the same altitudes.

Keywords: avocado, races, grafting, morphology, microsatellites, nuclear genes

1. Introduction

Avocado (*Persea americana* Mill.) is one of the most important subtropical crops of the Lauraceae family. The species is native to central Mexico [1–3] and belongs to the sub-genus *Persea* with two other species: *P. schiedeana* (Nees) and *P. parviflora* (Williams) [4]. *P. americana* is an evergreen tree that is heterogeneously branched from 40 to 80 ft. (12.9 to 24. 4 m) tall with elliptic leaves that are 3 to 10 inches (7.62 to 25.4 cm) long and has a dichogamous breeding system [5, 6]. First avocado movements from central Mexico occurred through big mammals (sloths and mammals of the family Gomphotheriidae) migrations to Mesoamerica [2]. Later on, *P. americana* was cultivated and domesticated by the first Mesoamerican cultures (The Mokayas) who transmitted their cultural practices to further civilizations such as the Mayas and Olmecs [7]. These two civilizations further originated three main avocado races recognized as Mexican [*P. americana* var. drymifolia (Schltdl. and Cham.) S.F.Blake], Guatemalan [*P. americana* var. guatemalensis L. Wms.], and West Indian [*P. americana* var. americana Mill.]. These races are distinguishable at morphological, physiological, and horticultural levels [8, 7] and adapted to different conditions. However, the first two are adapted to cooler (Mediterranean and subtropical) climates and medium elevations, while the third requires warmer (tropical) conditions and lowland humid tropics [2–11]. In Colombia, most commercial cultivars are interracial hybrids developed from chance seedlings [12] where propagation originated native or criollo trees and selected cultivars that are asexually reproduced and vary in flavor and nutritive traits [13] with unknown genetic origin.

Several studies based on molecular characterizations have been made in avocado, mainly with microsatellites [1–16]. They are advantageous since they are co-dominant, bi-parentally inherited, and easily standardized [17]. Additionally, sequencing of nuclear genes endo-1-4-D-glucanase (Cell), Chalcone synthase (CHS) and serine–threonine-kinase (STK) have also been made as they have been useful for studying the genetic origin of avocados [7, 18]. DNA sequencing represents the best method to infer gene genealogies, as historical events (coalescence) are better followed through chloroplast and nuclear DNA in plants and mitochondrial DNA in animals [17, 19]. On the other hand, landscape genetics provides relevant information on the interaction between landscape features and microevolutionary processes, such as gene flow, genetic drift and is a very useful tool for resolving genetic differentiation within a species across different geographical scales at fine taxonomic levels [20].

The chapter aimed to carry out a landscape genetic analysis together with a phylogeographic study on 90 criollo trees sampled in Antioquia (Northeast Colombia) based on molecular (microsatellites and nuclear genes sequencing) and morphological characterizations. Results obtained here are relevant for criollo avocado certification and for future grafting purposes of the species in this department of Colombia where importation of this fruit is highly demanded, particularly the variety Hass.

2. Study of landscape genetics and phytogeography of criollo avocadoes (*P. americana*) from Northeast Colombia

2.1 Materials and methods

2.1.1 Plant material and genomic DNA extraction

Sampling was performed in the Antioquia department (Northern Colombia) during 2008 and 2009. Young leaves were collected from 90 criollo avocado trees. Trees were chosen according to differences in Holdrige life-zones and climatic conditions. Collections included two municipalities from the western sub-region páramo [Sonsón (N = 17) and Abejorral (N = 9) at 2.475 and 2.275 m.a.s.l. respectively], five municipalities from the Western sub-region altiplano between 2.080 and 2.200 m.a.s.l. [El Retiro (N = 5), Marinilla (N = 6), Rio negro (N = 8), La Ceja (N = 6) and San Vicente (N = 6)], and 4 municipalities from the Southwestern sub-region [Caramanta (N = 13), Santa Bárbara (N = 2), Valparaiso (N = 3) and Montebello (N = 12) with elevations ranging from 1.375 in Valparaiso to 2.350 in Montebello]. Total genomic DNA was obtained from leaves based on Cañas-Gutierrez et al. (2015) from a study on an AFLP (Amplified fragment length polymorphism) characterization of the species made in 111 avocado trees.

2.1.2 Microsatellites analysis

Fourteen microsatellites were selected from those used by Alcaraz and Hormaza (2007) based on their high polymorphism. PCR amplifications were performed in 15 μL vol containing 16 mM (NH4)2SO4, 67 mM Tris–HCl pH 8.8, 0.01% Tween20, 2 mM $MgCl_2$, 0.1 mM each dNTP, 0.4 μM of each primer, 25 ng genomic DNA and 0.5 units of BioTaqTM DNA polymerase (Bioline, London, UK). Forward primers were labeled with WellRed fluorescent dyes on the 5′ end (Proligo, France). Reactions were performed in an I-cycler (Bio-Rad Laboratories, Hercules, CA, USA) thermocycler using the following temperature profile: an initial step of 1 min at 94°C, 35 cycles of 30 s at 94°C, 30 s at 50°C and 1 min at 72°C, and a final step of 5 min at 72°C. The PCR products were analyzed by capillary electrophoresis in a CEQTM 8000 capillary DNA analysis system (Beckman Coulter, Fullerton, CA, USA). Each reaction was repeated twice to minimize run-to-run variation.

2.1.3 Sequencing analysis

Amplification primers for the loci Cellulase (endo-1,4-D-Glucanase) (Cell), Chalcone synthase (CHS) and Serine–threonine-kinase (STK) were designed according to avocado sequences reported in the GenBank (Chen et al. 2008, 2009), using Primer 3 Plus software (http://primer3plus.com/). Amplification reactions contained 1 × buffer (20 mM Tris–HCl, pH 8.4, 50 mM KCl), 1.5 mM of magnesium chloride, 0.2 mM of each dNTP (Thermo Scientific, Waltham, MA), 2 μM of each primer, 1.5 U of Taq DNA polymerase (Thermo Scientific), 50 ng of DNA and water to complete a final reaction volume of 25 μL. Temperature cycles were made in a T100™ Thermal Cycler (Bio-Rad, Hercules, CA) with an initial temperature of 95°C preheating for 3 min, then 95°C for 45 s, annealing temperatures of 57°C for Cell primers and 58° for CHS and STK primers for 1 min, extension temperature was 72°C. Amplification included 35 cycles with a final extension of 72°C for 10 min. The amplification products (3 μl) were separated by electrophoresis in 1.0% agarose gels with GelRedTM and visualized in a UV transilluminator.

Polymerase chain reaction (PCR) conditions were as follows: 95°C preheating for 3 min, then 95°C for 45 s, annealing temperatures of 57°C for Cell primers and 58°C for CHS and STK primers for 1 min, extension temperature was 72°C. Amplification included 35 cycles with a final extension of 72°C for 10 min. Reactions were made in an I-cycler (Bio-Rad Laboratories, Hercules, CA, USA) thermocycler [7, 18]. Allele-specific PCR (AS-PCR) was not employed here [7, 18]. Instead, presence of heterozygous and homozygous genotypes was performed from individual chromatograms following the procedure given by 21 in Rhizophora genus (mangrove).

2.1.4 Morphological descriptors

Morphological characterization of 90 criollo trees was made according to 28 morphological descriptors that are specific for avocado [21]. These traits included: leaves measurements of length and width, leaves pubescence, branch and leaves colors, leaves margins, the number of primary and secondary veins found in leaves, leaves shape, branch insertions, trunk surface, and tree shape, amongst others.

2.1.5 Landscape variables

Elevation (altitude) was considered in this study according to the altitude above the sea level measured in meters. For the landscape analysis both topography and georeferentiation measurements were also considered as they were taken on each sampling site. Topographic analysis was associated to elevation profiles of the surface land, following FAO (1990) protocol as follows: 1) flat: elevation from 0 to 0.5%, 2) almost flat: elevation from 0.6 to 2.9%, 3) little wavy: elevation from 3 to 5.9%, 4) wavy: elevation from 6 to 10.9%, 5) broken: elevation from 11 to 15.9%, 6) hillside: elevation from 16 to 30%, 7) Strong undermined: > 30%, moderate variation of elevations, 8) mountainous: > 30%, large variations of high elevation range (˃300 m), 9) Other: specify other features.

2.1.6 Population genetics analysis

Microsatellite locus (N) diversity estimators were calculated with GENALEX 6.501 [22] for 90 criollo trees. Genepop 4.2 package [23] was used for exploring Hardy–Weinberg equilibrium and linkage disequilibrium per population. AMOVA test was performed to determine the genetic population differentiation amongst 11 sampling sites within Antioquia with the program GENALEX 6.501 [22]. The model-based clustering analysis STRUCTURE 2.3.4 was 1) first applied to 14 microsatellites and then 2) used with 3 concatenated nuclear loci in 90 individuals [24]. These two separated analyzes were carried out 1) to assess the most probable avocado cluster membership from Antioquia and 2) to determine the genetic assignment of criollo trees within the three avocado races according to GenBank data bases for these reasons, each molecular marker was analyzed separately. This program was run for 150.000 Markov chain Monte Carlo steps after a burn-in period of 15.000 interactions from K = 1–15 considering: a) no admixture model and b) model-independent allele frequency for each marker following the suggestions given by Chen et al. (2008) and Chen et al. (2009). Each K was calculated from 15 independent runs and 10 iterations. For DNA sequences, each nucleotide was numerically coded as follows: A = 1, T = 2, C = 3 and G = 4 and missing data as −9. A filter was applied to the three loci such that SNPSs were recognized from the total number of segregating sites (S) detected by the program DNAsp V5 [25]. A similar procedure was performed by Chen et al. (2009). The ad hoc estimated likelihood of K (delta K) [26] was obtained with STRUCTURE HARVESTER [27]. K mean values and their estimators were calculated with CLUMPP [28] and graphs were produced with Distruct 1.1. Finally, the Mantel test was estimated (correlation between genetic distances (from microsatellites) vs. geographic distances of collecting sites) with GENALEX 6.501.

2.1.7 Heterozygote detection in sequences and phylogeographic analysis

For sequencing analysis, DNA amplicons were purified and sequenced in Macrogen Inc., (South Korea with an ABI 3730XL sequencer (Perkin Elmer/Applied Biosystem, Foster City, CA). Sequences (forward and reverse) were edited by hand with Bioedit (Hall 1999). Chromatographs were examined to detect heterozygous (with double peaks in a polymorphic site) vs. homozygous genotypes (with single peaks in a polymorphic site) [29]. Heterozygous segregate into several haplotypes, for example: ATG/CGC/TA segregates into ATGGCAC, ATCGCTAC, ATGGTAC, etc. [30]. This variation was detected by Chen et al. (2008) from sylvester and domesticated avocados with the program POLIPHRED (Ewing and Green 1998). An alternative to this method is the use of chromatographs [29]. In this study, the highest peak for each heterozygous genotype (with two peaks per nucleotide position) in all sequence was selected per individual sample (collected specimen) to produce only one haplotype per avocado (**Figure 1**) and to simplify further population genetic analysis as Chen et al. (2008) and Chen et al. (2009)

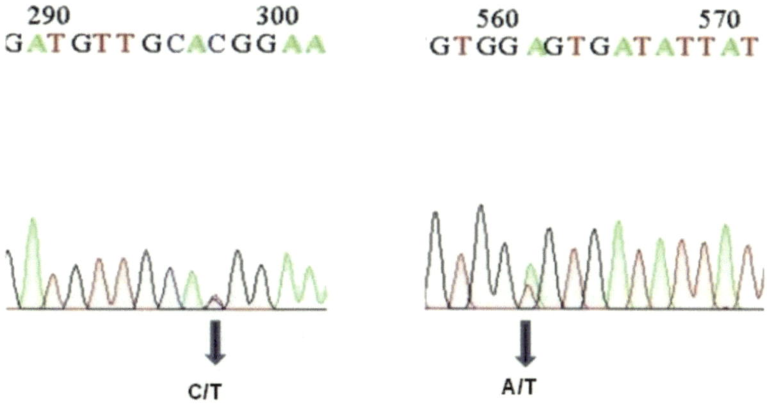

Figure 1.
Heterozygous genotypes were detected in the nuclear loci were considered in this study. Two double peaks at genotype AT.

mostly reported one haplotype per avocado (either domesticated or sylvester) in the GenBank.

Clustal W was used for all alignments in Mega 7.0 (Kumar et al. 2016). The variation in DNA sequence (extension of DNA polymorphism in a DNA sequence) for each locus was measured with the following parameters: 1) nucleotide polymorphism: q = s/a1, a1 = S 1/I and s = a number of polymorphic markers, 2) nucleotide divergence: $\pi = \Sigma$ dij/c, dij = number of nucleotide differences (substitutions) per site between the "ith" and "jth" alleles and c = a total number of sequences studied, 3) segregating sites: (S) where S = a nucleotide site with more than one nucleotide variation in "m" sequences comparisons, 4) a number of haplotypes (H), and 5) haplotype diversity = Hd [19]. All of them were computed on DNAsp V5 [25]. Sequenced loci were considered as independent genes based on the results of the linkage disequilibrium made by Chen et al. (2008). For a phylogenetic Bayesian analysis, jmodeltest [31, 32] was used to determine the model of evolution concatenated data set (Antioquian and Genebank sequences) obtaining the model General Time Reversal (GTR). The platform Beast 2.0 [33] was used to obtain the phylogeny. Data was analyzed in Beauti using an MCMC with 100 million generations. Each consensus tree obtained for the COI gene and ITS region was maintained every 1000 generations and preBurning was established for the first 10 million generations for the MCMC. The posterior probability was corroborated with Tracer v1.6 [34], the consensus trees were summarized with Treeannotator and graphed with Figtree [34]. No out-group was used in this phylogeny as no other reported sequences for Cell, CHS and STK are reported in the GenBank in other species.

2.1.8 Morphological data analysis

The program Past 1.2 [35] was used to carry out statistical analysis on morphological data. This analysis included a PCA (Principal Component Analysis) of 28 morphological descriptors used for the species together with the 14 microsatellites information obtained on each criollo tree. Finally, the average expected heterozygocity obtained for each criollo tree was correlated (Person correlation) to two environmental variables: topography and elevation according to 36 procedures. This analysis was made to determine if genetic diversity in criollo avocado varied in altitude and/or in topography [36].

3. Results

3.1 Population genetics results

Genetic diversity estimated with 14 microsatellites in 90 criollo trees showed that the total number of amplified alleles ranged from 248 to 28. The mean allele number was from 5.7 to 2, the number of effective alleles varied from 3.33 to 1.84. The Shannon Weber index ranged from 1.34 to 0.52. Maximum and minimum estimators obtained here were detected at the municipalities of Sonsón and Santa Bárbara. Observed heterozygocities were between 0.48 and 0.107 and expected heterozygocities were from 0.66 to 0.38. All microsatellites were in Hardy – Weinberg equilibria (**Table 1**). Linkage disequilibria were not significant after Bonferroni corrections (data not shown). Analysis of molecular variance (AMOVA) with 14 microsatellites generated FST = 0.05439 ($p < 0.0001$) suggesting population structure amongst 11 municipalities where avocado samples were taken. This test also showed that most genetic variation was within (94.5%) than between populations (5.44%). STRUCTURE HARVESTER estimated K = 2 according to mean Ln (P|D) = −1445.81;

Collecting site	Ni	Nt		N	Na	Ne	I	Ho	He	F	HWE (P)
Sonsón	18	248	Mean	17.710	5.780	3.336	1.344	0.486	0.663	0.258	1.000
			SDE	0.160	0.576	0.341	0.099	0.043	0.031	0.061	
Abejorral	9	125	Mean	8.920	4.143	2.820	1.120	0.523	0.603	0.120	1.000
			SDE	0.071	0.533	0.279	0.107	0.051	0.035	0.083	
El Retiro	5	67	Mean	4.780	3.143	2.683	0.984	0.560	0.570	−0.011	0.983
			SDE	0.155	0.275	0.257	0.105	0.092	0.052	0.146	
La Ceja	6	84	Mean	6.000	4.786	3.717	1.373	0.536	0.702	0.221	1.000
			SDE	0.000	0.422	0.345	0.087	0.398	0.025	0.081	
San Vicente	6	82	Mean	5.857	3.857	2.775	1.007	0.398	0.574	0.316	1.000
			SDE	0.097	0.443	0.294	0.122	0.073	0.054	0.095	
Marinilla	6	83	Mean	5.929	3.643	2.626	1.010	0.425	0.548	0.213	1.000
			SDE	0.071	0.341	0.307	0.109	0.081	0.051	0.118	
RioNegro	9	123	Mean	8.786	5.643	4.008	1.485	0.539	0.724	0.233	0.985
			SDE	0.155	0.401	0.340	0.084	0.051	0.026	0.084	
Santa Bárbara	2	28	Mean	2.000	2.000	1.814	0.582	0.393	0.384	−0.030	
			SDE	0.000	0.182	0.157	0.097	0.107	0.061	0.193	0.985
Valparaiso	3	42	Mean	3.000	3.214	2.655	1.021	0.571	0.587	0.028	
			SDE	0.000	0.214	0.223	0.077	0.089	0.035	0.137	1.000
Caramanta	14	191	Mean	13.643	6.429	3.739	1.485	0.435	0.703	0.365	
			SDE	0.133	0.510	0.345	0.086	0.052	0.028	0.078	1.000
Montebello	12	162	Mean	11.571	5.429	3.117	1.305	0.521	0.212		
			SDE	0.137	0.402	0.241	0.088	0.055	0.073		

(Ni = number of individuals sampled per population, Na = number of different alleles, Nt = total number of alleles, Ne = effective number of alleles, N = mean allele number, I = Shanon Index SDE = standard deviation), Ho = observed heterozygocity, He = expected heterozygocity, F = Wright F index, HWE = Hardy Weinberg equilibrium test (P = p value)).

Table 1.
Genetic diversity estimated in 90 criollo avocados.

Stdev = 18.08 and ΔK = 7.79 (**Figure 2a**). This result suggests that samples collected in 11 municipalities were assigned to two sub-populations where most criollo avocados were assigned to cluster 1 (orange) and a few individuals to cluster 2 (purple).

Pearson correlation showed no association between HE (expected heterozygocity) and topography as r = −0.2893 (p = 0.3886) (**Figure 3**) but a significant association between HE and elevation (altitude) as r = 0.7112 (p = 0.014029) (**Figure 4**). Additionally, Mantel test was significant (r = 0.0097, p = 0.015), suggesting isolation by distance (**Figure 5**).

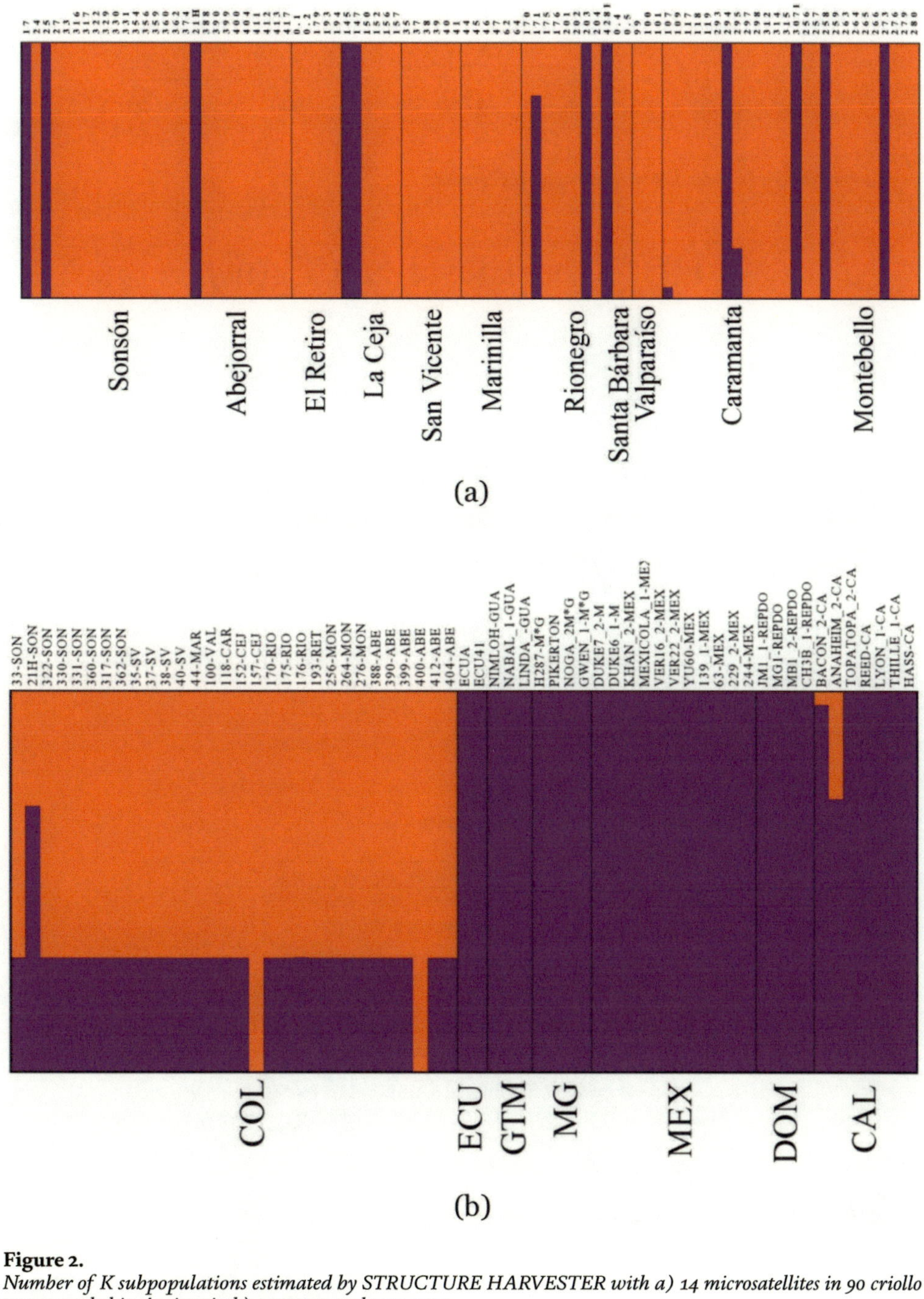

Figure 2.
Number of K subpopulations estimated by STRUCTURE HARVESTER with a) 14 microsatellites in 90 criollo trees sampled in Antioquia b) concatenated sequences.

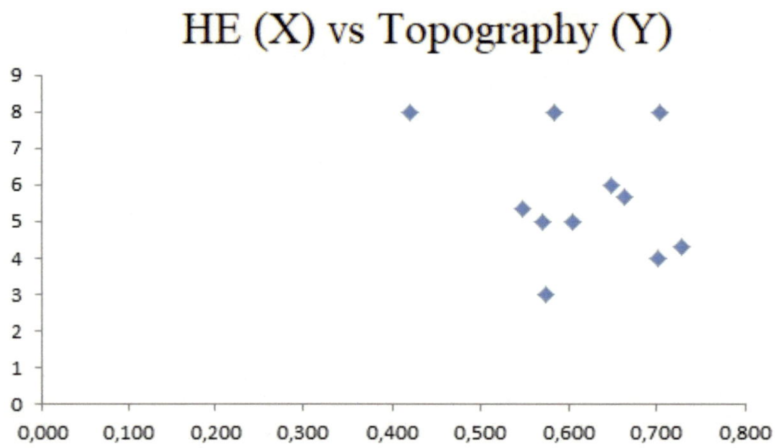

Figure 3.
Correlation between expected heterocigosity and topography.

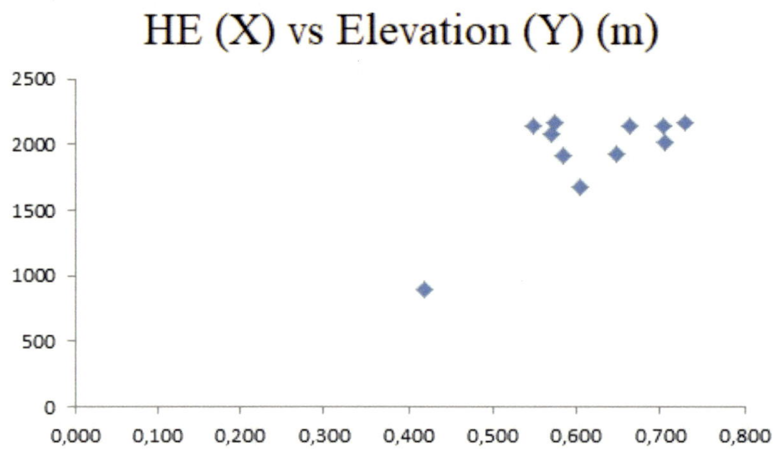

Figure 4.
Correlation between expected heterocigosity and elevation (m).

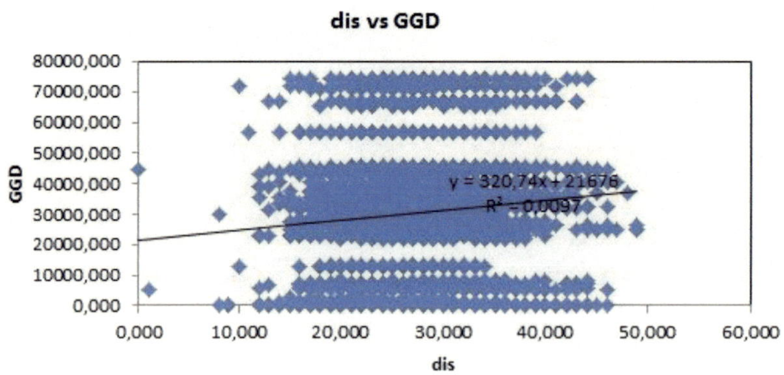

Figure 5.
Mantel test. Geographic vs. genetic distances measured in avocado.

3.2 Phylogeographic results

Amplification of the three nuclear genes was obtained from 87/90 criollo cultivars. Cell gene was sequenced in 49/87 cultivars, CHS gene in 80/87 and STK gene in 57/87. Amplification and final edition for Cell gene produced a final fragment of 997 bp, for the gene CHS of 827 bp and the gene STK of 1170 bp. These fragments were shorter than those obtained by Chen et al. (2008) and Chen et al. (2009). Chromatograph visualizations showed that locus Cell exhibited 10/49 cultivars with heterozygous genotypes in 3 positions: 68 (A/T), 75 (C/T) and 249 (C/G). Allele variation in position 75 was the most frequent. Locus CHS had 31/80 cultivars that were heterozygous in 14 positions: 9 (T/C), 19 (T/C), 58 (T/G), 75 (G/C), 96 (C/T), 121 (G/A), 126 (G/C), 201 (A/T), 214 (T/C), 218 (G/C), 257 (T/C), 258 (A/G), 302 (G/C) and 365 (T/C). The most abundant heterozygous variants were found in positions 121 and 201. Locus STK exhibited 14/57 cultivars with heterozygous genotypes in 25 positions: 17 (C/A), 88 (C/A), 89 (C/A), 104 (G/A), 193 (T/C), 194 (C/T), 209 (C/T), 215 (A/G), 229 (G/A), 297 (G/A), 307 (G/T), 314 (A/G), 379 (C/T), 399 (A/G), 414 (A/T), 418 (T/C), 428 (C/T), 468 (A/G), 472 (A/T), 485 (A/G), 631 (T/C), 648 (C/T) and 737 (A/G). The most frequent heterozygous were 88, 229, 379, and 399.

Genetic diversity obtained in the three loci showed that Cell locus exhibited high values for the criollo trees, and the high values between criollo trees vs. cultivars reported by Chen et al. (2009) from Costa Rica, Dominican Republic, Ecuador and México and the interracial hybrids between Guatemala x Mexico races (**Table 2**). D Tajima test was $D = -2.54$, $p < 0.001$ in Antioquia suggesting purifying selection or population avocado expansion [19]. Locus STK showed the highest gene diversity for all estimators in criollo avocados (even higher than Cell) and also between criollo trees vs. avocado cultivars reported by Chen et al. (2009). However, D Tajma test ($D = 0.237$, $p > 0.01$) was not significant suggesting no population expansion (positive or purifying selection pressure) or reduction (bottleneck effect) [19] in Antioquia. On the contrary, locus CHS presented the lowest gene diversities estimations within Antioquian avocados and between Antioquia vs. other cultivars [18]. Also, D Tajma test ($D = 0.796$, $p > 0.01$) was not significant. Genetic diversities increased when criollo sequences were combined with GenBank avocado sequences as more haplotypes were included in the analysis.

In locus Cell, the haplotype number was 20 (**Tables 2** and **3**). Haplotype 1 was composed by 31 sequences from Antioquia solely [Sonsón (9), Abejorral (5), Montebello (5), Marinilla (3), San Vicente (3), El Retiro (2), Rionegro (1), Santa Bárbara (1) Caramanta (1) and Valparaiso (1)]. Haplotype 2 was integrated by 30 cultivars: 7 from Antioquia [Sonsón (3), La Ceja (2), Marinilla (1) and Caramanta (1)], 12 from Guatemala, 5 to Mexico and 5 were hybrids between Guatemala x Mexico (7, 17). Haplotype 13 was integrated by 13 cultivars: 7 native to Mexico, 2 native to Ecuador and the varieties: Topa-Topa, Khan, Mexicola and Puebla. All of them were classified as Mexican races by Chen et al. (2009) except for Puebla (that is a hybrid between the three avocado races M × G × WI with assignment percentages of 6%, 82%, 12% respectively). Haplotype 16 included 5 cultivars: 4 of them from the Dominican Republic and one haplotype from the variety Arue. This haplotype was identified as the West Indian race [18]. Haplotype 15 was composed by sequences from the varieties Thomas and Duke 6 and the Mexican cultivar 63. All of them were assigned to Mexican races by Chen et al. (2009). Haplotype 14 included the commercial varieties: Zutano, Thille, Gwen, Esther, Bacon and Anaheim. All of them were assigned by Chen et al. (2008) and (2009) as Guatemalan races except

Gen	Aligned length	Population Studied	N	S	H	Θ × 10–3	π × 10–3	D Tajima	P
Cell	1540	Total	51	33	15	4.14	2.93	−0.89	>0.10
		Wild	20	30	12	4.6	3.46	−0.85	>0.10
		Cultivars	31	15	10	2.22	2.24	0.035	>0.10
	997	Total	101	89	20	4.29	4.27	2.52	<0.001
		Antioquia	49	80	11	5.89	5.43	2.54	<0.001
		Genebank	52	21	10	2.8	2.79	1.29	>0.10
CHS	1210	Total	43	35	27	6.09	4.92	−0.61	>0.10
		Wild	16	27	15	5.66	4.36	−0.81	>0.10
		Cultivars	27	29	19	5.67	5.02	−0.38	>0.10
	827	Total	115	26	26	3.6	3.58	−1.300	>0.10
		Antioquia	80	9	13	1.52	1.52	−0.79	>0.10
		Genebank	35	25	15	5.47	5.43	−1.09	>0.10
STK	1398	Total	53	48	22	6.56	5.51	−5.02	<0.001
		Wild	20	43	16	7.23	6.18	−5.51	<0.001
		Cultivars	33	27	11	4.06	4.98	−0.72	>0.10
	1170	Total	113	41	33	7.09	6.42	−0.24	>0.10
		Antioquia	57	19	19	3.68	3.66	−0.032	>0.10
		Genebank	56	31	15	4.87	4.89	−0.66	>0.10

Table 2.
Genetic diversity was estimated from three nuclear loci in criollo avocado trees sampled at the department of Antioquia. N = total number of genotyped individuals, S = segregant sites, Θ = nucleotide polymorphism, π = nucleotide diversity, D = Tajima test, P = p-value.

Haplotype	Cultivar	Haplotype	Cultivar
1	47-MAR –46-MAR –44-MAR –412-ABE –411-ABE –40-SV - 400-ABE –390-ABE –38-SV - 388-ABE - 362-SON -360-SON -35-SV -33-SON - 330-SON -322-SON -317-SON -29-H-SON - 265-MON -264-MON -263-MON -258-MON - 256-MON -24-SON -194-RET –193-RET –17-SON -170-RIO -118-CAR –100-VAL –04-SB	2	Yu60 - Reed- Pinkerton_2- Noga_2 - Nimlioh- -Nabal- Lyon- Linda- LeavenHass- - HX48 -Hass- -H670- -H287- Fuerte_2- Duke7_2 - Daily11- COSTRI - Ch35_2- 65_2 19-R-SON -145-CEJ –12-R-CEJ21-H-SON – MEX - 46_2-MEX - 445-R-MAR –307-H-CAR –25-SON -229_2- MEX - 229_1- MEX-
3	37-SV - 257-MON	4	147-CEJ
5	152-CEJ	6	157-CEJ
7	175-RIO -171-RIO	8	176-RIO
9	276-MON	10	404-ABE
11	428-H-RIO	12	446-F-MAR
13	Ver3-MEX -Ver22-MEX -TopaTopa-QRO1-MEX -Puebla_2- -Mexicola_2- Khan- -ECU - 65_1-MEX -63_1-MEX -46_1-MEX –41-ECU -139-MEX	14	Zutano_1 -Thille_1-Gwen_1- Esther_1-Bacon_1- Anaheim_1
15	Thomas_1 -Duke6_1–63_2_-MEX	16	MG1-REPD - MC1_1-REPD - MB1-REPD JM1_1-REPD Arue_1-SI
17	MC1_2-REPD -JM1_2_-REPD – Ch3B-REPD	18	244-MEX
19	Ver16_2-MEX	20	Ch35_1-MEX

Table 3.
Haplotype list of avocado cultivars produced by locus cell in 107 specimens.

for Zutano that is a hybrid between the Guatemalan x Mexican races. Haplotype 17 was composed of cultivars from the Dominican Republic that were assigned as West Indian race by Chen et al. (2008) and (2009). Finally, the rest of the haplotypes were composed of one or two sequences from Mexico or Antioquia.

For the gene CHS, the number of estimated haplotypes was 26 (**Tables 2** and **4**). Haplotype 5 (H5) included 48 avocado cultivars from Antioquia together with 2

Haplotype	Cultivar	Haplotype	Cultivar
1	CH-MB1-REPD CH-Ch3B-REPD CH-MG1_2-_REPD CH-JM1_2-REPD	2	CH-244-MEX CH-139-MEX CH-Ver16_2-MEX CH-KHAN-MEX CH-LYON_1-CA CH-BACON_2-CA
3	CH-Yu60_2-MEX	4	CH-Ver22_1-MEX CH-63-MEX CH-ZUTANO_1-CA
5	CH-1-ECU CH-41_1-ECU CH-262-MONT CH-256-MONT CH-193-RET CH-176-RIO CH-170-RIO CH-176-RIO(2) CH-179-RET CH-193-RET(2) CH-194-RET CH-201-RIO CH-202-RIO CH-259-MONT CH-276-MONT CH-279-MONT CH-280-MONT CH-293-CARA CH-331-SON CH-356-SON CH-359-SON CH-445-R-MARI CH-446-F-MARI CH-38-SV CH-45-MARI(2) CH-99-VAL CH-109-CARA CH-145-CEJ CH-156-CEJ CH-263-MONT CH-322-SON CH-362-SON CH-46-MARI CH-100-VAL CH-117-CARA CH-329-SON CH-374-SON CH-21-H-SON(2) CH-47-MARI CH-101-VAL CH-118-CARA CH-150-CEJ CH-390-ABEJ CH-05-SB CH-37-SV CH-44-MARI CH-64-MARI CH-152-CEJ CH-317-SON CH-399-ABEJ(2)	6	CH-2-COSTRI CH-NIMLIOH-GUAT
7	CH-Ch35_1-MEX CH-H287-MEXxGUAT CH-GWEN-MEXxGUAT CH-11-R-CEJ	8	CH-229_2-MEX CH-REED-CA CH-THILLE_1-CA CH-DUKE7-MEX
9	CH-THOMAS-MEX CH-ANHEIM_2-CA	10	CH-PIKERTON-MEXxGUAT CH-NOGA_2-MEXxGUAT CH-HASS_2-CA
11	CH-NABAL-GUAT	12	CH-DUKE6-MEX
13	CH-TOPATOPA-CA	14	CH-MEXICOLA-MEX
15	CH-LINDA-GUAT	16	CH-354-SON CH-354-SON(2) CH-27-SON CH-33-SON CH-40-SV(2) CH-256-MONT(2) CH-360-SON CH-400-ABEJ
17	CH-147-CEJ	18	CH-119-CARA CH-45-MARI CH-40-SV CH-35-SV CH-O4-SB CH-157-CEJ CH-175-RIO CH-388-ABEJ CH-399-ABEJ CH – 404-ABEJ CH-412-ABEJ CH-264-MONT CH-330-SON
19	CH-12-R-CEJ	20	CH-19-SON
21	CH-21-H-SON	22	CH-25-SON
23	CH-29-H-SON CH-147-CEJ(2)	24	CH-204-RIO
25	CH-307-H-CARA	26	CH-258-MONT

Table 4.
Haplotype list of avocado cultivars produced by locus CHS in 115 specimens.

sequences from Ecuador. These last two sequences were assigned to the Mexican race by Chen et al. (2008). H5 Antioquian cultivars were distributed as follows: Sonsón (9), Marinilla (6), Montebello (8), Rionegro (5), Caramanta (4), La Ceja (4), El Retiro (4), Valparaiso (3), Abejorral (2), San Vicente (2) and Santa Bárbara (1). H8 (N = 13) was also composed by several Antioquian samples distributed as follows: Abejorral (4), San Vicente (2), Santa Bárbara (1), Marinilla (1), Sonsón (1), Rionegro (1), La Ceja (1), Caramanta (1) and Montebello (1). H16 was composed by 8 haplotypes from Antioquia and the municipalities: Sonsón (5), San Vicente (1), Abejorral (1) and Montebello (1). Haplotype 2 was composed by 6 cultivars: Ver 16, Khan-CA, Lyon-CA, Bacon-CA, and two Mexican races, all these sequences were assigned to the Mexican race by Chen et al. (2008). H7 was integrated by 4 haplotypes: two from Mexico, the haplotypes H27 and Gwen that are hybrids between Guatemala x Mexico races, and one specimen from La Ceja (Antioquia). H8 was also integrated by 4 haplotypes from Mexico: one Mexican sequence, and the varieties Reed (California), Thille (California) and Duke 7 (Mexico). H3 was composed of 3 haplotypes: 2 Noga sequences (a Hass variety) and one cultivar named Pikerton (that represents a hybrid between a Guatemalan race x a Mexican race). Finally, the other haplotypes found for CHS were composed of one sequence from the varieties: Nabal (Guatemala), Duke 6 (Mexico) and Topa-Topa (California). Topa-Topa was classified as a Mexican race by Chen et al. (2008, 2009).

Amongst three loci, gene STK presented the highest number of haplotypes with N = 33 (**Tables 2** and **5**). This gene was analyzed in 113 cultivars (57 from Antioquia and 56 from the GenBank from native cultivars and commercial avocados). H12 represented most of the haplotypes identified for this locus with 24 cultivars, that included: 12 haplotypes of the Guatemalan race, 7 haplotypes that are hybrids between Guatemala x Mexico, a Mexican race, the cultivar 65 from Mexico according to Chen et al. (2008, 2009) and the Antioquian specimens: 19RSON, 12R-CEJ, 11R-CEJ. H3 was composed by 17 cultivars from Antioquia, distributed as follows: Sonsón (5), Abejorral (3), Montebello (3), Marinilla (1), San Vicente (1), El Retiro (1), Rionegro (1), La Ceja (1) and Valparaiso (1). H2 was integrated by 16 cultivars from Antioquia composed by individuals from: Abejorral (4), Sonsón (3), Rionegro (3), Montebello (2), San Vicente (1), Caramanta (1), El Retiro (1) y La Ceja (1). H24 included 9 cultivars, 4 from the Mexican race (commercial avocados), one specimen that is a hybrid between Guatemala x Mexico races and 4 native specimens collected in Mexico and assigned within the Mexican race [7]. H22 was composed of 8 cultivars: 2 native trees from Mexico, 3 from the Dominican Republic assigned within the West Indian race [18], the varieties Thomas and Khan assigned as Mexican races and a specimen from the variety Arue that was assigned within the West Indian race [18]. H33 was composed of haplotypes of the varieties Nabal, Bacon and Anaheim all assigned to the Guatemalan race by Chen et al. (2009). H23 was integrated by the sylvester specimens MG1 and Ch3B-1 and both are native to the Dominican Republic and cultivar COSTRI, from Costa Rica, all of them were assigned within the West Indian race. H4 was composed by Antioquian cultivars from Caramanta (2) and Sonsón (1). Other haplotypes were composed of one or two haplotypes from Mexico and some commercial varieties of avocado.

STRUCTURE HARVESTER produced K = 2 for the three concatenated loci according to the values: mean Ln P (D) = −1445.81, Stdev = 18.08 and Δ K = 7.79 estimated by the Evanno test implemented in STRUCTURE HARVESTER (**Figure 2b**). The first cluster was composed of most criollo samples. These criollo avocadoes were assigned together with Mexican and Guatemalan races whereas the second cluster was composed of only two specimens, one from Sonsón and the other from La Ceja.

Phylogeny obtained from concatenated sequences diverged in two main clusters, the first cluster was constituted by criollo avocados and the second by GenBank

Haplotype	Cultivar	Haplotype	Cultivar
1	33-SON	2	-404-ABE –399-ABE –388-ABE –360-SON -356-SON - 331-SON -293-CARA –276-MONT –256-MONT –202-RIO -201-RIO -193-RET –175-RIO -152-CEJ
3	44-MARI -412-ABE –400-ABE –390-ABE –362-SON - 35-SV -354-SON -322-SON - 280-MONT –27-SON - 279-MONT	4	359-SON - 118-CARA –117-CARA
5	329-SON - 259-MONT	6	445-R-MARI –21H-SON
7	05-SB	8	330-SON -119-CARA
9	37-SV	10	317-SON Zutano_2- Whitsell_1- Thille_1 -Teague_2-Reed -Pinkerton_2- Noga_2- Nimlioh -Nabal_1-
11	38-SV	12	Linda_1 -LHass -HX48 - Hass-H670 - H287-Gwen_1- -Esther- -Duke7_2 -Daily11_1- Andes3-65_2 -19RSON –
13	25-SON	14	157-CEJ
15	176-RIO	16	204-RIO
17	266-MONT	18	428H-RIO 411-ABE –40-SV
19	446F-MARI	20	Yu60
21	Ver3	22	Ver16_2- -Thomas_2- QRO1- MC1-MB1_2-Khan_2-JM1- Arue-SI
23	MG1 – COSTRI - Ch3B_1	24	Noga_1- Mexicola_1-Lyon_1 - Ganter-Fuerte_1 - 63-46-244 - 139_1
25	Ver22_2	26	Ver16_1
27	ECUA - Duke6 - 41_1	27	ECUA - Duke6 - 41_1
28	184_1	29	TopaTopa_2
30	Puebla_1	31	Ver22_1
32	229_2	33	Nabal_2 - Bacon_2 Anaheim_2

Table 5.
Haplotype list of avocado cultivars produced by locus STK in 113 specimens.

accessions (**Figure 6**). The closest varieties to criollo avocadoes were Bacon and Anaheim. These varieties are both commercial and classified as Guatemalan races by Chen et al. (2009).

3.3 Morphological results

Descriptive analysis of morphological data (N = 28) (**Table 6**) showed high Coefficient of Variation (CV) values for most traits demonstrating that morphological variation amongst criollo avocadoes is high. For this reason, these avocadoes were not differentiated amongst the three botanical races suggesting that morphological traits are not useful to distinguish criollo trees unless molecular data are included in the analysis. Since PCA analysis showed that the most important landscape variable was elevation (altitude) as 95% of the variation was explained by the first component and 5% by the second. Concatenation of this variable together with morphological and molecular data (microsatellites) produced three clusters distributed as follows: from 800 to 1900, from 2000 to 2100 and from 2200 to 2300 m. a. s. l. The first cluster was composed of avocadoes collected in the municipalities

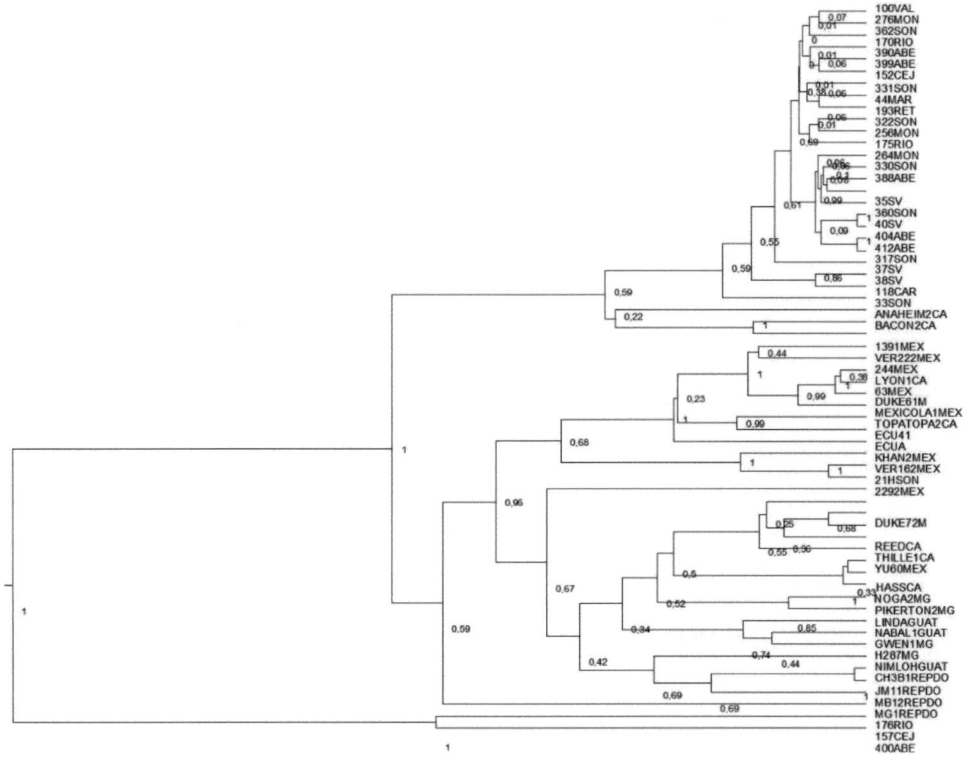

Figure 6.
Phylogeny of avocado cultivars from Colombia and world bank accessions.

Variable	Mean	Minimum	Maximum	SD	CV
Altitude (m)	2015.11	864	2281	242.79	12.05
Tree height (M)	13.05	1.8	25.7	4.66	35.72
Canopy with N-S	8.58	1	20.2	3.96	46.17
Canopy width N, W, E	8.66	1	22	4.18	48.29
Trunk circumference (cm)	117.79	19	234	45.45	38.59
Between stem lengths (cm)	0.62	0.234	1.428	0.25	39.62
Stems diameters (cm)	0.48	0.29	1.04	0.1	21.01
Leave length (cm)	18.58	9.8	30.8	4.12	22.15
Leave width (max) (cm)	8.78	4.7	14.5	2.08	23.64
Number of veins/ leave (Pairs)	7.44	1	16	3.68	49.42
Divergence between veins/leave (grades)	41.52	30	60	6.05	14.57

Table 6.
The descriptive analysis made on continuous traits measures in 90 criollo avocados.

of Abejorral, Montebello and Santa Bárbara found from 800 to 1900. The second avocadoes from Caramanta, Sonsón, El Retiro and Rionegro found from 2.000 to 2.200 and the third group by avocadoes from Valparaíso, La Ceja and Sonsón found from 2.300 to 2.400 m. a. s. l. (**Figure** 7).

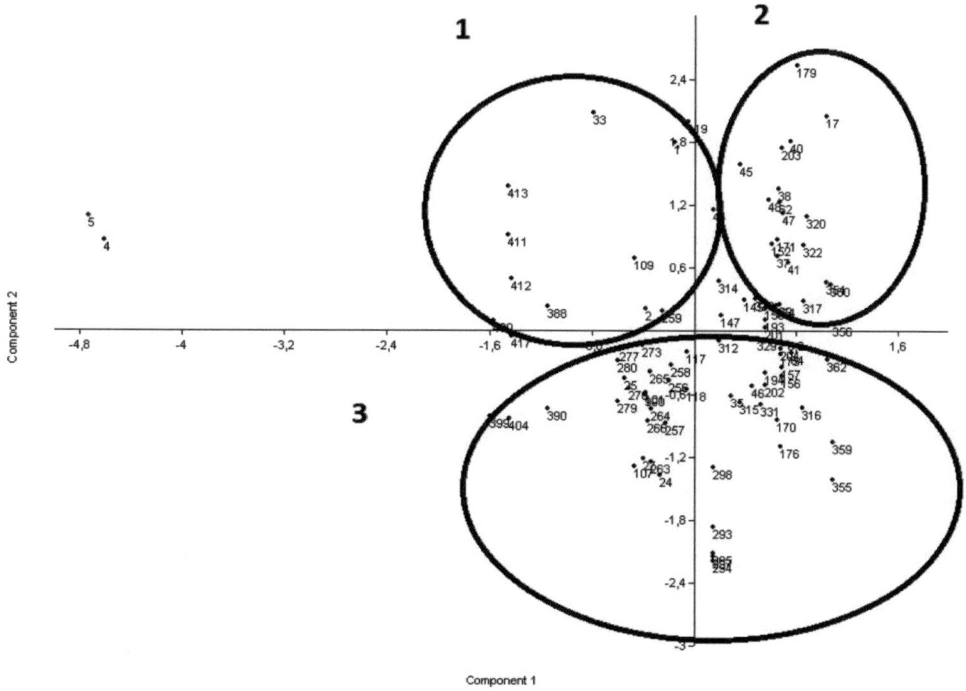

Figure 7.
PCA analysis made in criollo avocados with morphological traits.

4. Discussion

Molecular characterization was made here in 90 criollo trees from Northeast Colombia (Antioquia) by using 14 microsatellites previously standardized by Alcaraz and Hormaza (2007). These criollo trees presented similar genetic diversities to avocado germplasm banks from Spain [14] and US [1]. These germplasms are mainly composed by avocado cultivars from different countries such as Mexico, Guatemala, Israel, US, Chile, Ecuador, Canary Islands, amongst other countries. Also, genetic diversity found within criollo trees was higher than the diversity obtained with 6 microsatellites in avocados from Veracruz (Mexico) by Galindo-Tovar and Milagro (2011). Differences between these two studies could be due to criollo avocados from Antioquia being the product of multiple hybridizations through insect pollinators whereas Mexico samples were mostly composed of avocado cultivars. Cañas-Gutierrez et al. (2015) made a molecular analysis of 111 avocados from Antioquia with 38 AFLP and observed that criollo avocados were molecularly highly polymorphic. They also found that these trees shared AFLP bands with the varieties Fuerte, Hass and Reed demonstrating their hybrid origin.

Genetic population structure was also found amongst 11 Antioquian municipalities, where criollo avocadoes were collected, as $FST = 0.05439$ ($p < 0.001$). Similar results were obtained by Cañas-Gutiérrez et al. (2015) by using AFLP. The outcome obtained with these molecular markers could be explained due to differences in Holdridge zones amongst 11 municipalities. Avocadoes collected in these 11 sites were clustered in three groups by the PCA analysis. In contrast, only $K = 2$ groups were estimated with STRUCTURE HARVESTER based on molecular data only.

Person correlations between expected heterozygocities (estimated for 14 microsatellites) vs. elevation (altitude) and expected heterozygocities vs. topography were only significant for the former. This result suggests the importance of Holdridge zones (altitudes) in the genetic differentiation of criollo avocadoes from Colombia. Also, the Mantel test was significant suggesting isolation by distance in criollo Antioquian avocado. This outcome might be explained due to avocado farm producers sharing seedlings between neighbors, increasing genetic similarities between close orchards compared to distant orchards.

Also, a genetic characterization was made in 87 criollo trees based on Sanger DNA sequencing by using three nuclear genes: endo-1,4-D-glucanase (Cell), chalcone synthase (CHS), and serine–threonine-kinase (STK). These loci were previously standardized by Chen et al. (2008) in 21 sylvester avocado cultivars from Mexico, Ecuador, Costa Rica and the Dominican Republic and by Chen et al. (2009) in 33 domesticated cultivars. In these two separated studies, the authors estimated K = 3 avocado clusters with STRUCTURE and assigned their cultivars and domesticated avocadoes to the three botanical races.

Concerning the results related to the heterozygocities found in the criollo avocadoes. These results can be due to *P. americana* being a diploid species with 24 chromosomes (2n = 2x = 24) [37–39], meaning that during cell division the species segregates two alleles per locus producing homozygous and heterozygous genotypes. Chen et al. (2008) and Chen et al. (2009) re-sequenced four nuclear loci: endo-1,4-D-glucanase (Cell), chalcone synthase (CHS), flavanone-3-hydroxylase (F3H), and serine–threonine-kinase (STK) with AS (allele-specific)-PCR procedure. This last method facilitated the identification of heterozygous genotypes from each cultivar. In contrast, in this study, Sanger sequencing was implemented. This method required chromatograph observations to detect heterozygous genotypes [29, 40]. Chen et al. (2015) made a phylogeographic study of five Rhizophora species (mangrove) with five nuclear genes (22,454, 23,056, 22,274, 23,714 and C49) and sequenced four R. apiculate populations with seven nuclear genes (C22, 22,274, 23,056, 23,852, 22, 23,186 and 22,066). They found that the genus Rhizophora diverged into two clades: one composed by R. mangle and R. racemosa and the other by *R. apiculata*, *R. stylosa* and *R. mucronata*. In the case of *R. apiculate*, the presence of double peaks in nuclear genes allowed them to discover hybrids between *R. stylosa* and *R. apiculata* as they represented genotypes that were the product of the crosses between these two mangrove species. Likewise, heterozygous detection was made by Wei et al. (2017) with the nuclear gene PAL (phenylalanine ammonia-lyase) and the species *Camellia flavida*, an endangered population of yellow camellia from southwest China. Wei et al. (2017) used chloroplast and nuclear sequences to improve the taxonomic classification of *C. flavida* and found that the species was differentiated into three groups named: *C. flavida:* var. flavida 1, var. flavida 2, and var. patens. These two studies were useful to implement the same data analysis performed in criollo avocadoes made in this work.

In criollo avocadoes, results of the percentage of heterozygous genotypes showed that 20% of heterozygous were observed in locus Cell, 43% in locus CHS and 49% in locus STK. These outcomes were slightly lower to Chen et al. (2008) as they estimated heterozygocities of 40%, 70.6% and 50% for these three loci respectively from sylvester avocado cultivars and values of 48.4%, 69.2% and 68% from domesticated cultivars. Higher gene diversity estimators (number of haplotypes, Hd, π, Θ) were found in both Cell and CHS loci compared to haplotypes reported in the GenBank, but lower estimators were found for locus STK. These results might be due to differences between samples origins, sample sizes (sequence length) and selection pressures [17]. Also, it is important to mention that the Tajima test was only significant for locus Cell meaning that the gene might be under purifying selection pressure or the population is in expansion in Antioquia.

DNAsp results showed that some criollo haplotypes from Antioquia clustered with the Mexican and Guatemalan sequences (and therefore their races) reported from the three loci by Chen et al. (2008) and (2009). Our results also showed K = 2 for the concatenated sequenced loci with STRUCTURE HARVESTER. The first cluster was integrated by avocado sequences from Antioquia together by sequences reported by Chen et al. (2008) and (2009) from Ecuador, Guatemala, Mexico, California, and crosses between Mexico x Guatemala races and the second cluster by 3 samples from Antioquia that were genetically apart from the rest. In contrast, samples from Dominican Republic sequences deposited in the GenBank did not cluster with any criollo samples, demonstrating that they are not genetically similar to the West Indian race as these avocados were classified within this race by Chen et al. (2008) and (2009). Bayesian phylogeny obtained with concatenated sequences produced two main clades, one was composed of Antioquian avocados and the other by sequences reported previously [7, 18]. This outcome was similar to the results obtained with STRUCTURE HARVESTER as the tree clustered criollo avocados within one group that was closely related to two samples that were previously classified as Guatemalan races.

Concerning morphological data analyzed here, most continuous traits presented large Coefficient of Variation (CV) estimations suggesting that most morphological traits are highly variable, for this reason, they were not useful to classify criollo trees within the three botanical races [41]. According to Montes-Hernández et al. (2017), CV values superior to 20% indicate a high morphological variation within vegetal species. During avocado samplings, fruit collections were not carried out, for this reason; other morphological traits were used to classify criollo trees within the three botanical races. They were: stem form, leaves length and width and trunk surface. According to 44, the Mexican race exhibits small leaves, the Guatemalan race medium size leaves and the West Indian race large leaves. Leaves length mean was 18. 58 cm for criollo avocados and ranged from 9.80 cm to 30.80 cm. Leaves' width mean was 8.78 cm and ranged from 4.70 cm to 14. 50 cm. The trunk surface is usually smooth in both Mexican and Guatemalan avocado races, whereas in the West Indian race is rough [42]. Results on the trunk surface showed that 83.5% of criollo trees were rough and the rest were smooth. Finally, another trait that is relevant to distinguish the three races is elevation ranges as Mexican and Guatemalan races are usually found in high altitudes whereas West Indian races are found in low altitudes [7]. In this study, criollo avocados were distributed from 864 to 2.281 m. a. s. l. suggesting they are within Mexican and Guatemalan races ranges of distribution. The Mexican race is mainly sowed from 1.000 to 2.000 m. a. s. l. And the Guatemalan race above 2.000 m. a. s. l [43]. Furthermore, results from PCA analysis demonstrate that the elevation is the most relevant landscape variable discovered here as it plays an important role in the clustering pattern obtained for 90 criollo trees.

In sum, results obtained here showed that criollo avocados are highly diverse in genetics and morphology. Morphological traits failed to differentiate criollo trees within the 3 botanical races. Population genetic structure was found with Antioquia and also that elevation differences between sampling sites played an important role in the genetic differentiation of the criollo trees studied in this work. This result was also corroborated by PCA analysis. STRUCTURE HARVESTER showed K = 2. Criollo samples were genetically similar to Guatemalan and Mexican races. Further analyses are necessary for the species related to its genetic characterization, particularly next-generation sequencing studies.

5. Conclusions

This work has shown the importance of the genetic analysis of crops of economical importance based on diverse molecular markers where microsatellites are relevant to show the genetic variability of the species and current gene flow of a species while DNA sequencing generates information related to the genetic history of the species though comparisons of DNA sequences hold in databases vs. DNA sequences obtained in an investigation. Landscape analysis also requires the use of morphological traits as they can be used to explain the genotypic and phenotypic interactions in a species. According to the results obtained in this study, we found that Colombian avocado named "criollo" is the product of multiple hybridizations between natural trees enhanced by pollinators according to the results obtained from microsatellites and the high genetic diversity found in our avocado. Also, morphological variation showed that the species in Colombia is diverse and a product hybridization given the intermixed traits found on 90 criollo trees. Finally, DNA sequencing of three nuclear genes showed that Colombian avocados are genetically closer to the Mexican and the Guatemalan races of avocado and for this reason grafting between criollo trees and Hass variety is possible since both avocados are produced within the same altitudes.

Acknowledgements

This project was financially supported by Universidad Nacional de Colombia, sede Medellín to Clara Inés Saldamando Benjumea. Code No 29424. Collection and genetic access were provided for this study by ANLA (Autoridad Nacional de Licencias Ambientales) to UNAL (Universidad Nacional de Colombia) No 0255 given on March 14, 2014.

Authors contributions

GPC made the collections and molecular analysis. JM and CISB analyzed the data. RAI trained GPC, GPC and CISB wrote the manuscript.

Author details

Clara Inés Saldamando-Benjumea[1,2]*, Gloria Patricia Cañas-Gutiérrez[3], Jorge Muñoz[2] and Rafael Arango Isaza[1,2]

1 Escuela de Biociencias, Facultad de Ciencias, Universidad Nacional de Colombia, Medellín, Colombia

2 Plant Biotechnology Unit, Corporation for Biological Research, Medellín, Colombia

3 Plant Health and Biological Control Unit, Corporation for Biological Research, Medellín, Colombia

*Address all correspondence to: cisaldam@unal.edu.co

References

[1] Ashworth VETM, Clegg MT. Microsatellite markers in avocado (*Persea americana* Mill.): genealogical relationships among cultivated avocado genotypes. The Journal of Heredity. 2003;**94**(5):407-415. DOI: 10.1093/jhered/esg076

[2] Galindo-Tovar ME, Arzate -Fernández. Consideraciones sobre el origen y primeras dispersiones del aguacate (Persea americana, Lauraceae). Cuadernos de Biodiversidad. 2010;**33**:11-15

[3] Corona-Jacome EC, Galindo-Tovar ME, Lee-Espinosa HE, Landero-Torres I. Diversidad genética del aguacate (*Persea americana* Mill.) en cuatro zonas de su área de dispersión natural. Agro Productividad. 2016;**6**:80-85

[4] Bergh B, Ellstrand N. Taxonomy of the avocado. Calif Avocado Society Year Book. 1986;**70**:135-145

[5] Davenport TL. Avocado flowering. Horticultural Reviews. 1986;**8**:257-289

[6] Sharma CB, Martinez GC. The component triacylglycerols of avocado fruit-coat. Journal of American Oil Chemistry Society. 1972;**49**:229-232

[7] Chen H, Morrell PL, De la Cruz M, Clegg MT. Nucleotide diversity and linkage disequilibrium in wild avocado (Persea americana Mill.). The Journal of Heredity. 2008;**99**:382-389. DOI: 10.1093/jhered/esn016

[8] Bergh BO, Lahav E. Avocados. In: Janick J, Moore JN, editors. Fruit Breeding. Vol 1. Tree and Tropical Fruits. New York: Wiley; 1996. pp. 113-166

[9] Chanderbali AS, Albert VA, Ashworth VTEM, Clegg MT, Litz RE, Soltis DE, et al. *Persea americana* (avocado): bringing ancient flowers to fruit in the genomics era. Bioassays. 2008;**30**:386-396. DOI: 10.1002/bies.20721

[10] Galindo-Tovar M, Milagro P. Genetic relationships within avocado (*Persea americana* Mill.) in seven municipalities of central Veracruz characterized using microsatellites markers. Tropical and Subtropical Agroecosystems. 2011;**13**(3):339

[11] Schaffer B, Gil PM, Mickelbart MV, Whiley AW. In: Schaffer B et al., editors. Ecophysiology The Avocado: Botany. Production and Uses. 2nd ed. CAB International; 2013. pp. 168-199

[12] Cañas-Gutiérrez GP, Galindo-López LF, Arango- Isaza RE, Saldamando-Benjumea CI. Diversidad genética de aguacate (*Persea americana*) en Antioquia. Colombia. Agronomía Mesoamericana. 2015;**26**(1):129-143. DOI: 10.15517/am.v26i1.16936

[13] Mijares P, López L. Variedades de aguacate y su producción en México. México: Memoria Fundación Salvador Sánchez Colín CICTAMEX S.C; 1998

[14] Alcaraz M, Hormaza I. Molecular characterization and genetic diversity in an avocado collection of cultivars and local Spanish genotypes using SSRs. Hereditas. 2007;**144**:244-253. DOI: 10.1111/j.2007.0018-0661.02019x

[15] Gross-German E, Viruel MA. Molecular characterization of avocado germplasm with a new set of SSR and EST-SSR markers: genetic diversity population structure and identification of race-specific markers in a group of cultivated genotypes. Tree Genetics & Genomes. 2013;**9**:539-555. DOI: 10.1007/s11295-012-0577-5

[16] Corona-Jacome EC, Galindo-Tovar ME, Lee-Espinosa HE, Landero-Torres I. Diversidad genética del

aguacate (Persea americana Mill.) en cuatro zonas de su área de dispersión natural. Agro Productividad. 2016;**6**:80-85

[17] Freeland J. Molecular Ecology. Jonh Wiley & Sons: West Sussex; 2005, p. 388

[18] Chen H, Morrell PL, De la Cruz M, Clegg MT. Tracing the geographic origins of major avocado cultivars. The Journal of Heredity. 2009;**100**:56-65. DOI: 10.1093/jhered/esn068

[19] Nei M, Kummar S. Molecular Evolution and Phylogenetics. Oxford: Oxford University Press; 2000. p. 333

[20] Manel S, Schwartz MK, Luikart G, Taberlet P. Landscape genetics: combining landscape ecology and population genetics. Trends in Ecology & Evolution. 2003;**18**:189-197. DOI: 10.1016/S0169-5347(03)00008-9

[21] IPGRI (International Plant Genetic Resources Institute). Descriptors for Avocado (Persea spp.). Rome, Italy: International Plant Genetic Resources Institute; 1995. p. 58

[22] Peakall R, Smouse PE. GENALEX 6: Genetic analysis in Excel. Population genetic software for teaching and research. Molecular Ecology Notes. 2006;**6**:288-295. DOI: 10.1111/j.1471-8286.2005.01155.x

[23] Raymond M, Rousset F. GENEPOP (version 1.2): population genetics software for exact tests and ecumenicism. Journal of Heredity. 1995;**86**:248-249. DOI: 10.1093/oxfordjournals.jhered.a111573

[24] Pritchard JK, Stephens M, Donnelly P. Inference of population structure using multilocus genotype data. Genetics. 2000;**155**:945-959

[25] Librado P, Rozas J. dnaSP v5: A software for comprehensive analysis of DNA polymorphism data. Bioinformatics. 2009;**25**:1451-1452. DOI: 10.1093/bioinformatics/btp187

[26] Evanno G, Regnaut S, Goudet J. Detecting the number of clusters of individuals using the software STRUCTURE: A simulation study. Molecular Ecology. 2005;**14**:2611-2620. DOI: 10.1111/j.1365-294X.2005.02553.x

[27] Earl DA, von Holdt BM. STRUCTURE HARVESTER: A website and program for visualizing STRUCTURE output and implementing the Evanno method. Conservation Genetics Resources. 2012;**4**:359-361. DOI: 10.1007/s12686-011-9548-7

[28] Jakobsson M, Rosenberg NA. CLUMPP: A cluster matching and permutation program for dealing with label switching and multimodality in analysis of population structure. Bioinformatics. 2007;**23**:1801-1806. DOI: 10.1093/bioinformatics/btm233

[29] Chen Y, Hou Y, Guo Z, Wang W, Zhong C, Zhou R, et al. Applications of multiple nuclear genes to the molecular phylogeny, population genetics and hybrid identification in the mangrove genus Rhizophora. PLoS One. 2015;**10**(12):e0145058. DOI: 10.1371/journal.pone.0145058

[30] Clark AG. Inference of haplotypes from PCR-amplified samples of diploid populations. Molecular Biology and Evolution. 1990;**7**:111-122. DOI: 10.1093/oxfordjournals.molbev.a040591

[31] Nei M, Kummar S. Molecular Evolution and Phylogenetics. Oxford: Oxford University Press; 2000. p. 333

[32] Posada D. jModeltest: Phylogenetic model averaging. Molecular Biology and Evolution. 2008;**25**:1253-1256. DOI: 10.1093/molbev/msn083

[33] Bouckaert R, Heled J, Kühnert D, Vaughan T, Wu C-H, Xie D, et al. BEAST 2: A software platform for bayesian evolutionary analysis. PLoS

Computational Biology. 2014;**10**: e1003537. DOI: 10.1371/journal.pcbi.1003537

[34] Rambaut A. FigTree v1.4.2. [WWW Document]. 2014. Available from: http://tree.bio.ed.ac.uk/software/figtree/ [Accessed: June 9, 2019]

[35] Hammer OD, Harper AT, Ryan PD. PAST: Palaeontological statistics software package for education and data analysis. Palaeontologia Electronica. 2001;**4**:1-9 http://palaeo-electronica.org/2001_1/past/issue1_01.htm

[36] OhsawaT TY, Saito Y, SawadaY IHY. Altitudinal genetic diversity and differentiation of Quercus crispula in the Chichibu mountains, Central Japan. International Journal of Plant Sciences. 2007;**168**:333-340

[37] Garcia A. Cytogenetic studies in the genus Persea (Lauraceae). I. Karyology of seven species. Canadian Journal of Genetics and Cytology. 1975;**17**:173-180

[38] Ibarra-laclette E, Méndez-Bravo A, Pérez-Torres CA, Albert VA, Mockaitis K, López-Gómez R, et al. Deep sequencing of the Mexican avocado transcriptome, an ancient angiosperm with a high content of fatty acids. BMC Genomics. 2015;**16**:599. DOI: 10.1186/s12864-015-1775-y

[39] Boza EJ, Tondo CL, Ledesma N, Campbell RJ, Bost J, Schnell RJA, et al. Genetic differentiation, races and interracial admixture in avocado (*Persea americana* Mill.), and Persea spp. evaluated using SSR markers. Genetic Resources and Crop Evolution. 2018;**65**: 195-1215

[40] Wei SJ, Lu YB, Ye QQ, Tang SQ. Population genetic structure and phylogeography of camellia flavida (Theaceae) based on chloroplast and nuclear DNA sequences. Frontiers in Plant Science. 2017;**8**:718. DOI: 10.3389/fpls.2017.00718

[41] Montes-Hernández SJ, de la Torre-Vizcaino D, Heredia-García E, Hernández-Martínez M, Camarena-Hernández MG. Caracterización morfológica de germoplasma de aguacate MEXICANO (Persea americana var. drymifolia. LAURACEAE). Interciencia. 2017;**42**(3):175-180

[42] Barrientos-Priego AF, López-López L. Historia y genética del aguacate. In: Téliz D, González H, Rodríguez J, Dromundo R, editors. El aguacate y su manejo integrado. México: Mundi-Prensa; 2000. pp. 19-31

[43] Barrientos PAF, López LL. Historia y genética del aguacate. In: Memoria 2002 de la Fundación Salvador Sánchez Colín. Coatepec de Harinas, México: Centro de Investigaciones Científicas y Tecnológicas del Aguacate en el Estado de México; 2002. pp. 100-121

Chapter 6

The Use of NDVI and NDBI to Provide Subsidies to Public Manager's Decision Making on Maintaining the Thermal Comfort in Urban Areas

Arthur Santos, Fernando Santil and Claudionor Silva

Abstract

The use of physical indexes such as NDVI (Normalized Difference Vegetacion Index) and NDBI (Normalized Difference Built-up Index), related to the variation of Surface Temperature (LST), have been widely used as support for mapping and monitoring land use and occupation, mainly in urban centers, due to, among other factors, changes in the energy balance and, consequently, increase heat of cities. Thus, this study approaches the urban space of the municipality of Paracatu, Minas Gerais (MG) and aims to verify urban growth, through the variation of NDVI, NDBI and LST, between the years 1990 and 2019 by using images of the LANDSAT-5 and LANDSAT-8 satellites. As a final result, an urbanization map of the municipality was obtained, and it was possible to verify that these indexes were adequate to size the environmental impact caused by disordered urbanization, since the degradation of vegetation caused in the area was responsible for reducing and/or increasing the values recorded by the indexes. In addition, the results made it possible to identify areas with higher and lower temperature variations, causing the agility of decision-making and the development of projects that meet the peculiarities of each sector of the city.

Keywords: NDVI, NDBI, LST, urban growth, spectral indexes

1. Introduction

There is an increasing trend, by public and private administrations, on seeking a renewal on the way in which the data representing its territorial aspects are managed. This change is mainly due to the strong need for basic, and reliable, information to be used as base of decision-making for society.

Thus, and given its ability to observe the earth's surface at local, regional and global scales, Remote Sensing [SR] presents itself as an important tool to perform monitoring effectively, in addition to improving the systems already available therefore support managers in their decisions [1]. Moreover, because they are obtained, on most occasions, free of charge, these data have been used for science, education and technology purposes in many countries, which are fundamental resources to identify problems, visualize panoramas and propose viable political alternatives in territorial management [2].

It should be noted that, regarding the anthropic activities that cause negative environmental impacts and that require monitoring, the growth of urban centers in a disorderly manner, especially in developing countries, has been causing gradual changes in land use and cover [3] due to the change in vegetation by materials that have the capacity to store heat [4, 5] and, as a result, increase the thermal temperature of urbanized areas and causes changes in the local, regional microclimate and in the well-being of the population through thermal discomfort, according to related studies [3, 6].

Historically, urbanization, begun around 1950, initiated changes in natural landscapes. The disorderly growth of large cities and their growing interference in the environment has made them, and keep on making them, increasingly less sustainable [7]. Therefore, what has been happening in these areas, is the emergence of a specific climate on the site, which is related to the impact of urban development on the surface heat balance [8], causing the formation of the Urban Heat Island (ICU) phenomenon and which can be considered the most evident example of climate change caused by anthropic action [8, 9].

In order to analyze and monitor these variations, the results of research that analyze, through data from SR - spectral indices and variations in Surface Temperature [LST] - is gradual, the impact of these changes on the multiscale landscape, particularly with the use of vegetation indices, due to their ability to detect the presence and absence of vegetative, with emphasis on the Normalized Difference Vegetation Index [NDVI] and also for the Normalized Difference Built-Up Index [NDBI], which assesses the urban development of the built area.

In Brazil, there are few studies that address the impacts of urban expansion in a disorderly manner. As a result, cities grow in disharmony to their ecosystem aspects. In this context, there is the municipality of Paracatu, Minas Gerais (MG), an area of study of the present study and that very little is known, so far, about the impacts caused by its urban expansion. The present work is justified by the fact that, according to the Atlas of Human Development in Brazil [ADHB], in the last decade, the population of Paracatu grew at an average annual rate of 1.20%, while in Brazil, this rate was 1.17% for the same period [10] and aims to analyze, between 1990 and 2019, the urbanization of the municipality of Paracatu through the NDVI and NDBI, as well as its consequences on the well-being of the population.

2. Material and method

2.1 Study area

Located 220 km from the Brazil's capital, Brasília - DF, Paracatu - MG (**Figure 1**) is the only historical city at the northwest of the mesoregion of Minas Gerais. Founded due to its mineral wealth, the municipality is a national reference in the exploration of gold and zinc [11]. According to the latest IBGE estimate, Paracatu has a current population of approximately 94,000 inhabitants [12].

2.2 Data acquisition

Data from LANDSAT-5 and LANDSAT-8 satellites were obtained from the United States Geological Survey (USGS) website. The years in which el niño and

Figure 1.
Study area. Source: Authors (2021).

La Niña phenomena did not occur were analyzed. Subsequently, autumn was chosen in the southern hemisphere as the period of the year to be analyzed.

The choice of this season was for 2 reasons: a) due to the low incidence of clouds, which guarantees the better perceptibility of the surface and the greater attenuation of atmospheric effects and; b) because it is a transitional period between summer and winter [13]. At this point, it is worth noting about the influence of the presence/scarcity of water on LST records, because the moisture of the material tends to alter the albedo, which represents the part of the incident solar radiation that is reflected by the material, and, as a result, greater/lower is its ability to absorb and reemit energy later, also increasing the tendency of temperature elevation/decrease [14–16].

The images of the following years were requested: 1990, 1995, 2005, 2014, 2019 and then cut to the study area. All data were standardized for the same reference system: SIRGAS 2000 in UTM Time Zone 23S. It should be noted that, for each year, 2 images were chosen that could represent the autumn season, being: 05-June-1990; 20-May-1990; 03-June-1995; 18-May-1995; 11-April-2005; 13-May-2005; 04-June-2014; 22-May-2014; 02-May-2019 e; 21-June-2019.

For the multitemporal analysis and production of the urbanization map, we used the 6 band of the LANDSAT-5 satellite and the 10th band of the LANDSAT-8 satellite, which correspond to their respective thermal ranges in the electromagnetic spectrum, and the vector files in shapefile format (.shp), which represent the country, state, municipality and area of interest of the research.

The vector files used were: a) limit of the territory of Brazil; b) limit of the state of Minas Gerais; c) boundary of the municipality of Paracatu - MG; d) urban mesh. It is notable here that items (a), (b), and (c) are available free of charge on the platform of the Brazilian Institute of Geography and Statistics (IBGE) and on the scale 1:250,000 [17]. Regarding the mask of the urban mesh (d), this was obtained from the Planning Secretariat of the municipality of Paracatu and with a scale of 1:50,000.

2.3 LST extraction

For the calculation of the LST of the areas under study, Eq. (1) was applied using the software QGis 3.2.12 [18] in each image obtained.

Eq. 1 - LST calculation.

$$T = \left(\frac{K2}{\ln\left(\left(\frac{K1}{ML * Qcal + AL} \right) + 1 \right)} \right) - 273.15 \qquad (1)$$

Source: [19].

Being: ML = Multiplicative factor of resizing the thermal band*, AL = Thermal band-specific additive resizing factor*, Qcal = Quantized value calibrated by pixel in DN, T = temperature in Celsius, K2 = constant of calibration 2* e K1 = constant of calibration 1*.

*Values used in the processing of images and taken from the metadata file.

Subsequently, the average was performed between the two images in **Table 1**. For the landsat-8 satellite, the value of −0.29 was added for each pixel of the image resulted from the mean, according to the recommendation of [19], because the 10 and 11 thermal bands receive scattered light interference from areas adjacent to the imaged scene and therefore require this adjustment.

Afterwards, in each year analyzed, the shape of the urban mesh and the mining area of the last year of the time scale was applied, which represents the current situation of the study area. Finally, the maps containing the LST of each place of interest were elaborated. Regarding the elaboration of the layout, these images were separated into classes, and their values were expressed in degrees Celsius (°C): <18; 18.1 to 20; 20.1 to 22; 22.1 to 24; 28.1 to 30. The values were classified by the standard deviation technique.

2.4 Extraction of shape pixel values and data analysis

Regarding the extraction of the pixel values of the shapes used, the ENVI software was used in version 5.2 [20]. First, the image representing the year was exported from Qgis to ENVI after the application of the method used. Once completed, the shape elaborated in Qgis was applied in the image, and was performed the pixels conversion into the shape using the Region Of Interest (ROI) plug-in

Year	Image 1	Image 2
1990	05/06/1990	20/05/1990
1995	03/06/1995	18/05/1995
2005	11/04/2005	13/05/2005
2014	04/06/2014	22/05/2014
2019	02/05/2019	21/06/2019

Table 1.
Years and dates of the images chosen for analysis.

available in the software. With the LST values of all shapes and all years of the series, these were statistically analyzed.

It is worth mentioning that, at this point, the validation of the data obtained was performed, based on the first and last year of the analyzed series, by means of Atmospheric Temperature (AT) measured in a conventional station of the National Institute of Meteorology (INMET) (code 83479; altitude 711.41 m; latitude: −17.244166 and longitude: −46.881666), accessed from its official website.

2.5 Analysis of the NDVI, NDBI and mapa de urbanização

In order to verify a possible cause of the largest and smallest variation between the analyzed neighborhoods, the images were corrected of the effect of the atmosphere and, later, the NDVI (Eq. (2)) was applied in the first and last images of the analyzed series.

Eq. 2 - NDVI calculation.

$$\mathrm{NDVI} = \frac{\left(\mathrm{NIR}\,\rho_{(830\mu m)} - \mathrm{RED}\,\rho_{(660\mu m)}\right)}{\left(\mathrm{NIR}\,\rho_{(830\mu m)} + \mathrm{RED}\,\rho_{(660\mu m)}\right)} \tag{2}$$

Source: [21].

Where: NIR corresponds to the near infrared band and RED corresponds to the band located in the red region.

For the calculation of NDBI, the Near Infrared (0.76–0.90 μm) and Mid-Infrared (1.55–1.75 μm) bands were used. Due to the fact that the NDBI is based on the fact that constructed lands have a higher infrared reflectance of medium waves than in shortwave infrared, it is expected, with this index (Eq. (2)), that it presents higher values in more densely urbanized areas.

Eq. 3 - NDBI calculation.

$$\mathrm{NDBI} = \frac{\left(\mathrm{NIR}\,\rho_{(0,76-0,90\mu m)} - \mathrm{MID}-\mathrm{IR}\,\rho_{(1,55-1,75\mu m)}\right)}{\left(\mathrm{NIR}\,\rho_{(0,76-0,90\mu m)} + \mathrm{MID}-\mathrm{IR}\,\rho_{(1,55-1,75\mu m)}\right)} \tag{3}$$

Source: [22].

Where: NIR corresponds to the near infrared band and MID-R corresponds to the band located in medium infrared.

In order to verify the effect of urbanization between the first and last year of analysis, the mean LST was performed among the 10 images analyzed in the last 29 years.

3. Results

Regarding the variation in spectral indices, between 1990 and 2019, these are arranged, in relation to the NDBI, in **Figure 2** and, in relation to the NDVI, in **Figure 3**.

Regarding the urbanization of the study area, between 1990 and 2019, this is shown in **Figure 4**.

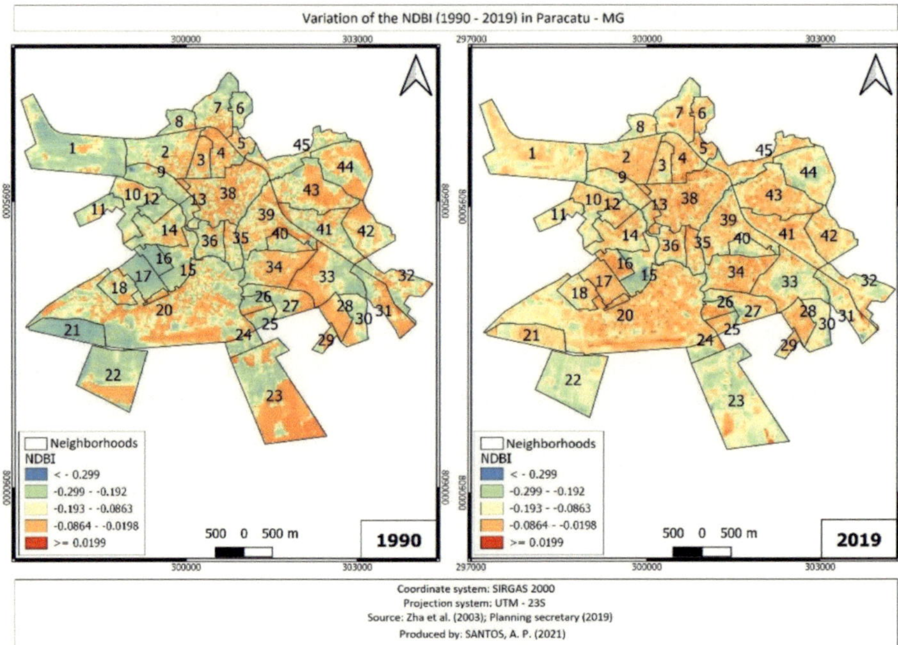

Figure 2.
Variation of the NDBI in the urban mesh of Paracatu - MG. Source: Authors (2021).

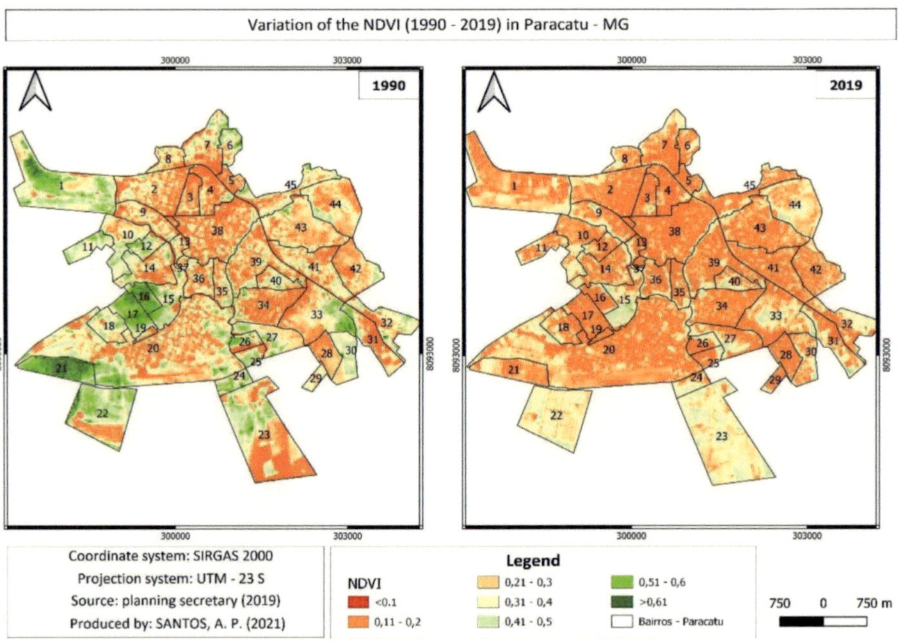

Figure 3.
Variation of NDVI in the urban mesh of Paracatu - MG. Source: Authors (2021).

4. Discussion

In relation to the multitemporal variation of spectral indices, it was possible to observe the reduction, practically total, of vegetation during expansion in some

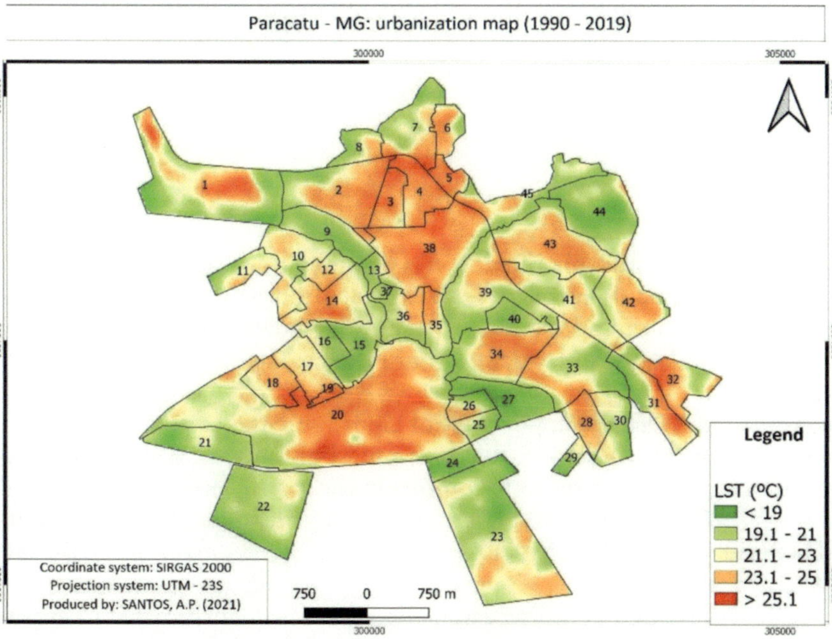

Figure 4.
Map of urbanization of the municipality of Paracatu - MG. Source: Authors (2021).

parts of the urban mesh of the municipality, especially in its central region. This reduction is possibly related to the increase in constructed areas, as presented by the NDBI and, consequently, suppression of vegetation, considering that vegetation degradation is noticeable in the visual comparison between the years analyzed.

It is worth mentioning that, according to [23, 24], urban meshes have a scenario in which urban areas built, or central, have low NDVI values and low vegetation. The authors also point out that the spatial distribution of NDVI in urban heterogeneity, with different uses and land cover, presents values opposed to LST.

Through the results obtained it is possible to verify that, currently, the municipality of Paracatu needs to have some areas - here called strategical - to be replanned environmentally. Neighborhood 20 is one of the examples of this situation. Located in the peripheral area of Paracatu, it is considered the largest neighborhood, in its population and extension, of the municipality, a fact evidenced by the change in NDVI and NDBI over the years and also with the urbanization of the neighborhood in this period.

In relation to neighborhood 38, considered as the center of the municipality of Paracatu, it presented a scenario of growth in the variation of spectral indices. This situation may be related to the high urban growth that the municipality presented until 2010.

Finally, it is worth noting that some neighborhoods located in a peripheral area, distant from the central region, have never had a high peak of urbanization, which may be related to the geographical distance between these neighborhoods and the city center. Currently, the municipality does not have many services (banks, hospitals, pharmacies, schools, supermarkets, sports centers and leisure areas) outside its central region, which ends up driving residents away from these most distant neighborhoods.

Finally, it is important to point out that some neighborhoods, that are closer to the municipal center, presented variation in spectral indices and also high values of urbanization, a fact that may be related to the choice of the population to prioritize the occupation of this zone closer to the central region. It should be noted that, in

2019, the public management of the municipality began to move all the services of the city to an administrative center located in a peripheral area of the municipality to, among other factors, relieve the concentration of vehicles and services at just one area of the municipality.

5. Conclusions

Considering the results obtained, it was possible to conclude that NDVI and NDBI, tied to LST data, is adequate to size the environmental impact caused by urbanization in a multitemporal way, since the degradation of vegetation caused in these areas is responsible for reducing and/or increasing the values recorded by the indexes.

The results reinforce the importance of environmental parks and urban afforestation, not only as landscaping, that is as a mere architectural element, but as an element of maintenance of thermal comfort in the urban space, especially in the area of study, being possible, in this way, to modify the microclimate and bring several benefits to the well-being of the population, as it contributes to the quality of life and health.

The results presented in this research also allow the identification of areas with greater and lower temperature variations, causing the agility on the decision-making process and the development of projects that meet the peculiarities of each sector of the city, thus enabling the creation of green areas in strategic points of the municipality, in order to, among other factors, reduce the possible formation of heat islands.

Author details

Arthur Santos*, Fernando Santil and Claudionor Silva
Federal University of Uberlândia (UFU), Uberlândia, Brazil

*Address all correspondence to: arthursantos@ufu.br

References

[1] Bégué, A., Leroux, L., Soumaré, M., Faure, J. F., Diouf, A. A., Augusseau, X., TOURÉ, L., Tonneau, J. P. (2020). Remote Sensing products and services in support of agricultural public policies in Africa: overview and challenges. Frontiers in Sustainable Food Systems, 4, 58.

[2] REGINATO, Vivian da Silva Celestino. Remote Sensing Applied to Regional-Scale Mapping of Solar Potential—Case Study on Florianopolis Island. Journal of Geographic Information System, v. 12, n. 5, p. 432-450, 2020.

[3] Li, G., Zhang, X., Mirzaei, P.A., Zhang, J., Zhao, Z. Urban heat island effect of a typical valley city in China: responds to the global warming and rapid urbanization. Sustain. Cities Soc. 38, 736-745. 2018.

[4] Santiago, D.B., Gomes, H.B., 2016. Estudo de Ilhas de Calor no Município de Maceió/AL, por meio de Dados Orbitais do Landsat 5. Revista Brasileira de Geografia Física 9, 793-803.

[5] ALVES, L. E. R. et al. Space-temporal evaluation of changes in soil use and soil cover and temperature in the metropolitan region of Baixada Santista. Biosci J, 2019.

[6] Guha, S., Govil, H., Dey, A., Gill, N., 2018. Analytical study of land surface temperature with NDVI and NDBI using Landsat 8 OLI and TIRS data in Florence and Naples city, Italy. Eur. J. Remote. Sens. 51, 667-678. https://doi.org/10.1080/22797254.2018.1474494

[7] HAUGHTER, G.; HUNTER, C. Sustainable Cities. London: J. Kingsley Publishers, 1994.

[8] OKE, T. R. et al. The energy balance of central Mexico City during the dry season. Atmosphe-ric Environment, [s.l.] v. 33, n. 24/25, p. 3919-3930,1999.

[9] BARROS, H. R.; LOMBARDO, M. A. A ilha de calor urbana e o uso e cobertura do solo no município de São Paulo-SP. GEOUSP Espaço e Tempo (Online), [S. l.], v. 20, n. 1, p. 160-177, 2016.

[10] ATLAS DO DESENVOLVIMENTO HUMANO NO BRASIL. Perfil - Paracatu, MG. Disponível em: http://www.atlasbrasil.org.br/acervo/biblioteca. Acesso em: 27 mai. 2021.

[11] PIMENTEL, H. U. A história de Paracatu. Paracatu, 2020. Disponível em: http://paracatu.mg.gov.br/cidade.

[12] – IBGE. Instituto Brasileiro de Geografia e Estatística. Censo demográfico. https://cidades.ibge.gov.br/brasil/mg/paracatu/panorama

[13] INSTITUTO NACIONAL DE METEOROLOGIA (Brasil). Base de dados históricos. Disponível em: http://www.inmet.gov.br/portal/. Acesso em: 20 jun. 2021.

[14] AMORIM, M. C. C. T. Detecção remota de ilhas de calor superficiais: Exemplos de cidades de porte médio e pequeno do ambiente tropical, Brasil. Finisterra-Revista Portuguesa de Geografia, Lisboa, n. 105, p. 111-133, 2017.

[15] QUERIN, C. A. S.; BENEDITTI, C. A.; MACHADO, N. G.; SILVA, M. J. G.; QUERINO, J. K. A. S.; NETO, L. A. S.; BIUDES, M. S. Spatiotemporal NDVI, LAI, albedo, and surface temperature dynamics in the southwest of the Brazilian Amazon forest. Journal of Applied Remote Sensing, v. 10, n. 2, p. 026007, 2016.

[16] TRLICA, A.; HUTYRA, L. R.; SCHAAF, A. ERB, A.; WANG, A. land

cover, and daytime surface temperature variation across an urbanized landscape. Earth's Future, v. 5, n. 11, p. 1084-1101, 2017.

[17] Instituto Brasileiro de Geografia e Estatística. IBGE. 2018. Banco de dados geográficos.http://www.metadados.geo.ibge.gov.br/geonetwork_ibge/srv/por/main.home. Acesso em: 18 fev. 2020.

[18] QUANTUM GIS. Geographic Information System. Open Source Geospatial Foundation Project. Disponível em: http://www.qgis.org/it/site/.

[19] USGS. EarthExplorer. 2016. Disponível em: https://earthexplorer.usgs.gov/.

[20] ENVI. Environment for Visualizing Images (ENVI).

[21] ROUSE, J.W.; HAAS, R.H.; SCHELL, J.A.; DEERING, D.W., 1973. Monitoring vegetation systems in the Great Plains with ERTS. Third Symposium of ERTS, Greenbelt, Maryland, USA. NASA SP-351, V1:309-317.

[22] ZHA, Yong; GAO, Jay; NI, Shaoxiang. Use of normalized difference built-up index in automatically mapping urban areas from TM imagery. International journal of remote sensing, v. 24, n. 3, p. 583-594, 2003.

[23] PESSI, D.D.; PIMENTEL, C.F.; CÂNDIDO, A.K.A.A.; JUNIOR, P.L.M.; SILVA, N.M. Análise da relação entre NDVI e a temperatura da superfície terrestre como técnica no planejamento urbano dos municípios, Terr@Plural, Ponta Grossa, v.13, n.3, p. 237-251, 2019.

[24] YUE, W., MENEVEAU, C., PARLANGE, M.B., ZHU, W., VAN HOUT, R.; KATZ, J. A comparative quadrant analysis of turbulence in a plant canopy. Water resources research, [s.l.], v, 43, n. 5, 2007.

Chapter 7

Detailed Investigation of Spectral Vegetation Indices for Fine Field-Scale Phenotyping

Maria Polivova and Anna Brook

Abstract

Spectral vegetation indices (VIs) are a well-known and widely used method for crop state estimation. These technologies have great importance for plant state monitoring, especially for agriculture. The main aim is to assess the performance level of the selected VIs calculated from space-borne multispectral imagery and point-based field spectroscopy in application to crop state estimation. The results obtained indicate that space-borne VIs react on phenology. This feature makes it an appropriate data source for monitoring crop development, crop water needs and yield prediction. Field spectrometer VIs were sensitive for estimating pigment concentration and photosynthesis rate. Yet, a hypersensitivity of field spectral measures might lead to a very high variability of the calculated values. The results obtained in the second part of the presented study were reported on crop state estimated by 17 VIs known as sensitive to plant drought. An alternative approach for identification early stress by VIs proposed in this study is Principal Component Analysis (PCA). The results show that PCA has identified the degree of similarity of the different states and together with reference stress states from the control plot clearly estimated stress in the actual irrigated field, which was hard to detect by VIs values only.

Keywords: vegetation indices; irrigated crops, agriculture management, field spectroscopy, space-borne spectral imagery, spatial variability early stress detection, principal component analysis

1. Introduction

The 2030 Agenda represents an agreement between all 193 UN Member States to introduce a set of common strategies to achieve 17 goals (the Sustainable Development Goals, or SDGs) and 169 targets before the year 2030. SDGs are a collection of global goals to attain a better future. Sustainable agriculture is at the heart of this agenda. This goal is responsible for ensuring food production systems and implementing resilient agricultural methods, influencing the increase in production and productivity, assisting in maintaining ecosystems, adjusting to climate change and extreme weather [1–4]. Thus, simultaneously taking into consideration improvement of land and soil quality. One of the ways that can help achieve sustainable agriculture is with Precision Agriculture (PA), a method to accurately apply the right treatment, in the right place, at the right time.

Timely detected crop stress allows rapid correspondence and adaptation of planned agricultural activities and preventing negative effects on the yield. Special attention is paid to water stress due to its effect on plant growth and yields [5]. Moving towards PA that stands for concept of managing crop fields considering spatial variation and local field requirements, involves data collection to characterize field spatial variability, mapping, decision-making, and management practice implementation. The growth number of precision agriculture applications has influenced the development of remote sensing technology owing to its ability to conduct higher spatial, spectral and temporal resolutions capabilities and cost-effectiveness. Remote sensing at visible and near-infrared wavelengths (vis–NIR) has been used to formulate many spectral indices for estimating crop properties (e.g. [6]). In irrigation management, crop water state can be estimated by water content or water potential in soil and plants [7]. These parameters can be measured directly and indirectly. Direct methods are gravimetric soil moisture, relative water content [8], energy status of soil water and plant water potential. Indirect estimation crop-water state performed by microwave or radar techniques [9], soil moisture balance calculations [10], and air-plant temperature differences [11]. Indirect methods have many advantages: they do not damage plants and soil structure; do not require laboratory conditions, expensive equipment, complex measurement protocols, and specific technical knowledge. Regardless to the measurements type, the prevalent method for crop stress detection in practice is a comparison of actual crop parameters with reference values for the normal state [12]. Visual agronomic field inspection for assessment wilting, morphometric changes (plant organ shrinkage), and growth rate [13] to estimate crop-water state is still the widespread approach in practice because of its traditional origins, low cost and ease application. The essential drawback of this visual inspection method is ability to detect stress at the obvious visual stage when a crop is already undergoing significant damage. In general any manually assessed plant trait in the traditional way is time-consuming, laborious; introducing errors and sometimes destructive [14, 15]. From 2000s, the era of non-destructive plant-phenotyping platforms, based on image-based techniques, has begun [16–18].

Spectroscopy can improve current agronomic inspection methods. The physical properties of plants directly influence its reflectivity at different spectral ranges [19]. When crop stress is clearly pronounced in visual spectrum range, it can be detected by traditional field inspection. Before this, stress has already been caused by crop properties and has influenced its reflectivity. Spectral tools provide measurements of plant reflectance with higher sensitivity and in a wider range than the human eye's capability. Therefore, spectroscopy allows for detecting stress at earlier stages than traditional visual methods [20, 21].

Simple Vegetation Indices (VI) have significantly improved the ability and sensibility of the detection of green vegetation [22]. Vegetation indices are widely used to estimate crop growth status and crop parameters, such as biomass, yield, photosynthesis, Leaf Area Index, etc. [23–26], and many studies have analyzed the potential of using spectral reflectance indices in wheat starting from the early years of the century (e.g., [27]) and up to the present day (e.g., [28]). High-resolution VIs may detect changes of wheat crop status and it might help to improve crop monitoring [29], nitrogen management [30], and crop yield estimation [31, 32]. Furthermore, it is indeed known that yield prediction while using VIs in wheat can be accurate two months prior to harvest, because yield estimates stabilize and especially during the flowering period, significant correlations between UAV-VIs and yield components were found [33, 34].

VIs estimate plant state by calculating ratios or more complicated mathematical models of reflectance measurements using different spectral wavelengths. Its development began with discovering a strong linear correlation between plant green biomass and the ratio of 2 spectral bands obtained by satellite imagery [35]. The concept of applying spectral data to assess plant parameters was introduced in the 1970s. The VIs era began with the 1972 launch of the first ERTS satellite (Landsat 1) with its MultiSpectral Scanner. The first index was the Normalized Difference Vegetation Index (NDVI) developed as quantitative measurement of vegetation conditions by calculating the ratio between visible and infrared (VIS/IR) spectral bands. Further on, NDVI was applied to assess plant health and estimate other physical properties presenting sufficiently good linear correlation with plant height [36] and asymptotic relationships with Green LAI [37], and has shown its ability to indicate different phenological stages [38]. Further high correlations of NDVI with crop biomass were discovered, Leaf-Area Index (LAI), absorbed photosynthetically active radiation, and canopy photosynthetic capacity [39]. However, NDVI does not have sufficient sensitivity to all crop features [40]. The next developed VIs were sensitive to crop photosynthesis and plant-water state [41, 42], there were several indices sensitive to crop pigmentation; e.g. chlorophyll, carotenoid, anthocyanin [43, 44].

As NDVI includes chlorophyll absorption band, it finds application in estimating chlorophyll and used as health state parameter as well [45, 46]. Likewise, NDVI's response to physical crop characteristics and chlorophyll content makes it suitable tool to predict yield and detect N deficiency [47], estimate actual evaporation rates [48], assess fraction of Absorbed Photosynthetically Active Radiation [49], indicate soil salinity [50] and calculating crop coefficients for irrigation needs [51]. In addition, there were developed many modified VIs applied under detailed/dedicated conditions for estimating physical parameters and other specific tasks [24–26, 44]. Other approaches were associated with the ability for collecting spectral data at high spectral resolutions (1 nm). These approaches allowed obtaining changes in plant pigmentation, nutrient content, and chemical properties [32, 52, 53].

Nowadays, there are several dozen VIs and models for different spectral data types, which can be applied for estimating crop state, determining stress, salinity, diseases, and hazardous substances. Irrigation management is one of the most promising directions where VIs could be applied. For this task, there were already developed VIs that are directly sensitive to water absorption; e.g., Water Band index [54] and indirectly by estimating changes in pigmentation and photosynthesis rates [51]. It is important mentioning soil moisture indices such as Temperature Vegetation Dryness Index (TVDI) [55], on account of its potential applicability for estimation water state of agricultural fields. Besides estimation soil moisture TVDI provides information about groundwater depression cone which improves remote monitoring and allows to reduce in-situ measurement [56].

Despite the great importance and potential benefits of these studies, they do not consider the informativeness of indices without co-core physical measurements described above, such as water content, soil water balance and etc. Most of the indices based on water absorption bands require high-resolution spectral data in the short-wave infrared range: Normalized Difference Water Index (NDWI), Moisture Stress Index, Normalized Difference Infrared Index, and Normalized Multiband Drought Index [57–60]. Spectral data in near-infrared range 900–970 nm provides an opportunity to estimate water content by Water Band Index [61]. Due to the high cost of obtaining required data, the above indices were not widely applied in agricultural management. Consequently, a number of studies were conducted to find a method of applying the current VIs from visible and near-infrared spectral ranges for water stress detection.

Several studies were devoted to discovering a direct correlation between widespread VIs (NDVI, Simple Ratio, Photochemical Reflectance Index, and Structure

Insensitive Pigment Index) and leaf water-content, but did not find a significant correlation [62, 63]. Nevertheless, NDVI and Soil Adjusted Vegetation Index (SAVI) found application in irrigation management, because they were proved to be a strong approach for estimating crop coefficients (Kc) and predicting crop evapotranspiration [64–68]. Further investigation showed that combination between VIs and physical parameters might increase accuracy (R2 = 0.5–0.7) of estimated crop-water state, than single VIs analysis [69, 70]. Accordingly, considering water deficit by VIs, canopy temperature, air temperature, stomatal conductance, or stem water potential in one application might contribute to better detection. The PRI alone showed good correlation with plant water-content (R2 = 0.8) with no additional physical measurements in the model [71, 72]. However, this method also has a disadvantage, as a good PRI correlation can be found only at specific plant stages and when Photosynthetically Active Radiation was above 700 µmol/m2/s.

Estimating plant state only by VIs consists of calculating values and deciphering results. Some of the indices provide a range of optimal values for plant health; e.g., Structure Insensitive Pigment Index, Carotenoid Reflectance Index, Modified Red Edge Simple Ratio, etc. [25, 42, 44]. Other VIs only provide interpretations of physical processes based on increasing or reducing values; e.g., Anthocyanin Reflectance Index, Chlorophyll Absorption Ratio Index, Triangular Vegetation Index, etc. [26, 44, 73]. Since agricultural management requires a common design for a large area, a series of measurements (pixels from spectral images or points by a portable field spectrometer) should be converted to one field parameter, which can lead to loss of essential information. Likewise, the analysis could be complicated since the variations in the VIs from date to date are confounded with crop phenology and, possibly, atmospheric conditions. Consequently, field state analysis by VIs should include calculating values, merging them into a single field parameter, assessment variety, interpretation value according to the vegetation index response, specific environment and plant conditions.

In addition to the complications described above, the fact that the same VIs can be assessed by these two fundamentally different sources of spectral data, namely, space-borne and airborne imagery, and point-based spectral measurements, raises questions regarding its universality and accuracy. Several studies confirm that models based on field spectroscopy need direct comparisons, modification, and validation for application of space-borne and airborne spectral imagery [74]. The imagery data acquires spectral data from the whole scene, including soil, plants, and atmospherics [75]. Field point-based spectral measures could minimize the atmosphere's effect and reduce signal noise from the soil by simply excluding it from the observed scene (ground footprint). Thus, it is important to remember that the VIs originally developed for one type of spectral data might lose estimation accuracy once applied to another data source without any additional validation or modification.

Sensitivity to plant stress at early stages, estimation of crop parameters, non-destructive method of observations, and existing low-cost equipment for spectral data collection, make VIs the prospective approach for monitoring crop-water state and implementing the optimization of an irrigation schedule. Despite all the advantages and attractiveness, the absence of clearly interpreted VIs' behavior, relatively low-correlation of individual index with real-field water state, limits spectroscopic application approach in irrigation planning.

The main aim of this study is to assess the performance level of the selected VIs in application to different spectral data sources on actual irrigated agricultural fields without co-core physical control measurements. There were three specific objectives:

1. To determine suitability of different spatial and spectral (e.g. spaceborne multispectral imagery and point-based field spectroscopy) data sources for early detecting stress in the irrigated crops;

2. Define VIs corresponding to early water stress in the irrigated crops.

3. The aim of this study is to identify water stress on actual irrigated field by VIs.

For this propose, three crops are investigated: cotton, tomato and chickpea. The third objective was tested based on chickpea crop, one under actual agricultural management and other field with limited water treatments were studied. Stress identification was implemented using several methods. The first method is interpretation obtained VIs values according description from the original papers and their temporal behavior. The second is validation the correlation between physical field parameters and VIs stated in existing studies. The last is novel approach for estimation early crop stresses by principal component analysis (PCA) proposed in this study. PCA is introduced for processing the matrix of VIs, and further identification uniqueness and similarity of crop states and estimation stress by introduction reference VIs values of crop suffered from dryness.

2. Materials and methods

2.1 Study area

This study was carried out in Kibbutz Hazorea, in northern Israel (32°38′42.0"N 35°07′17.6″E). The fields are located in the typical Mediterranean climate with mild winters and dry summers. The average daily mean annual temperature is approximately 18°C and relative humidity is about 68%. Our observations were conducted during the 2015 summer growing season, from March 1 through August 31. The average daily temperature in spring is 18°C and in summer reaches 25°C. The total precipitation during the study period was 111.1 mm of which 59.9 fell in two days (April 10–11).

The soil is classified as Vertisol according to the USDA Soil Taxonomy [76]. The soil characteristics were as follows: heavy-textured soil, bulk density 1.8–2.0 g cm − 3, pH value 8.0–8.6, organic matter 5–10%, clay 50–60%, and CaCO3 9%. Vertisol is a widespread soil type which is prominent on almost every continent: Africa, India, Australia, southwestern USA (Texas), Uruguay, Paraguay and Argentina [77]. The main feature of this soil type is its high level of clay content (between 50 and 60%). As a consequence, it has a high moisture-holding capacity. Vertisol is known as suitable soil for agriculture, but the level of clay content requires very careful irrigation management [78]. Dry Vertisol soils are conglomerated and cracked. The specificity of the soil in Kibbutz Hazorea required great effort by the farmers: drainage, land reclamation, and finding appropriate agricultural crops. Initially, attempts were made with dry farming. In early years, vegetable gardens and grains were the leading sources of income. Gradually the high watering cost of vegetables melons and fruit trees were changed to more profitable irrigated crops [79].

The following three types of crops were chosen for this study: tomatoes, cotton, and chickpeas. These crops belong to various families of plants: tomatoes are Nightshades, cotton is Malvaceae, and chickpea is Fabaceae. The crops have different sensitivity to water scarcity, the strategy of development, and stress resistance. Chickpea (*Cicer arietinum* L. 'Yarden') was planted on 1st of January 2015 with density 190 kg seeds per hectare. The cover area was nearly 12 hectares. The rows were located at a distance of 90 cm from each other and had East/West direction. The crop was irrigated by a sprinkler on mobile irrigation systems. The irrigation period was 3 weeks since 18th of May. The total water amount was 2400 cubic meters per hectare. Watering was carried out in equal parts daily. Determination of actual phenological stage and assessment plant state by field inspections were conducted based on guideline WATERpak [80]. Tomatoes (*Solanum lycopersicum* '4107') was planted on open ground on 15th of March 2015 with density 25000 seedlings per hectare. The cover area was nearly 13 hectares. The crop was irrigated by surface drip. Cotton (*Gossypium hirsutum* L. 'HA – 195') was planted on 28th of March 2015 with density 15000 seedlings per hectare. The cover area was nearly 10 hectares. The crop was irrigated by a sprinkler on mobile irrigation systems. Both crops have distance between rows 90 cm and rows had North/South direction. Criteria for estimation actual phenological stage and for assessment crop state by farmers field examination in tomatoes and cotton was provided in the relevant guidelines developed by NaanDanJain Irrigation Ltd. The daily irrigation amount to the crops was calculated by multiplying evapotranspiration to crop coefficient [81]. The obtained value was corrected according to recommendations based on field inspections.

Due to the reported impact of environmental characteristics on correlation strength between VIs and crop-water state, the study was conducted on actual irrigated field. To monitor a crop mainly suffering from water lack, a monitored area preferably should have risks of high temperatures and drought. Thus, crops grown on open fields in Israel are an appropriate object for the proposed study.

Chickpea was grown in open fields (*Cicer arietinum* L. 'Yarden') was chosen for investigation. The monitored chickpea was grown from January 25 until June 28, 2016, on open ground. Planting density was 190 kg seeds per hectare. The planted area was nearly 5.3 hectares. The crop was irrigated by surface drip. The plants have a 90 cm distance between rows, in a North/South configuration. Planting distance on the row was 10 cm. Limit of irrigation water amount for the whole growing season on the chickpea field was 120 mm and actual water consumption amounted 123.62 mm. The potential chickpea yield at the given seeding density declared by the seed producers is 0.55 t/ha (Nir Agricultural Works, Ltd). In practice, a yield of 0.5 t/ha is considered by farmers as good due to the low planting density proposed to prevent fungus development. In 2016, the chickpea yield from the monitored field was 0.47 t/ha.

Irrigation schedule was developed based on potential water consumption [10] using weekly weather forecast and crop coefficients provided by chickpea growth guide [80]. In addition, the chickpea field was equipped with a tensiometer, which was used to determine the metric water potential (soil moisture tension). The field feedback data obtained by tensiometer was applied to correcting and improving the weekly irrigation schedule. The irrigation starts on May 2 and continues for four weeks. The water was distributed unevenly: 26.5, 34, 15.5 and 46 mm per week according to the sequence of the weeks.

For the purposes of the study, control plot of chickpea was planted jointly with main chickpea field on the distance 5 meters. Plot size is 2x2 meter. All activities and treatments on the control plot match to the management on the main monitored field till flowering stage on April 25. Since this date the control plot was not irrigated.

2.2 Data collection

Field trips were conducted from January 2015 through August 2015. During these trips, the crops underwent agronomic inspection and spectral measurements.

Crop phenological stages were determined according to crop guides [82]. Agronomic examination confirmed the state of the tomatoes' and cotton's normal development and health during the entire growth period. The actual tomato yield was 145 ton per hectare versus 160–170 ton per hectare maximum [82]. The actual cotton yield was 5.2 ton per hectare versus a potential 6.5–7 ton per hectare maximum [81].

The chickpeas developed under limited water treatments. Irrigation started in mid-May 2015 (at yield formation stage) and lasted two weeks. The agronomic inspection noted visual marks of plant dryness at the end of May 2015 [81]. Under-irrigation during vegetative and flowering stages leads to a slowdown in phenological development and a decrease in yield. The actual growth cycle of the chickpeas was 202 days compared to theoretical 150–170 days. The actual yield was 0.45 ton per hectare in contrast to 0.55 ton per hectare expected by farmers.

Additional chickpea tested field was observed during the growth period from February to June 2016. In total, there were 8 observations on the main field. Each campaign consisted of visual crop inspection and health state assessment [13], identifying the phenological stage, height measures, and spectral data-collection. Crop health state was assessed by leaf color, dryness, and plant wilting. The phenological stage was identified according to the description provided by crop growth guides [81]. Crop height was defined as the average of 5 random measures from ground to top crop leaves.

Field spectral reflectance was measured directly on the leaves using a portable field spectrometer (USB4000, Ocean Optics Inc., USA [83]) acquiring data across the VIS and NIR range from 350 to 1100 nm with a resolution of 0.5 nm and an accuracy of 1 nm. The spectrometer was calibrated according to protocol [84] against a white Spectralon plate (Labsphere Inc., North Sutton, NH). The detailed spectra were collected at a nadir view angle approximately 5–10 cm above the crop's leaf-scale by a bare fiber optic with 25° field-of-view positions. All measurements were carried out between 11:00 to 13:00 o'clock on sunny days when cloud cover was less than 10%. For observation, upper leaves orientated approximately perpendicularly relative to the sun were chosen. During the field trip, about 10 spectra were collected for each crop with 10-spectra repetition for each measurement. The plants were chosen randomly on the same 10x10 meter area. Selecting the control area size is conditioned to ensure uniform water distribution to crops due to homogeneous landscape, which as a complex with other identical parameters (crop and soil type, weather conditions and fertilizers) provides the expectation of the close water treatment for the tested crops. The spectral data from the chickpea field were collected from 5 x 5 meter control patch with a grid layout, and nearly 50–60 spectral signatures were obtained.

The plant water concentration (PWC) was introduced for validation results of agronomic inspections and providing accurate detection water stress in the crops. The plant water concentration is estimated by reflectance Water Index [85]. The ground-based reflectance measurements required for this method were obtained during the field campaigns according to the Penauelas protocol and processed according to Eq.(1).

$$\mathbf{PWC} = \mathbf{684} * (\mathbf{R900}/\mathbf{R970}) - \mathbf{620} \tag{1}$$

The plant water concentration obtained from the tested samples is performed in **Table 1**. In addition to visibly health plants tested in this study, the plant suffered from the dryness stress was detected in the cotton field (on 22nd of May) by PWC and introduced to the analysis as a reference of drought.

For remote field observation, RapidEye data was kindly provided by the RESA project 597 [86]. The data was presented at preprocessing level 3A [87].

	Chickpea	Tomatoes	Cotton
22.01	94.17%	—	—
15.03	81.51%	—	—
19.04	89.15%	—	—
22.05	80.45%	87.33%	85.08% (63.85% stressed)
07.06	77.60%	84.17%	83.29%
28.07	—	78.59%	75.93%
31.08	—	—	68.10%

Table 1.
Plant water concentration (PWC) by Penauelas method.

Atmospheric correction was conducted with the generic processing chain CATENA developed at the German Aerospace Center (DLR) [88]. RapidEye images have a spatial resolution of 5x5 meter per pixel and 5 spectral bands (440-510 nm; 520-590 nm; 630–685; 690-730 nm; 760-850 nm). Images were obtained from February 2015 through July 2015, with an average frequency of twice a month. The spatial resolution covered about 5 pixels that matched the study area 10x10 meter chosen for field spectral measures.

Soil moisture tension was measured by tensiometer ("Mottes Tensiometers" Company). The tensiometer consists of plastic tubes filled with distilled water and porous ceramic caps. The tubes were installed in the tomato field at depths of 30, 60, and 90 cm. High tensiometer values means dry soil, while low values indicate high soil moisture. The tensiometer logger recorded data every 20 minutes from April 11 until May 29 (**Figure 1**).

Meteorological data for this study was provided by the Israel Meteorological Service from Haifa University meteorology station, which is situated 16 km from the observed fields.

Field data for the further analysis consists of weekly irrigation amount records, tensiometers date, measured crop height, and estimated phenological stage. "Irrigation week amount" is an amount of water applied to the field during the week before a field campaign including precipitation. Tensiometer data considered the

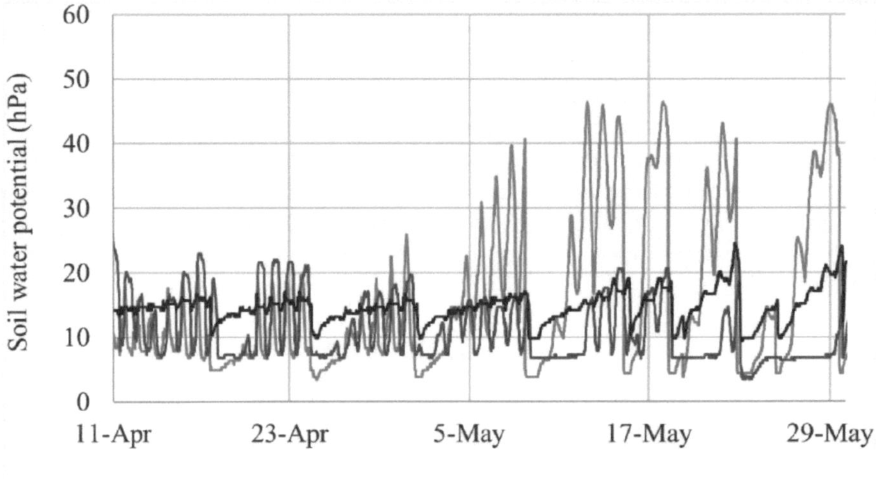

Figure 1.
Tensiometers records on depth 30, 60 and 90 cm from the chickpea field in 2016.

maximum and minimum recorded values (amplitude) during the last week before a field campaign ("Tens_max" and "Tens_min" respectively), and the actual value at the moment of field spectral measurements "Tens_actual" (at a depth of 60 cm). In this study, plant development was divided into 6 phenological stages ("Stage"): emergence, early vegetative, late vegetative, flowering, fruit formation, and ripening. Field input data for analysis is presented in **Table 2**.

According to tensiometer records, at the week before observation the soil water content on the depth 60 cm turns to the lowest values. It corresponds to the irrigation schedule: at that week the chickpea got the lowest amount of water. Field inspection on May 21 noted a relatively high concentration of yellow leaves not typical for this phenological stage.

Chickpea control plot has no irrigation since April 25. During the month since May 1, the control plot was observed every five days. At this period there was three rainy days from May 23 to May 28 and 8.1 mm fell. The plot monitoring consists of visual inspection, spectral measurements and collection leaf samples. The spectral data was collected according the same protocol like on the main field with a grid layout. Nearly 20 spectral signatures were obtained from the plot at each date. Also, at every observation five representative leaf samples were collected and sent to the lab measurements. In the lab, the leaves were weighted and dried by the oven at 70° C for 24 hours. Leaf water content (LWC) was defined (Eq.(2)) as the difference between initial weight and weight after drying divided to initial weight and converted to the percent [89].

$$\mathbf{LWC} = (\mathbf{initial\ leaf\ weight} - \mathbf{weight\ of\ dried\ leaf})/(\mathbf{initial\ leaf\ weight}) \times \mathbf{10} \quad (2)$$

The dried samples were ground into powder and mixed with 100% acetone at the rate of 0.1 g of dry matter per 10 ml of acetone. The mixtures were centrifuged for 30 min in glass tubes to make the extract fully transparent. The resulting extracts were immediately measured by spectrometer USB4000. Specific absorption coefficients of Chl a and Chl b reported by Lichtenthaler [90] were used to estimated chlorophyll concentration in studied leaves. The results of measured leaves' parameters are presented in **Table 3**.

ID	Date	Stage	Irrigation week amount (mm)	Height (cm)	Tens_max (hPa)	Tens_min (hPa)	Tens_actual (hPa)
C1	19.02	emergence	0	2	—	—	—
C2	16.03	early vegetative/ late vegetative	15	20	—	—	—
C3	03.04	late vegetative	0.4	32	—	—	—
C4	18.04	flowering	21.1	40	19.1	6.8	12.2
C5	06.05	flowering/ fruit formation	26.5	61	20.6	3.4	6.8
C6	11.05	fruit formation	34	66	21.6	6.4	11.3
C7	21.05	fruit formation	15.5	68	44.1	3.9	34.3
C8	30.05	fruit formation/ ripening	54.1	72	32.8	3.9	20.6

Table 2.
Crop parameters and watering conditions on the chickpea field in 2016.

ID	Date	LWC	Chlorophyll content mg/g
S1	May 1	95.25 ± 2.17%	1.33 ± 0.25
S2	May 6	95.12 ± 2.09%	1.01 ± 0.26
S3	May 11	95.51 ± 2.14%	1.04 ± 0.39
S4	May 16	96.33 ± 2.35%	0.83 ± 0.37
S5	May 21	86.27 ± 6.72%	0.76 ± 0.21
S6	May 26	69.01 ± 13.41%	0.68 ± 0.16
S7	May 30	25.48 ± 13.72%	0.49 ± 0.06

Table 3.
Leaf water content and chlorophyll concentration estimated in control stressed chickpea leaves.

2.3 Data analysis

VIs for this study was selected according to the following criteria: 1) the considered indices should have application in irrigation management; 2) different types of VIs original response should be presented in this study; 3) required spectral data for VI calculation should match the spectral ranges obtained from the spectrometer and space-borne images. The last one, the considered VI should be developed by different technologies: remote sensing images (satellite and aerial) and field spectrometer. This requirement allows considering the influence of data source on processing results.

Statistical analysis was used to bring all the data to a single format, convenient for interpretation and comparative analysis. This goal was achieved by estimation VI's average mean and internal variability. Measures of each crop at the same time by the same sensor (spectrometer or satellite) were merged into one dataset. From space-borne images, values of 5 pixels coinciding with the location of ground-based measurements were extracted. For further comparative assessment of variability, the number of field spectral measures was reduced from 10 to 5 by random sampling to match the number of values from spaceborne images.

The heterogeneity of the data (space-borne images and spectral plots) complicates the choice of analytic methods due to their applicability for both data sources. Sophisticated methods such as semivariogram have already proved their effectiveness for geostatistical analysis of remote sensing imagery [91]. However, the specificity of field spectral measures limits the suitable methods for assessment variety of VIs. For the purpose of further comparative analysis, the common statistical method was chosen: a coefficient of variability was applied to the quantification measured variability (Eq.(3)).

$$\mathbf{Cv} = \mathbf{s}/|\,\bar{\mathbf{x}}| \tag{3}$$

where Cv is a coefficient of variation, s is the standard deviation, and $\bar{x}$ is the average mean.

Cv values close to zero were excluded from the analysis. Additionally, the temporal variability of each VI was estimated. The temporal variability was defined as the coefficient of variation between average means of datasets during the whole season. The absolute average means of VI also was calculated from average means of datasets during the whole season. The temporal variability and absolute average mean are presented in the result section and named "Total".

The suitability of application the coefficient of variation was examined by normality Shapiro–Wilk W test in the SPSS environment [92]. The VIs corresponding to the requirements of the normal distribution was rescaled (Eq. (4)).

$$\mathbf{X\_new} = (\mathbf{X} - \mathbf{X\_min})/(\mathbf{X\_max} - \mathbf{X\_min}) \quad (4)$$

where Xnew is a rescaled VIs value, X is the original value, Xmin and Xmax are minimum and maximum permissible VIs values respectively. The minimum and maximum values from "Range of values" in **Table 3** are used for rescaling.

The sensitivity limits of VI's, spatial and temporal variables were assessed by statistical methods for outlining extreme values (outliers). For this examination, the following methods were used: histogram of distribution, percentiles (Tukey's Hinges), tests for normality (Kolmogorov–Smirnov, Shapiro–Wilk, Normal Q-Q plot), outlier labeling by quarters with g = 1.5 [93] and stem-and-leaf plot [94] to examine VIs spatial and temporal variables on data collected with both portable spectrometer and satellite.

The obtained spectral range allows calculating indices responding to visible and NIR diapason and excludes indices related to SWIR. To find the stress manifestation on different levels (physical and pigment), VIs related to different plant parameters and responded to drought stress are considered: biomass (NDVI, RENDVI, MRNDVI, MRESR), canopy coverage (MCARI2, MTVI, and MTVI2),

VI	Original response	Application for stress detection	Range of values	Reference
NDVI	Green biomass	Drought stress and nutrient uptake	0 to 1	[35]
WBI	Relative water content	Drought stress	0.8 to 1.2	[61]
ARI1	Anthocyanin	Rust infection, drought stress, low-temperature stress	0 to 3	[22]
ARI2	Anthocyanin	Rust infection, drought stress, low-temperature stress	0 to 2	
CRI1	Carotenoid	Drought stress, salinity level	1 to 20	[44]
CRI2	Carotenoid	Drought stress, salinity level	1 to 20	
PSRI	Carotenoid/ Chlorophyll	Leaf senescence and fruit ripening, salinity stress and water use efficiency	-1 to 1	[43]
PRI	Photosynthesis rate	Water stress	-1 to 1	[41]
SIPI	Carotenoid/ Chlorophyll a	Leaf senescence and unhealth plant, rust infection and aphid	0 to 2	[42]
MCARI	Chlorophyll	Stress detection, nitrogen availability and water stress	0 to 0.7	[95]
MCARI2	LAI	Heavy metals stress, water and nitrogen stresses	0 to 1	[26]
MRENDVI	Chlorophyll	Water and stress	0.2 to 0.8	[24]
MRESR	Chlorophyll	Drought stress	2 to 8	[25]
MTVI	LAI	Drought and salinity stressed	0 to 1	[26]
MTVI2	LAI	Water and nitrogen stresses	0 to 1	[26]
RENDVI	Green biomass	Salinity stress, early stress, stem water potential	0.2 to 0.9	[25]
TCARI	Chlorophyll	Nitrogen stresses	0 to 0.4	[26]

Table 4.
The list of considered vegetation indices and their stress response.

photosynthesis rate (PRI), chlorophyll (MCARI, TCARI), carotenoid (CRI, PSRI, and SIPI), anthocyanin (ARI), and leaf water content (WBI). **Table 4** shows VIs chosen for this study.

The estimation of VIs applicability for water-stress detection was performed separately for each data set. This estimation is based on a comparative analysis between VI behavior in health and presumably early drought-stressed crop. Tomatoes and cotton were considered as healthy crops. The chickpea lacked water during the month since the last rain (60 mm) on April 11, 2015, until the first irrigation on May 15, 2015. The chickpea state during this period was assumed as water lack suffered, and the absence of visible signs is perceived as the initial stage of (early) stress.

2.4 Regression analysis

Linear regression analysis was applied for estimation VIs' response to chickpea irrigation. The analysis is designed to detect linear correlation between physical parameters (stage, height, weekly irrigation amounts, and tensiometer data) and calculated VIs averages. Input chickpea states for this analysis are C4-C8 (**Table 2**) when tensiometers measurements were obtained. The strength of correlation was assessed by Spearman and Pearson coefficients and two-tailed significance, since among the parameters there are both: scale and ordinal. Correlation analyses were performed in the SPSS environment.

2.5 Principle component analysis

The proposed method for estimation crop water stress by VIs is principal component analysis (PCA) implemented in MatLab environment [96]. This approach is a multivariate technique that analyzes a matrix of numerous intercorrelated quantitative dependent variables. PCA extracts the dominant patterns to new orthogonal variables called principal components and plotting the variables in new multi-demotion, where each dimension presents estimated principle component [97, 98]. The purpose of applying this method is consideration VIs values combination as single characteristic for estimating a pattern in crop behavior under stress. The VIs values from control chickpea plot were used as reference crops stress to the identification pattern corresponded to crop-water lack. The analysis was conducted on 3 input datasets: VIs averages with their variability, VIs averages and variability separately. For this analysis VIs values were rescaled by Eq. (4).

The first test of PCA determines each crop state as a point in multidimensional space, where every dimension is an individual parameter estimated from VIs averages and variability and plotting it in 2-dimensional space by determining 2 principal components. In this test the similarity of crop state assessed by mutual arrangement of points on the plot. The second PCA test was implemented in the reverse: combinations of indices and variability were plotted by 2 principal components and crop states became variables and were presented on the biplot as vectors showing the contribution to these components. The manner of vector presentation calls a correlation circle: if vectors with the same length are close to each other, they are significantly positively correlated (R close to 1); if they are orthogonal, they are not correlated (R close to 0); if these vectors are on opposite sides, they are negatively correlated (R close to −1). When the variables are close to the center, some information is carried on other axes, and then any interpretation might be hazardous.

3. Results

The results of VI variability assessment showed strong differences between space-borne imagery and portable field spectrometer data. Consequently, this section is subdivided by the data sources. For each data source, the VIs and their variability, sensitivity limits of variability were considered, and comparative analyses between healthy and stressed crops were performed.

3.1 Space-borne imagery data-source

The chickpea field was observed by space-borne imagery 11 times from February 1 until July 15, 2015, with a frequency of every 2–3 weeks. The first three observations in February and March characterized relatively low values of NDVI: 0.239, 0.325 and 0.323. Since NDVI ranges from 0.05 on bare soil to 0.90 on dense vegetation, the observed low values corresponded to the low density of crop seedlings. On May 4, NDVI considerable increased to 0.717, which corresponds to crop growth. Then, NDVI declined to a value of 0.493 on April 20, without obvious reasons. All three observations in May recorded high NDVI values (0.861, 0.840 and 0.835). The last three observations during the summer coincided with the crop ripening and show an NDVI decline: 0.606, 0.328 and 0.329.

The majority of the examined indices (NDVI, EVI, Green Atmospherically Resistant Index (GARI), Green Normalized Difference Vegetation Index (GNDVI), LAIwp, LAIc, Pigment Specific Simple Ratio (PSSR), TVI, MCARI2, Modified Triangular Vegetation Index (MTVI), MTVI2 and Modified Chlorophyll Absorption Ratio Index (MCARI)) display similar behavior, presenting low values during the first three observations; a significant increase on April 4; an inexplicable decline for the next observation; high values in May, and a decline during the summer. Slight differences in the behavior of these indices could be detected only with the first three observations. Some of the indices gradually increased during this period; e.g., EVI, GARI, LAIwp, TVI, MCARI2, MTVI, and MTVI2. An additional group had approximately equal values for the second and third observation: NDVI, LAIc, PSSR, and MCARI. Only the GNDVI had equal values for the first and third observations and a higher value in the second. Anyhow, these differences are insignificant in comparison with changes during the season.

Two indices related to carotenoid and anthocyanins display unique behavior in chickpeas. The Plant Senescence Reflectance Index (PSRI) behaves in an inverse manner with respect to health indices. At the beginning of the season, it had high values: 0.442 and 0.474. In April, it declined to 0.192, and then rose to 0.221. During May, the PSRI turned to its lowest values: 0.111, 0.061, and 0.155. In summer, it gradually increased: 0.327, 0.436, and 0.532. Despite the PSRI, a behavior pattern or correlation with the crop's physical development in Anthocyanin Reflectance Index (ARI) was not found. At first observation, ARI was 4.617, the value then rose to 6.249. At a third observation, it dropped to 2.854. In April, ARI behavior matched the health indices: on April 4, it rose to 7.956 and on April 20, it declined to 3.034. During May, the ARI consistently gave values of 8.824, 6.907, and 7.914. During the summer, ARI also behaved without a pattern and displayed values of 4.979, 3.715, and 6.122. During the whole observation period, ARI values were approximately in the same range.

Observations of tomato fields were conducted from April 4 until July 16, 2015. In total, 9 surveys were conducted with a frequency of 2–3 weeks. NDVI gradually increased from the beginning of observation until June 4: 0.201, 0.209, 0.465, 0.648, and 0.810. This behavior is a response to crop growth and an increase in green biomass. On June 9, a small decline of NDVI (0.777) was detected.

Nevertheless, on the June 22 observation, the value rose and started to decline as related to the crop's ripening (0.789, 0.597, and 0.585).

Similar to the chickpea field, the majority of indices calculated for tomatoes displayed close behavior: NDVI, EVI, GAVI, LAIc, PSSR, TVI, MCARI2, MTVI, MTVI2, and MCARI. These indices showed increments since June 4, a recession on June 9, a small rise in value at the next observation, and further decline until the end of the growing session. The recession on June 9, pronounced with different strength in these VIs. In some of them the decline was significantly pronounced: EVI (0.751–0.635-0.738), LAIc (1.734–1.422-1.528), TVI (31.649–26.600-32.476), MTVI (0.769–0.641-0.791), MCARI (0.307–0.158-0.344), MCARI2 (0.696–0.604-0.689) and MTVI2 (0.696–0.604-0.689). In other indices, the differences were barely noticeable: NDVI (0.810–0.777-0.789), GAVI (0.634–0.562-0.584), and PSSR (814.036–619.157-678.157).

On April 9, some indices displayed behavior close to NDVI, but without a decline. The GNDVI and LAIwp rose from April through June, and then gradually declined during the summer. Except for an increment in value at the last observation, ARI also matched this group of parabolic behavior. The PSRI in tomatoes did not have a behavior pattern. During the season, 4 cycles of increment and decline were observed, with the values ranging from 0.163 to 0.327.

The cotton field is located next to the tomato field, thus, its observations were carried out on the same dates: 9 surveys with a 2–3 week frequency from April 4 until July 16, 2015. The study was conducted from cotton seedlings until flowering. This explains NDVI behavior: low values in April–May (0.205, 0.219, 0.212, and 0.266), an increase in June (0.498, 0.570 and 0.858), and consistently high values during July (0.785 and 0.871). As in the other crops, NDVI behavior is repeated in other indices: EVI, GARI, GNDVI, LAIwp, LAIc, PSSR, TVI, MCARI2, MTVI, and MTVI2. On May 6 and July 2, NDVI showed a slight decline. The same reaction on crop state was pronounced in other indices with various intensities. LAIc, PSSR, MCARI2, and MTVI2, repeat NDVI's behavior and displayed a decline on both dates. LAIwp, TVI, and MTVI declined only on May 6. GARI and GNDVI declined only on July 2.

All three pigment indices developed under laboratory conditions showed unanticipated behavior in cotton. MCARI grew unevenly during the two months from April 4 until June 4: 0.011, 0.014, 0.016, 0.016, and 0.082. The next 3 measurements during the summer show a decrease: 0.054, 0.047, and 0.028. The last measurement unexpectedly increased to 0.035. The PSRI also increased at the beginning of the season: 0.273, 0.302, 0.398, and 0.499. Since summer, PSRI behavior had not displayed a trend: 0.3, 0.393, 0.116, 0.108, and 0.182. Also, ARI did not exhibit a behavior trend during the whole season. The index growth changed to a regression and then reversed: 3.012, 2.844, 3.419, 4.898, 4.739, 5.878, 9.463, 6.889, and 11.286.

Abnormal behavior can be observed in chickpea: on April 20, increased VIs that are linked to carotenoid and sharply decreased VIs that are linked to greenness and physical parameters. During May, the VI recovers and returns back to expected behavior. A similar situation was observed in tomatoes during June 2015 and in cotton during July 2015. Unlike chickpeas, the VI decline in tomatoes and cotton was slight and not pronounced in all the VIs. Among the declined indices in tomatoes and cotton, there were NDVI, GARI, LAIc, PSSR, MCARI, MCARI2, and MTVI2.

The range of VI variabilities calculated from satellite images is from 0.3 to 34%. It was not possible to calculate variability for GARI because of mixed (negative and positive) values in datasets: the mean average was close to zero. A strong correlation between values or phenological stages and variability was not found.

The normality tests show the nonnormal distribution of measurement variability from the satellite data source. The outlines were determined in two ways. Steam-and-Leaf Plot defined extreme values above 16.1%. Labeling by quarters defined the limits of the acceptable range from −5.52% to 15.64%. According to this outlining, LAIwp, LAIc, and GARI should be excluded because of their high variability. Despite MCARI2; MTVI, and MTVI2 also showed outline variability. They will await further consideration because the extreme variability occurs only once – at the late ripening stage of chickpeas.

The temporal variability of VI as calculated by satellite images ranges from 22.95% to 137.82%. In the study case, the high temporal variability indicates a significant change in VIs during the growth period. Statistical methods did not define outlines. The smallest temporal variability (less 40%) was found in GNDVI, ARI, and PSRI. The highest variability (more than 100%) was observed in GARI, LAIc, PSSR, and MCARI.

Pattern behavior of all considered VI (except ARI) in healthy crops is linked to phenological development (as has already been described earlier). Both healthy crops had a small decline in greenness and physical parameters, and an increase in carotenoid during the flowering period (in cotton) and yield formation (in tomatoes). The chickpea's VIs were also linked to phenological development and had a behavior similar to healthy crops. An exception is an abnormal behavior during April 2015 at the flowering stage. Despite healthy crop behavior in the middle of the growth period, a decline in chickpea greenness VIs was evident and pronounced in all VIs.

3.2 Spectrometer data-source

The chickpea field was examined by spectrometer during the growth period 5 times from the end of January until the beginning of June with an average frequency of once a month. NDVI calculated from spectrometer shows high values and increments during the season: 0.671, 0.740, 0.743 and 0.763. In this growth trend, the decline to 0.654 on May 22 seems abnormal. Similarly, high values and increases during the season with an abnormal decline in the end of May were shown by LAIc, PSSR, MCARI2, and MTVI2. Most of the health indices duplicate NDVI behavior but show a slight decline at the last observation: EVI, GARI, GNDVI, LAIwp, TVI, and MTVI.

ARI, PSRI, and MCARI, developed in laboratory conditions, display a common behavior trend in chickpeas: high value at the beginning, a decline during March, and an increase during April and May, and then a decrease at the end of observation in June.

The tomato field was observed 3 times during the irrigation period: May 22, June 7, and July 28. Analogous to the previous crop, spectrometer measurements recorded high values for the indices during crop growth: NDVI (0.905 and 0.915), EVI (0.797 and 0.983), GARI (0.764 and 0.770), GNDVI (0.758 and 0.761), LAIwp (5.005 and 5.251), LAIc (2.802 and 2.936), PSSR (2501 and 4367), TVI (27.445 and 37.120), MCARI2 (0.845 and 0.913), MTVI (0.735 and 1.002), and MTVI2 (0.845 and 0.913).

All indices, except those developed in laboratory conditions, veer to the lowest values for the last measurements: NDVI (0.822), EVI (0.680), GARI (0.557), GNDVI (0.615), LAIwp (2.830), LAIc (1.849), PSSR (1794), TVI (24.368), MCARI2 (0.723), MTVI (0.682), MTVI2 (0.723). This could be explained by the fact that the tomatoes were at their ripening stage. Thus, the difference in indices behavior is when the values were higher: on vegetative stage at May 22, or June 7. The NDVI, LAIc, and PSSR rise during this period. The EVI, LAIwp, TVI, MCARI2, MTVI, and

MTVI2 declined from the vegetative to the flowering stage. GARI and GNDVI remained constant. Laboratory indices behave differently from each other. MCARI increased during the whole period of observation: 0.200, 0.221, and 0.324. The ARI displayed an increase and then decrease (1.210, 2.246, and 2.135); where the first value is significantly lower than the next two. PSRI values were close to zero: −0.003, −0.002, 0.010.

During the irrigation period, the cotton field was examined 4 times, monthly from May through August. Additionally, during the first field trip, a few plants were found with visual signs of leaf dehydration. These plants were measured separately from healthy plants. Thus, consideration of indices for cotton includes behavior studies and comparative analyses.

NDVI in cotton showed growth from May until the end of July (0.757, 0.815, and 0.837) and a decrease at the fruit formation stage at the end of August (0.780). As in the previous analysis, most of the indices match the NDVI trend. GARI, GNDVI, LAIwp, LAIc, PSSR, MCARI2, and MTVI2 also increased from May to July and declined at the end of August. However, some of the health indices did not react to crop changes at the fruit formation and growth stage during the whole observation period: EVI, TVI, and MTVI.

In laboratory indices, there was no trend corresponding to crop development. The ARI reached its highest value (0.549) when first measured, and then during the growth period values remained approximately the same: 0.301, 0.326, and 0.323. The MCARI behaved in a wavelike manner: 0.170, 0.247, 0.141 and 0.204. PSRI values were too low for detection of changes: 0.006, −0.009, −0.003 and − 0.001.

Comparison of cotton health and stress on May 22 shows a reaction of all the indices on visually detected leaf dehydration. Health indices developed in the field conditions have values in stressed crops that were 1.5–2 times lower than in healthy ones: NDVI (0.537–0.757), EVI 0.545–0.916), GARI (0.350–0.587), GNDVI (0.464–0.601), LAIwp (1.111–3.044), LAIc (0.542–1.207), PSSR (306.455–528.177), TVI (35.787–24.459), MCARI2 (0.488–0.746), MTVI (0.640–0.945), and MTVI2 (0.488–0.746). Indices developed under laboratory conditions in contrast with field conditions, have higher values in stressed crops: ARI (2.823–0.549), MCARI (0.170–0.305) and PSRI (0.151–0.006).

Variability in spectrometer measurements is also rather different from satellite images. The common range of variance in spectrometers' VIs is from 1.9% to 127.94%. PSRI for all crops and ARI for chickpeas have negative values. Therefore, estimating their variability is not possible.

The normality tests show the non-normal distribution of variability from the spectrometer. The outliers were determined by two methods. Steam-and-Leaf Plot defined extreme values after 50.28%. Labeling by quarters defined the limits of the acceptable range from −18.30% to 52.26%. Outlining variables belong to GARI (1 value), LAIwp (1 value), LAIc (3 values), PSSR (3 values), and MCARI (3 values). All these indices had extreme variability in chickpeas on May 22, 2015.The temporal variability of VI, calculated by the spectrometer, ranging from 3.84% to 46.01%. Both statistical methods, Stem-and-Leaf Plot and labeling by quarters determined only one extreme value – 46.01%. According to outlining, the optimal range of temporal variability is from 3.84% to 33.29%, and PSSR had an extreme value.

VIs calculated by field spectrometry could be grouped by behavior trends. The first group of variability showed a gradual increase in cotton and tomatoes during yield formation and ripening (**Figure 2a**). NDVI, LAIc, PSSR, MCARI2 (and MTVI2 – had the same values as MCARI2) belong to this group. In chickpeas, these indices displayed common behavior as well, but differ from cotton and tomatoes: high value at the establishment, lowest values at vegetative stage, a slight increase at flowering and yield formation, and an abnormally high value on May 22, 2015. In

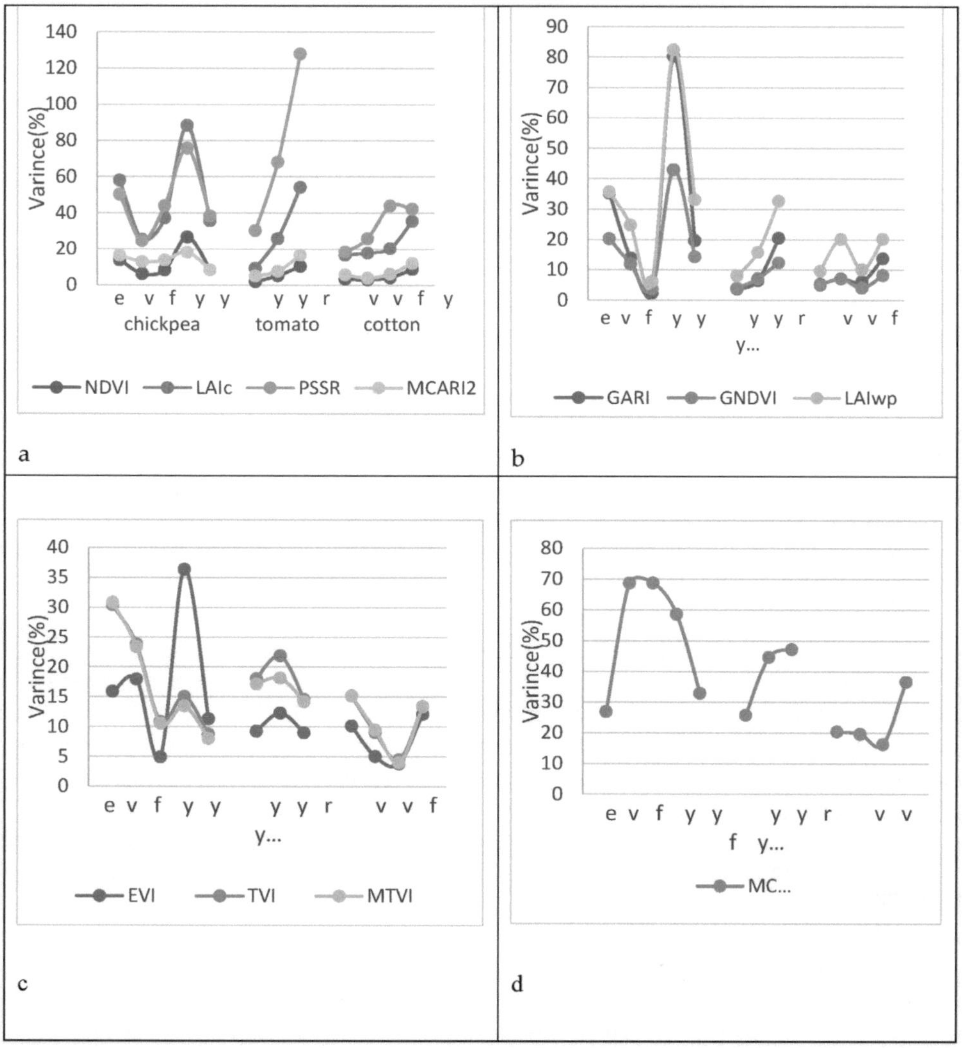

Figure 2.
Variability trends of VIs by spectrometer in different crop types, a is the first behavior group (NDVI, LAIc, PSSR, MCARI2), b is the second behavior group (GARI, GNDVI, LAIwp), c is the third behavior group (EVI, TVI, MTVI),d is indices not falling into the trends (MCARI). The x-axis labels: e – establishment, v – vegetative stage, f – flowering, y – yield formation, r – ripening.

MCARI2 and MTVI2, the increase was expressed slightly for this date. The next group includes LAIwp, GARI, and GNDVI (**Figure 2b**). These indices behaved in tomatoes in a manner similar to the previous group. In cotton and chickpeas, it also matched behavior in the previous group, except for relatively high values at the late vegetative stage. The third group consists of EVI, TVI, and MTVI (**Figure 2c**). These VIs have a common unique type of behavior for each crop type. MCARI did not match any group (**Figure 2d**).

Summarizing the above, VIs calculated by a spectrometer had no common behavior pattern for all crops. For chickpeas, all VIs related to chlorophyll, green biomass, and LAI have comparable behavior during growth season: increasing since establishment until flowering, and declining after yield formation until ripening. What is unusual is that the values of the indices for May 2015 were lower than those for June 2015. In tomatoes and cotton, VIs related to health can be divided into two groups according to their behavior. The first group includes NDVI, GARI, GNDVI,

LAIpw, LAIc, PSSR, MCARI2, and MTVI2. These indices have a direct correlation with crop development. The second group consists of indices from EVI, TVI, and MTVI. These indices show abnormal behavior for healthy crops: in tomatoes, the decrease begins at the late vegetative stage, while cotton has low VI values during the growth period and then begins to rise after the flowering stage. Indices developed in laboratory conditions (ARI, MCARI, and PSRI) show unique behavior for each crop and there is no possibility to distinguish any trends.

Behavior patterns for healthy crops (cotton and tomatoes) could be clearly defined in VIs related to phenological development: NDVI, GARI, GNDVI, LAIpw, LAIc, PSSR, MCARI2, and MTVI2. VIs grew until flowering and decreased during yield formation and ripening. In chickpeas, the same VI behavior was observed except at season end. At yield formation during May 2015, this decrease was pronounced in the VIs related to phenology. Regardless of expected continuing decline at yield formation in June 2015, all the indices increased. Thus, the decline in May 2015 during the drought period could be considered as a reaction to stress. In other indices, behavior patterns for healthy crops were not defined and consequently cannot reveal abnormal behavior in chickpea stress.

3.3 Validation

The first results show that health crops have an approximately common value of VIs obtained by the field spectroscopy. In the same time, results give the right to assume that crops under stress that situated in the same area have a high spatial variability of indices. Anyhow, a limited dataset of field spectral data for comparative analysis is not suitable for the approval of variability as a stress indicator. Thus, to verify the hypothesis about indication stress by variability, all field spectral measures obtained in the study was merged in "health" and "stress" groups. Control "stress" group consists of 18 point spectral measurements from the stressed cotton on 22.05.15. Control "health" groups are presented by 40 values of cotton (10 measures X 4 dates) and 20 (10 measures X 2 dates) of tomatoes field spectral measurements collected during the vegetative and fruit formation stages. Tested groups consist of chickpea field spectral measures. The "health" tested group consists of 20 (10 measures X 2 dates) spectral measures from the vegetative and flowering stage in chickpea. The proposed "stress" group is presented by 10 spectrums on 22.05.15 when the variability in the first results was the highest. For this test, indices with higher (more than 40%) variability detected in the first analysis were applied: GARI, GNDVI, LAIwp, LAIc, and PSSR.

The test shows (**Figure 3**), that variability of GARI, GNDVI, LAIwp and LAIc in the group of "health" cops merged from several dates is lower than in the "stress" dataset obtained on the one date. Also, the results display that level of variety of the control "stress" cotton and tasted "stress" chickpea are approximately the same and rather far from the "health" groups (especially in GARI, LAIwp, and LAIc). Only PSSR shows high variability in all groups. This fact was already noticed at the first stage of the analysis. The absence of a relationship with the stress in crops and high variability of the values indicates the non-occurrence of the use of the PSSR algorithm for field spectroscopy.

3.4 Early water stress detection by VIs

The first stage of analysis is calculation VIs averages and consideration their temporal behavior (**Figure 4**). The W value of the Shapiro–Wilk test is significant for all considered VIs. The VIs informativeness assessment for stress estimation based on their response. All graphs are presented in the range of allowable values

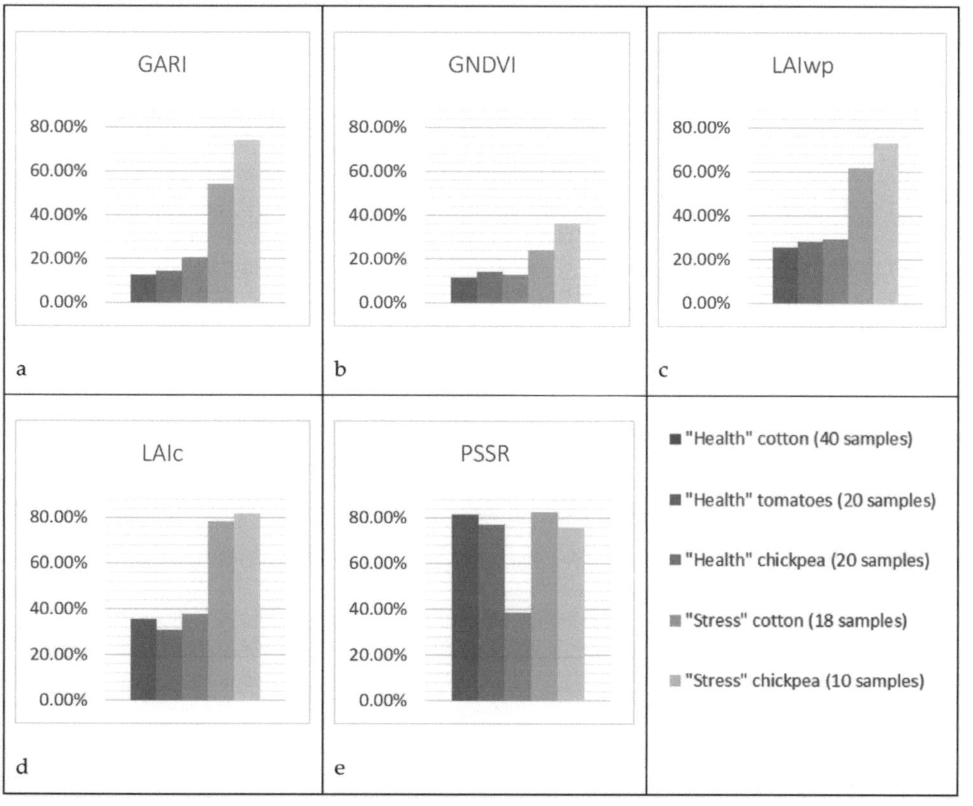

Figure 3.
Variability of GARI, GNDVI, LAIwp, LAIc and PSSR obtained by the spectrometer in "health" and "stress" crop groups order as follow: "Health" cotton, "Health" tomato, "Health" chickpea, "Stress" cotton, "Stress" chickpea, from dark gray to light gray.

from the original papers or in the range of optimal values for green vegetation where it was possible (CRIs, PRI, PSRI, SIPI).

Interpretation of VIs responded to anthocyanin and carotenoids (**Figure 2a, b**) is ambiguous. On one side, in the stress conditions, the concentration of anthocyanin and carotenoids should increase. This principle lays in interpretation of ARI and CRI behavior from original papers: weakening vegetation contains higher concentration of these pigments, thus increase indices values indicates stress. However, the content of these pigments also relates to chlorophyll level [99]. Subsequently, decline of anthocyanin and carotenoids can be consequences of chlorophyll reduction, which is also a sign of stress. This contradiction complicates the interpretation of Anthocyanin and Carotenoid Reflectance Indices. The highest values of ARIs (**Figure 2a**) were obtained at the beginning and at the end of season. The lowest values of this index correspond to the chickpea vegetative period, after which the anthocyanin concentration began to grow smoothly and slowed down on May 21 only, when chickpea get least watering. The high values of ARI at the beginning and end of observation corresponds to chickpea extreme stages: emergence and ripening. Detected sensitivity of ARI to critical stages coincides with the results of previous studies of this pigment [100, 101]. The decline on May 21 is abnormal and can be reaction on stress, but the existing information is not enough for an unambiguous interpretation. Like ARIs, CRIs have smooth graph with high values at the edges. But unlike anthocyanin, the level of carotenoids on May 21 increase.

Indices developed for estimating green LAI and biomass, chlorophyll content and health (**Figure 4b–f**) have common temporal trend: the lowest values at

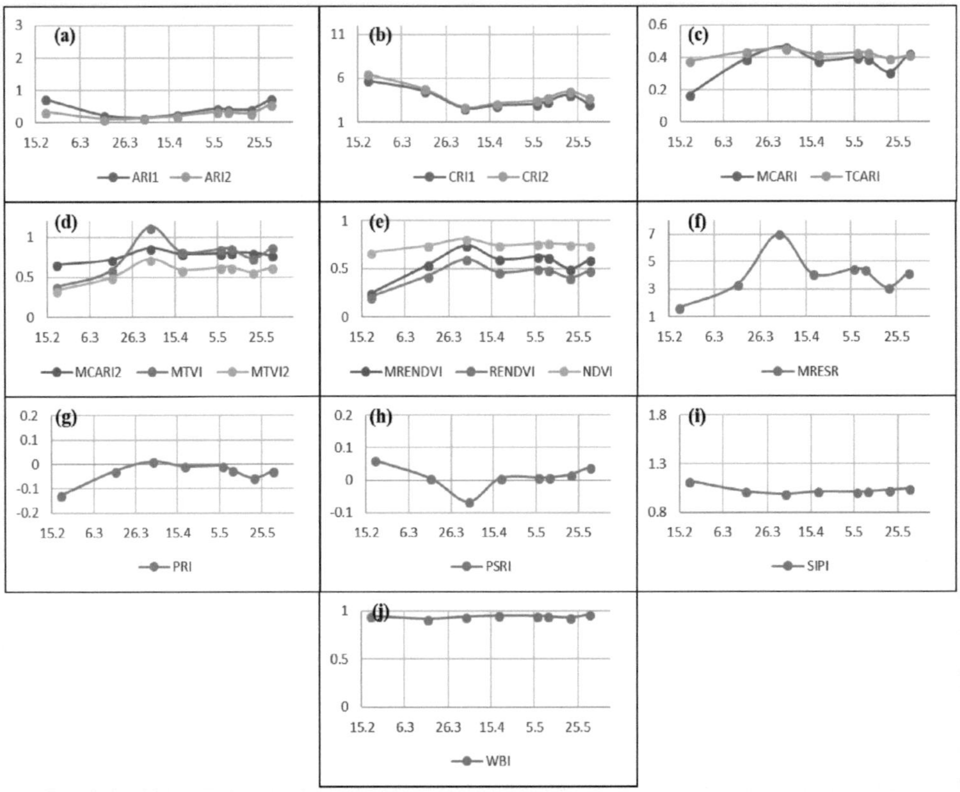

Figure 4.
VIs' changes over time on the chickpea field in 2016: a) anthocyanin indices; b) carotenoid indices; c) chlorophyll indices; d) canopy cover indices; e) NDVI group; f) MRESR; g) WBI; h) SIPI; i) PRI; j) PSRI.

emergence, increase and highest values on April 3, and notable decline on May 21. At other dates these indices have stable values and did not manifest any crop changes. Even deviation of these indices is almost inessential concerning the declared common range for green vegetation because of data obtaining on leaf scale [102], it is enough to identify chickpea at late vegetative stage as most stable and health and detect signs of weakness on May 21. However, the absence of a reference range for leaf scale spectral measurements does not allow to assess the significance of weakness and define it as stress. PRI (**Figure 4g**) related to photosynthesis rate match the trend of health indices behavior, but with less intensity of expression. On the smooth graph of PRI, the most pronounced is decline on May 21. Based on these graphs it can be concluded that chickpea experienced weakness and decreased activity at the week with the least watering.

PSRI (**Figure 4h**) shows carotenoid/chlorophyll ratio and originally proposed for estimation canopy senescence. Increase of the index value is interpreted as stress. In this study the index behaves opposite to indices of green LAI (**Figure 4d–f**), except May 21 when did not increased as expected. Thus, PSRI does not confirm stress estimated by previous group of health indices. Last two graphs on the Figure (**Figure 4i, j**) presents relative water content by WBI and carotenoid/chlorophyll ratio estimated by SIPI. Both graphs represent straight lines that do not reveal any significant changes during the time. On closer examination it can be seen that SIPI is similar to PSRI but with less expressed values.

As a result of the VIs temporal behavior analysis, one can assume the chickpea stress on May 21. This assumption is confirmed only by some of the examined indices and it is difficult to determine the degree of stress significance. The assumption of stress was made based on VIs temporal changes and information of irrigation amount. The absolute values of the indices on May 21 belongs to range of optimal values for green vegetation.

3.5 VIs variability

Variability is a complicating factor in interpreting VIs behavior and identifying stress. Since at one observation there was a measured series of spectral signatures, the VIs can vary throughout the field. Thus, consideration of VIs as the average of the dataset, as at the previous stage of analysis, should be combined with a VIs' variability assessment. The high level of variety (more than 50%) makes the average of VIs uninformative for estimating field crop state. As was mentioned above, the results of the Shapiro–Wilk test is significant for all applied VIs that allow to consider their variability. The results of calculation variability are presented in the **Table 5** and marked according to intensity.

The highest level of variability was detected in indices related to anthocyanin and carotenoid. Also, MCARI has variability more than 50% since chickpea flowering. Other indices were stable and has variability near or les 25%. WBI and SIPI that did show temporal deviation have the least variability from 0.4 to 5.7%.

To date, the role of the variability of leaf-scale spectral measurements for crop state estimation has been poorly studied, but already found that high variability indicates different stresses in leaves [103, 104]. In the next section of PCA, the VIs averages will be studied commonly and separately with their variability to identify its effect on crop state estimation.

3.6 Regression analysis

Considered in this study VIs did not reveal linear correlation with irrigation amount and crop height. CRIs, PRI, MRENDVI, TCARI show strong relationship with maximal drought obtained by tensiometers (R2 = 0.75–0.95) and the same indices together with MRESR and RENDVI fit the actual tensiometer data with accuracy R2 = 0.81–0.96). At the same time no relation stronger than R2 = 0.59 was found with minimum values obtained by tensiometers. Also, several indices reacted to the phenological stage. Nevertheless, these relationships were weak (R2 = 0.5–0.7). In this case of study, only 6 of 17 considered indices reacts on soil water content on chickpea field. These results highlight, the complexity of application VIs as estimator of physical parameters. Such approach requires validation for each specific crop and field to identify the most suitable and informative VIs. Also, incomplete correlation effects on crop state estimation accuracy that can reduces the possibility of detecting water stress at early stages.

3.7 Principal component analysis

The advantage of implementation PCA is the ability to find unique behavior patterns by combination of indices, despite their relatively close values in the irrigated chickpea that limits such approaches like interpretation temporal behavior and regression analysis. The field parameters were excluded from this analysis and crop water stress is identified by spectral data only. The detection of stress is based on comparison of the similarity considered and reference stressed crops. Despite the close values of the indices, their combination turns to the unique signature that describes the singularity of crop state at each campaign.

The variability and regression analysis results were used to define the input VIs for PCA: the indices with variability more than 50% were excluded (ARIs, CRIs and

	19.02	16.03	3.04	18.04	6.05	11.05	21.05	30.05
Cv of NDVI	◍ 11.3%	○ 4.8%	○ 1.9%	○ 6.2%	○ 4.7%	○ 7.5%	○ 7.5%	◍ 11.1%
Cv of WBI	○ 2.3%	○ 0.7%	○ 0.7%	○ 1.0%	○ 0.4%	○ 1.3%	○ 1.2%	○ 1.0%
Cv of ARI1	⊗ 62.5%	⊗ 92.4%	⊗ 171.1%	● 32.4%	● 44.8%	⊗ 73.5%	● 38.3%	◍ 18.1%
Cv of ARI2	⊗ 60.2%	⊗ 72.6%	⊗ 177.6%	● 36.0%	● 40.6%	⊗ 70.7%	● 35.9%	◍ 16.2%
Cv of CRI1	⊗ 58.1%	● 30.1%	◍ 21.6%	● 40.9%	● 29.9%	● 41.3%	◍ 23.7%	● 26.5%
Cv of CRI2	⊗ 57.1%	● 32.4%	● 28.5%	● 37.3%	◍ 24.9%	● 40.7%	◍ 24.0%	◍ 24.0%
Cv of SIPI	○ 3.8%	○ 2.3%	○ 1.1%	○ 0.4%	○ 0.5%	○ 0.7%	○ 2.4%	○ 5.7%
Cv of MCARI	◍ 22.7%	◍ 23.8%	◍ 10.0%	⊗ 52.2%	⊗ 56.4%	⊗ 56.4%	● 25.4%	◍ 19.6%
Cv of MCARI2	◍ 13.2%	◍ 13.4%	○ 2.4%	◍ 10.1%	○ 9.0%	◍ 10.3%	○ 7.4%	◍ 10.3%
Cv of MRENDVI	◍ 16.9%	○ 6.4%	○ 3.6%	○ 8.6%	○ 9.7%	◍ 11.1%	◍ 17.5%	◍ 18.4%
Cv of MRESR	○ 9.1%	○ 9.7%	◍ 14.0%	◍ 16.6%	◍ 18.8%	◍ 23.2%	● 26.6%	● 30.2%
Cv of MTVI	◍ 22.5%	◍ 22.6%	○ 3.6%	○ 4.9%	○ 6.9%	○ 6.5%	◍ 13.5%	◍ 11.2%
Cv of MTVI2	◍ 18.8%	◍ 14.0%	○ 2.2%	○ 3.6%	○ 3.8%	○ 8.7%	◍ 13.6%	◍ 13.9%
Cv of RENDVI	◍ 19.9%	○ 6.5%	○ 2.2%	○ 3.6%	○ 7.2%	◍ 10.9%	◍ 19.0%	◍ 20.2%
Cv of TCARI	● 40.7%	◍ 14.0%	○ 5.2%	◍ 11.1%	◍ 17.9%	◍ 20.2%	◍ 17.1%	● 25.8%

Table 5.
The variability (Cv) of the considered VIs on the tomato field in 2016: ○ *– less than 10%;* ◍ *– from 10% to 25%;* ● *– from 25% to 50%;* ⊗ *– more than 50%.*

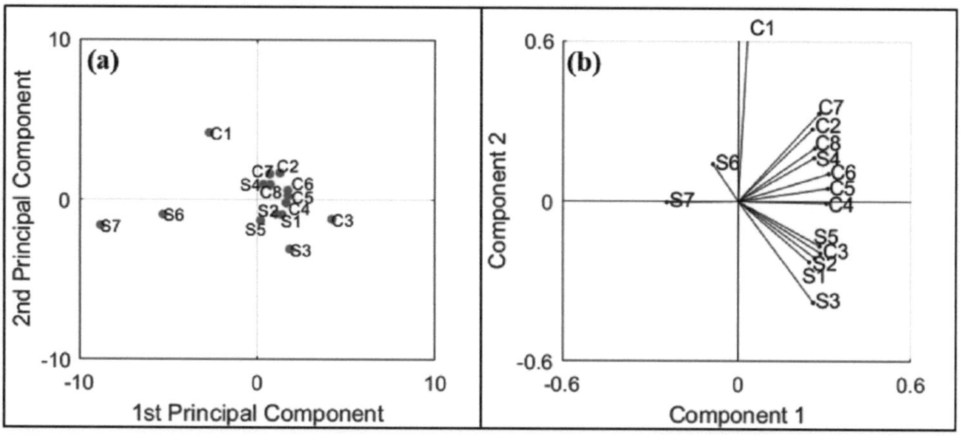

Figure 5.
Principal component analysis for the chickpea states from the field and control plot: a) first PCA test for VIs; b) second PCA test for VIs.

MCARI) as non-informative. The first runs of PCA were failed due to significant differences in crop states at specific development stages (**Figure 5**). The initial chickpea stage C1 (**Table 2**), and late stress stages from the control site S6 and S7 (**Table 3**) were plotted by PCA far from other states and make deviation of other states insignificant and hard identifiable. The second reason of closeness the plotted states in the first results of PCA is slight VIs deviation because of too wide range of values used for normalization.

Based on the experience of the first results, new configurations for PCA have been developed. The extreme chickpea states (C1, C2, S6, S7) when watering is inappropriate were excluded from the analysis. To increase the sensitivity to changes in crop states, the considered VIs were normalized (Eq. (4)) by minimum and maximum values obtained in these states. After these changes, the analysis became more accurate and allowed to distinguish slight variation in chickpea (**Figure 5**). In the first PCA test, the crop states are plotting according to estimated 2 principal components makes possible to estimate their similarity based on their closeness in space (**Figure 6a, c, e**). In the second PCA test, the crop state is characterized by vectors (**Figure 6b, d, f**). The similar crop states have the same length and angle of the vector in the second PCA test. In the all performed tests, the first and second principal components together explained 70–80% of the variance.

In the first tests, C3, C8 and S5 are most distant from the plot center and can be called extreme states. C3 is situated at the opposite part of plot from C8 and S5 and has opposite value of the first principal component. States C8 and S5 are located on the same plot side by the first principle component, but at a considerable distance from each other. While states C7 and S4 have almost same location on the plots. States C4-C6 and S1-S3 lay in the middle of the plot. Among the first PCA tests, variability and VIs with their variability plotted the chickpea states such way that it can be seen the dynamics of state changes. At the beginning of the experiment the state of the chickpea on the control plot is close to the chickpea on the field (**Figure 6c, e**). During first weeks of the experiment, when stress was not significant, the states C5, C6, S2 and S3 still has similar location, but they still located on the different sides of the second principle component axis.

The second PCA tests were less informative. The vectors of chickpea states were distributed in several groups and have different length that limits interpretation and identification correlations. The most graphic is the plot of second PCA test for VIs

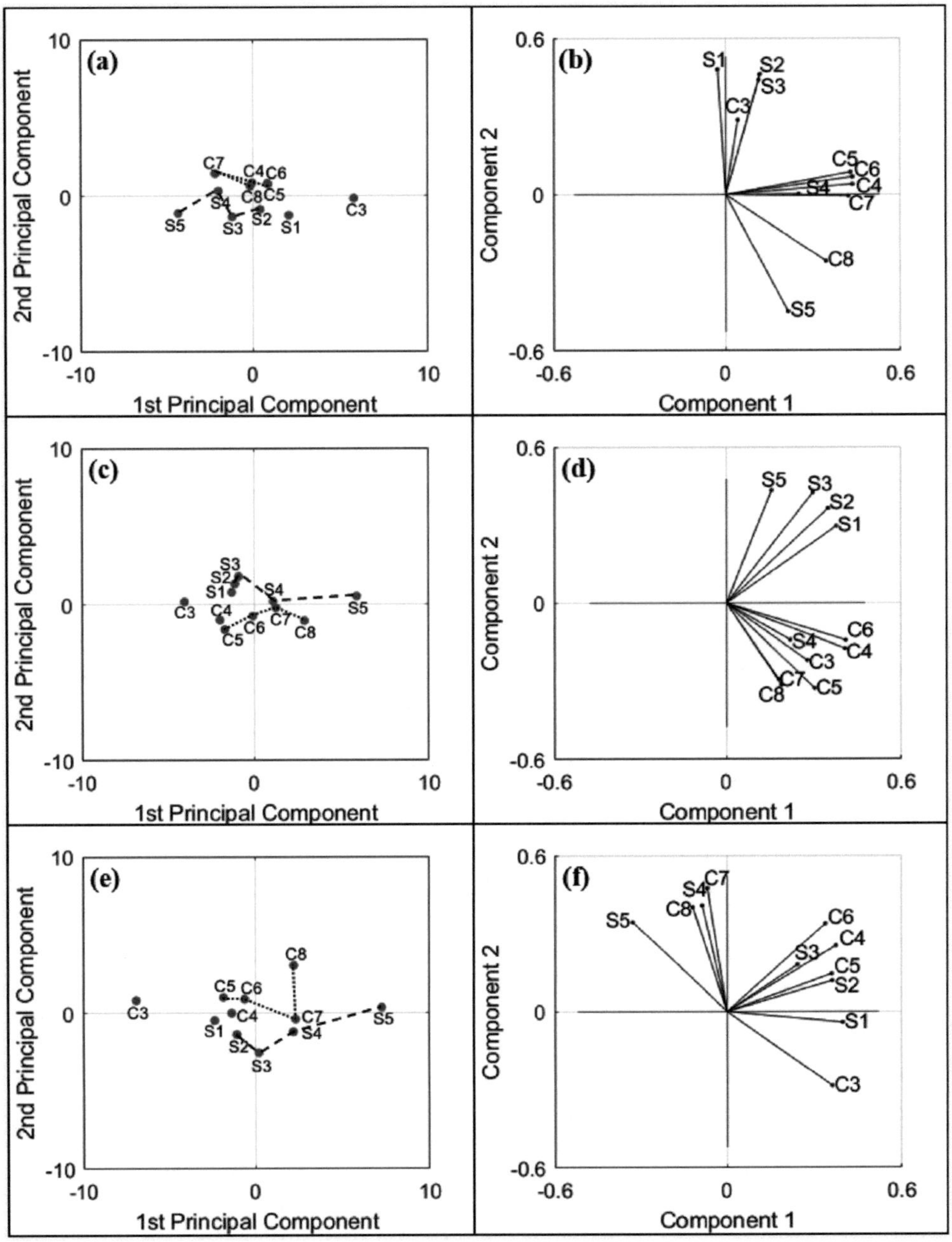

Figure 6.
Principal component analysis for the chickpea states C3-C8 and S1-S5: a) first PCA test for VIs; b) second PCA test for VIs; c) first PCA test for VIs' variability; d) second PCA test for VIs' variability; e) first PCA test for VIs and their variability; f) second PCA test for VIs and their variability.

and their variability (**Figure 6f**). All chickpea states are distributed between two opposite conditions: active growth at vegetative stage C3 and suffered from drought control chickpea S5. Stressed and ripening states C7, C8, S4, S5 have negative values of the first component, while other states have positive.

The results of PCA tests indicates similarity of chickpea state on May 21 with crop suffered from drought. The assumption of stress in chickpea field on May 21 corresponds to the low soil moisture at week before the spectral measurement. The implemented PCA test indicates the early water stress in chickpea by spectral data only, while other considered in this study traditional methods of VIs analysis could not uniquely detect lack of water in chickpea field. The obtained results confirm already discovered PCA's applicability for precise analysis in agricultural management [105] like estimation nitrogen concentration [106] or detecting diseases [107].

4. Discussion

The discussion section presents a recommendation for applying VI's for determining water stress in irrigated crops. Differences in the variance of intensity, temporal changes, and correlation to phenological development were mainly dependent on the data source. VIs from satellite data sources are strongly related to phenological stages and significantly changes with time. This correlation is caused by the common measuring of vegetation and soil background. Since crop growth and leaf development, the noise from soil is reducing and greenness rises. Thus, satellite imagery collects not only spectral plant characteristics but also records physical development. On the one hand, it is a good approach to estimate crop physical parameters: LAI and biomass. On the other hand, it reduces sensitivity to changes in crop pigmentation.

In the satellite data source, the variability was estimated by few image pixels (due to the field size). Therefore, the variance in measurements could be interpreted as differences in crop characteristics on adjacent 5 x 5 m areas. All indices showed low variability (less 20%). This can be explained by crops conditions that should not be significant different across the field. Nevertheless, the crop characteristics cannot be absolutely the same and variability shows up in satellite's sensitivity to the differences. The lowest variability was observed in indices developed for satellite data sources. The indices developed by field and laboratory spectroscopy, and modified for airborne detection provide high variance regardless the proposed response (physical or pigmentation). Variability in TVI with its modifications reacts on crop type. In chickpeas, these indices have lower variance than in tomatoes and cotton.

To unequivocally estimate water stress in chickpeas by VIs using satellite images is complicated. The abnormal VI behavior for April 2015, in chickpeas, could be considered a reaction to water stress. There were already 9 days of drought since the last intensive precipitation (60 mm). The recommended level of irrigation at this growth stage is nearly 5 mm per day [80]. It is logical to assume that water stress would result and reach a maximum at the beginning of the irrigation period, in May 2015. However, VIs calculated from May 1, 2015, toMay 16, 2015, did not detect stress. Moreover, all VIs testified to the maximally a healthy and strong chickpea state on May 1, 2015, and May 6, 2015. Thus, abnormal VI behavior on April 20, 2015, is not related to water crop stress.

Measurements by field spectrometer have typically high variability and slightly pronounced changes over time. Despite satellite data, direct leaf measurements reduced soil background noise and minimized the influence of physical plant parameters. It can explain weekly correlation with phenological development and lower temporal variability than in VIs calculated by satellite images. Measurements on leaf-scale by high-resolution spectral equipment provided pigmentation sensitivity.

Crop pigments are a suitable stress indicator. Stressors affecting plants primarily lead to impaired or perturbed metabolism and photosynthesis [108]. Chlorophyll level is the basic parameter of photosynthesis. Thus, an abnormal chlorophyll level points to stress. Apart from metabolism, pigments play a protective role against stress. The increase of carotenoid and anthocyanin indicate crop weakness [22, 44].

That is why pigments are considered to be the optimal approach for early, in vivo, detection of abiotic plant stress [86].

Particular attention should be paid to the variance of measurements during one field trip; as stress leads to changes in pigment production, the adjustment is not evenly implemented. Therefore, the increase in pigmentation variability could be a sign of stress. The highest values of VI variances was detected at emergence, ripening stage, and at water deficit in chickpeas on May 22, 2015. Emergence, ripening, and water lack could be considered as stress periods. At emergence stage, the crop is a weak, unformed plant that is practically non-resistant to external conditions. At ripening stage, the plant stops supplying nutrients to leaves and begins dying. Prolonged water deficiency leads to crop drought and a limitation of nutrients for leaves. The unstable state of the plant due to these stresses leads to uneven leaf nourishment resulting in pigmentation variability. Likewise, high variability of point-by-point field spectral measures can be an indicator of unstable and stressed crop state.

Almost all VIs, apart from the highest variability during emergency, ripening, and stress, additionally showed increases in variance at other times. These defined a few behavior trends in VIs variance. The presence of trends in variability may mean that common increase of VI variances was a reaction to some unstable or stressed crop state that was not noticeable during visual inspection. This assumption gives grounds to study variability as a parameter for stress estimation during early stages in irrigated crops. Hence, the high temporal frequency of field spectral measurement could clarify variability trend in different phenological stages under normal conditions that allow accurate and timely highlight fluctuations associated with stress.

VIs calculated from spectrometers were more sensitive to stress than VI from satellites. Most of VI calculated by spectrometer displayed an abnormal behavior in chickpeas on May 22, 2015, when there was an expected lack of water due to a long drought. The reaction to stress was estimated only in VIs related to phenological development. VIs with high sensitivity to pigmentation displayed complex behavior during the growth period when a behavior pattern could not be observed in healthy crops requiring further stress detection by comparative analysis.

Determining lack of water in irrigated crops is a complicated task. The goal is to define stress before water lack causes problems in physical development. Consequently, stress should be detected at an early stage when it only affects metabolism and related pigmentation. Thereafter, VIs linked to crop pigmentation are expected to be more applicable than VIs related to LAI and green biomass. However, this study has shown that method and scale of spectral data collection from the irrigated crops significantly influence VI correlation with various parameters.

VIs calculated from satellite images are strongly related to physical crop development regardless of proposed physical or pigmentation sensitivity. Type of inputs applicated to develop VIs (satellite and airborne images, field and lab spectroscopy) not effect on VI sensitivity as was expected. Likewise, satellite spectral images could be considered as a useful source for estimating crop physical parameters and other tasks related to it, such as crop classification. The low variance of VIs on the same date and high temporal variability allow for clearly estimating the crop state for the whole field and observe crop development. At the same time, these reasons lead to low-pigmentation sensitivity and make satellite images unsuitable for detecting water stress in irrigated crops at early stages.

Point-by-point field spectroscopy measurements of the irrigated crops provide results with high sensitivity to crop pigmentation. Measuring reflectance with an accuracy of 1–2 nm allows identifying and quantifying leaf pigments [109, 110]. However, successful early stress detection depends not only on the degree of sensors'sensitivity to pigmentation but also the approach for estimating the crop state. Among the considered VIs, a few (NDVI, GARI, GNDVI, LAIpw, LAIc, PSSR, MCARI2, and MTVI2) were found exhibiting a correlation with phenological development. This correlation was weaker than in VIs calculated from satellite images. However, it was enough to define pattern behavior by crop development stages in healthy crops and detect abnormal behavior in drought-stricken chickpeas. The accurate estimating crop stress by VIs responded to the leaf pigmentation requires a comprehensive analysis that includes the specifics of the development strategy to each crop species and the metabolism features. Without these specific parameters, there could be identified only global trends of change in pigment composition that limits the opportunity to detect stress at the early stage.

Another feature of spectrometer measurement is hypersensitivity that leads to high variability of values on the relatively small and homogenous area. The variance of some VIs reached 127% High variability casts doubt on the measurements' validity. Outlining extreme values defined the optimal range of variability in this study - from 18.30% to 52.26%. Such high variability rates in field spectrometer measurements can significantly reduce accuracy and clarity of crop state estimation. Since the variability can reach extreme values, this parameter should be tested when the agriculture field is monitored by VIs from point spectral measures. Despite to this shortcoming, the variability itself can provide information about plant state: the study has shown that variability has certain behavior trend; e.g., high variability rates were associated with periods of crop stress (GARI, LAIwp, and LAIc). Since the level of variability is responded to uneven changes in pigment production, the increase of Vis variations indicates alteration of behavior strategy: the transition of phenological stage or reaction to uncomfortable environmental conditions (i.e. stress). Thus, VIs variability could be a suitable approach to detecting stress at the early stages.

In practical agriculture, crops should be in optimal conditions, since stresses due to imperfections in the management can lead to lower yield. Spectral measurements on leaf scale allows for tracking numerous slight changes in the plant associated with both early stresses and natural processes characteristic of each phenological stage [111]. On the one hand, this level of sensitivity is necessary to detect early stress in irrigated crops [112, 113]. On the other hand, it requires an approach for isolating external stress from natural processes. Also, the downside of high-precision measurements is high variability of spectra, that doubts the results [114], which was confirmed in this study. However, the fact that variability of leaf scale spectral measurements is also sign of stress forces to leave this parameter in the analysis. For this purpose, it is necessary to consider a dataset of parameters (VIs averages and their variability) that characterize different plant processes. The advantage of PCA is ability to process commonly the VIs and their variability and results of this study proves that inclusion of variability improves accuracy of stress estimation by PCA together with reference stress states.

5. Conclusion

To date, there are many scientific works claiming the advantages of using spectral indices in agricultural needs, including the detection of stress. Unfortunately, the results of this study discover: the fact that the index showed a response to stress is not enough to clear stress detection by VIs in practice. The stress detection task is complicated by the fact that an agricultural field requires a unified management approach and hence one parameter value of crop state. It means that the dataset of obtained spectral data (point measures by field spectrometer or a number of pixels from spectral imagery) should be transformed into a single unified value, which will be used in further decision-making about field treatments. Exactly this requirement has been met at the first stage of the analysis: calculation of the VIs average for each dataset. Meanwhile, results of the variability estimation indicate that field spectroscopy could provide high spatial variability that depreciates the significance of the VIs average itself. The intensity of variation obtained in this study was in the range from 0.4–177% on the relatively small test area (5 x 5 meter).

The next stage of analysis was dedicated to identifying in practice response of VIs to field parameters related to irrigation. 6 from 17 considered VIs show strong correlation with soil moisture on the chickpea field, while other 11 VIs also proposed for water state estimation did not show significant correlation. Based on these results it can be argued, monitoring crop state by VIs based on their correlation with physical parameters need to be supported by validation from each specific crop and field to define the informative VIs in the specific case. Such requirement can significantly reduce the attractiveness of this approach for practical applications.

The novel method considered in this paper for early stress detection by VIs, was developed based on the experience that the joint consideration of parameters is more effective than a separate one, which was stated in the "Introduction." The proposed consideration of common VIs and their variability was implemented through PCA. The results met expectations: the joint examination of the parameters turned out to be much more precise in determining crop state than the individual arguments in the first stages of analysis. Proposed in this study VIs and their variability processing by PCA in common with spectral library of reference crop states allow detect stress on early stages. An additional advantage of PCA for estimating crop state is the presentation of results as a grid of vectors, where extreme positions define opposite crop states (active growth and death), while other states within this grid can be estimated from the proximity to one or the other extreme state.

To conclude, field spectroscopy is a promising technology for detecting water stress in irrigated crops at an early stage. Nevertheless, existing remote spectral methods should be supported by additional physically measured ground-based data. VIs as a separate independent approach for crop monitoring is limited by the difficulty of unambiguous interpreting. Hypersensitivity of spectral measurements could provide early stress detection on the one hand, and on the other hand, it requires a modified approach for accurate analysis and interpretation. Determining crop stress by spectroscopy as an independent method should be implemented through common consideration of VI behavior patterns in healthy crops, the intensity of correlation with physical development, and rate of measurement variance on a particular date.

Acknowledgements

The authors want to thank the German Aerospace Center (DLR), RESA Science Team, Neustrelitz for the support by providing the satellite data of the RapidEye Science Archive (proposal no. 597) through funding from the German Federal Ministry for Economic Affairs and Energy. We also thank the staff of the DLR (Oberpfaffenhofen), namely Thomas Krauss and Peter Fischer, for the atmospheric correction using the generic processing chain CATENA.

This work was supported by the Israel Water Authority [Grant #41683, 2014], and by Israel Water Authority (grant number 41683); and the Israel Innovation Authority, Israel-Italy Joint Innovation Program for Industrial, Scientific and Technological Cooperation (grant number 60283).

Author details

Maria Polivova and Anna Brook*
Department of Geography and Environmental Studies, Spectroscopy and Remote Sensing Laboratory, Center for Spatial Information Systems Research, University of Haifa, Haifa, Israel

*Address all correspondence to: abrook@geo.haifa.ac.il

References

[1] Cano, E., Cano-Ortiz, A., Musarella, C. M., Fuentes, J. P., Ighbareyeh, J. M. H., Gea, F. L., & del Río, S., 2019. Mitigating climate change through bioclimatic applications and cultivation techniques in agriculture (Andalusia, Spain). In Sustainable Agriculture, Forest and Environmental Management (pp. 31-69). Springer, Singapore.

[2] Jhariya, M. K., Banerjee, A., Meena, R. S., & Yadav, D. K. (Eds.)., 2019. Sustainable agriculture, forest and environmental management. Springer.

[3] Cano-Ortiz, A., Fuentes, J. C. P., Canas, R. J. Q., Gomes, C. J. P., & Cano, E., 2020. Analysis of the Relationship Between Bioclimatology and Sustainable Development. In INTERNATIONAL SYMPOSIUM: New Metropolitan Perspectives (pp. 1291-1301). Springer, Cham.

[4] Bevilacqua, C., Calabrò, F., & Della Spina, L. (2020). New Metropolitan Perspectives. Springer International Publishing, SIST 178, pp. 1319–1329.

[5] Kramer, P. J. ,1983. Problems in water relations of plants and cells. International review of cytology,85, 253–286. https://doi.org/10.1016/s0074-7696(08)62375-x

[6] Chlingaryan, A., Sukkarieh, S., & Whelan, B. 2018. Machine learning approaches for crop yield prediction and nitrogen status estimation in precision agriculture: A review. Computers and electronics in agriculture, 151, 61–69.

[7] Jones, H. G., 2006. Monitoring plant and soil water status: established and novel methods revisited and their relevance to studies of drought tolerance. Journal of experimental botany, 58(2), 119–130. https://doi.org/10.1093/jxb/erl118

[8] Barrs, H. D.,1968. Determination of water deficits in plant tissues. In: Kozlowski TT, ed. Water deficits and plant growth. New York, NY: Academic Press, 235–368.

[9] Gardner, C. M., Robinson, D. A., Blyth, K., & Cooper, J. D., 2000. Soil water content. Soil analysis: Physical methods, 1–74. https://doi.org/10.1201/9780203908600.ch1

[10] Allen, R. G., Pereira, L. S., Raes, D., & Smith, M.,1998. Crop evapotranspiration-Guidelines for computing crop water requirements-FAO Irrigation and drainage paper 56. FAO, Rome, 300(9), D05109.

[11] Idso, S. B., Jackson, R. D., Pinter, P. J., Reginato, R. J., & Hatfield, J. L., 1981. Normalizing the stress-degree-day parameter for environmental variability. Agricultural Meteorology, 24, 45–55. https://doi.org/10.1016/0002-1571(81)90032-7

[12] Delimini L., 2012. Seed production and training manual. Berlin (Germany): FAO/Germany Cooperative Programe for Ministry of Agriculture, Forestry and Food Security.

[13] Jones, H. G., 2004. Irrigation scheduling: advantages and pitfalls of plant-based methods. Journal of experimental botany, 55(407), 2427–2436. https://doi.org/10.1093/jxb/erh213

[14] Preuksakarn, C., Boudon, F., Ferraro, P., Durand, J.B., Nikinmaa, E. and Godin, C., 2010, September. Reconstructing plant architecture from 3D laser scanner data. In 6th International Workshop on Functional-Structural Plant Models (pp. 12-17).

[15] Neilson, Elizabeth Heather, Aaron Mark Edwards, C. K. Blomstedt, Bettina Berger, B. Lindberg Møller, and Roslyn M., 2015. Gleadow. "Utilization of a high-throughput shoot imaging system to examine the dynamic

phenotypic responses of a C4 cereal crop plant to nitrogen and water deficiency over time." Journal of experimental botany 66, no. 7: 1817–1832

[16] Granier, C. and Vile, D., 2014. Phenotyping and beyond: modelling the relationships between traits. Current opinion in plant biology, 18, pp.96–102

[17] Yousfi, S., Gracia-Romero, A., Kellas, N., Kaddour, M., Chadouli, A., Karrou, M., Araus, J.L. and Serret, M.D., 2019. Combined use of low-cost remote sensing techniques and δ13C to assess bread wheat grain yield under different water and nitrogen conditions. Agronomy, 9(6), p.285.

[18] Nikolaou, G., Neocleous, D., Christou, A., Kitta, E. and Katsoulas, N., 2020. Implementing sustainable irrigation in water-scarce regions under the impact of climate change. Agronomy, 10(8), p.1120.

[19] Gates, D. M, 1970. Physical and physiological properties of plants. Remote Sensing with Special Reference to Agriculture and Forestry: Washington D.C., National Academy of Sciences , 224–252.

[20] Behmann, J., Steinrucken, J., & Plumer, L., 2014 . Detection of early plant stress responses in hyperspectral images. ISPRS Journal of Photogrammetry and Remote Sensing, 93, 98–111. https://doi.org/10.1016/j.isprsjprs.2014.03.016

[21] Belgiu, M., Bijker, W., Csillik, O. and Stein, A., 2021. Phenology-based sample generation for supervised crop type classification. International Journal of Applied Earth Observation and Geoinformation, 95, p.102264.

[22] Gitelson, A. A., Merzlyak, M. N., & Chivkunova, O. B., 2001. Optical properties and nondestructive estimation of anthocyanin content in plant leaves. Photochemistry and photobiology 74(1), 38–45.

[23] Rouse, J. W., Hass, R. H., Schell, J. A., Deering, D. W., 1973. Monitoring vegetation systems in the great plains with ERTS. Third Earth Resources Technology Satellite (ERTS) Symposium 1, 309–317.

[24] Datt, B., 1999. A new reflectance index for remote sensing of chlorophyll content in higher plants: tests using Eucalyptus leaves. Journal of Plant Physiology 154(1), 30–36.

[25] Sims, D. A., & Gamon, J. A., 2002. Relationships between leaf pigment content and spectral reflectance across a wide range of species, leaf structures and developmental stages. Remote sensing of environment 81(2), 337–354.

[26] Haboudane, D., Miller, J. R., Pattey, E., Zarco-Tejada, P. J., & Strachan, I. B., 2004. Hyperspectral vegetation indices and novel algorithms for predicting green LAI of crop canopies: Modeling and validation in the context of precision agriculture. Remote sensing of environment 90(3), 337–352.

[27] Aparicio, N., Villegas, D., Casadesus, J., Araus, J.L. and Royo, C., 2000. Spectral vegetation indices as nondestructive tools for determining durum wheat yield. Agronomy Journal, 92(1), pp.83–91.

[28] Anderegg, J., Yu, K., Aasen, H., Walter, A., Liebisch, F. and Hund, A., 2020. Spectral vegetation indices to track senescence dynamics in diverse wheat germplasm. Frontiers in plant science, 10, p.1749.

[29] Brook, A., De Micco, V., Battipaglia, G., Erbaggio, A., Ludeno, G., Catapano, I. and Bonfante, A., 2020. A smart multiple spatial and temporal resolution system to support precision agriculture from satellite images: Proof of concept on Aglianico vineyard. Remote Sensing of Environment, 240, p.111679.

[30] Baret, F., Houlès, V. and Guerif, M., 2007. Quantification of plant stress

using remote sensing observations and crop models: the case of nitrogen management. Journal of Experimental Botany, 58(4), pp.869–880.

[31] Polinova, M., Salinas, K., Bonfante, A. and Brook, A., 2019. Irrigation Optimization under a Limited Water Supply by the Integration of Modern Approaches into Traditional Water Management on the Cotton Fields. Remote Sensing, 11(18), p.2127.

[32] Kochubey, S. M., & Kazantsev, T. A., 2012. Derivative vegetation indices as a new approach in remote sensing of vegetation. Frontiers of Earth Science 1–8.

[33] Tsouros, D.C., Bibi, S. and Sarigiannidis, P.G., 2019. A review on UAV-based applications for precision agriculture. Information, 10(11), p.349.

[34] Radoglou-Grammatikis, P., Sarigiannidis, P., Lagkas, T. and Moscholios, I., 2020. A compilation of UAV applications for precision agriculture. Computer Networks, 172, p.107148.

[35] Rouse Jr, J., Haas, R. H., Schell, J. A., & Deering, D. W.,1974. Monitoring vegetation systems in the Great Plains with ERTS. Third ERTS Symposium, NASA SP-351 I, 309–317.

[36] Payero, J. O., Neale, C. M. U., & Wright, J. L., 2004. Comparison of eleven vegetation indices for estimating plant height of alfalfa and grass. Applied Engineering in Agriculture, 20, 385–393. https://doi.org/10.13031/2013.16057

[37] Vina, A., Gitelson, A. A., Nguy-Robertson, A. L., & Peng, Y., 2011. Comparison of different vegetation indices for the remote assessment of green leaf area index of crops. Remote Sensing of Environment, 115(12), 3468–3478.

[38] Motohka, T., Nasahara, K. N., Oguma, H., & Tsuchida, S., 2010. Applicability of green-red vegetation index for remote sensing of vegetation phenology. Remote Sensing, 2(10), 2369–2387.

[39] Sellers, P. J., 1987. Canopy reflectance, photosynthesis, and transpiration, II. The role of biophysics in the linearity of their interdependence. Remote sensing of Environment 21(2), 143–183.

[40] Schreiber, U. and Bilger, W., 1987. Rapid Assessment of Stress Effects on Plant Leaves by Chlorophyll Fluoreecence Measurements. In: Plant Response to Stress: Functional Analysis in Mediterranean Ecosystems; Tenhunen, J.D., Catarino, F.M., Lange, O.L., Oechel, W.C.; Springer-Verlag, Berlin, 27-53.

[41] Gamon, J. A., Penuelas, J., Field, C. B., 1992. A narrow-waveband spectral index that tracks diurnal changes in photosynthetic efficiency. Remote Sensing of environment 41(1), 35–44.

[42] Penuelas, J., Filella, I., & Gamon, J. A., 1995. Assessment of photosynthetic radiation-use efficiency with spectral reflectance. New Phytologist 131(3), 291–296.

[43] Merzlyak, M. N., Gitelson, A. A., Chivkunova, O. B., & Rakitin, V. Y., 1999. Non-destructive optical detection of pigment changes during leaf senescence and fruit ripening. Physiologia plantarum 106(1), 135–141.

[44] Gitelson, A. A., Zur, Y., Chivkunova, O. B., & Merzlyak, M. N., 2002. Assessing Carotenoid Content in Plant Leaves with Reflectance Spectroscopy. Photochemistry and photobiology 75(3), 272–281.

[45] Oyundari, B., 2008. Spectral indicators for assessing the effect of hydrocarbon leakage on vegetation: International Institute for Geo-information Science and Earth

Observation. Enschede: the Netherlands (ITC).

[46] Jarocinska A and Zagajewski B., 2006. Remote sensing tools for analysis of vegetation condition in extensively used agricultural areas. University of Warsaw, K Krakowskie Przedmiescie. 30: 00–927

[47] Basso, B., Cammarano, D., & De Vita, P., 2004. Remotely sensed vegetation indices: Theory and applications for crop management. Rivista Italiana di Agrometeorologia, 1, 36–53.

[48] M.S. Moran M.S., S.J. Maas S.J., V.C. Vanderbilt V.C., M. Barnes M., S.N. Miller S.N., T.R. Clarke T.R., 2004. Application of image-based remote sensing to irrigated agriculture. S.L. Ustin (Ed.), Remote sensing for natural resource management and environmental monitoring, Manual of remote sensing, John Wiley & sons, Hoboken, pp. 617–676

[49] Atzberger, C., 2013. Advances in remote sensing of agriculture: Context description, existing operational monitoring systems and major information needs. Remote Sensing, 5 (2), 949–981. https://doi.org/10.3390/rs5084124

[50] Hamzeh, S., Naseri, A. A., AlaviPanah, S. K., Mojaradi, B., Bartholomeus, H. M., Clevers, J. G., & Behzad, M., 2013. Estimating salinity stress in sugarcane fields with spaceborne hyperspectral vegetation indices. International Journal of Applied Earth Observation and Geoinformation, 21, 282–290. https://doi.org/10.1016/j.jag.2012.07.002

[51] Oliveira, T. C. D., Ferreira, E., & Dantas, A. A. A., 2016. Temporal variation of normalized difference vegetation index (NDVI) and calculation of the crop coefficient (Kc) from NDVI in areas cultivated with irrigated soybean. Ciência Rural, 46(9), 1683–1688. https://doi.org/10.1590/0103-8478cr20150318

[52] Hernández, E. I., Melendez-Pastor, I., Navarro-Pedreño, J., & Gómez, I., 2014. Spectral indices for the detection of salinity effects in melon plants. Scientia Agricola 71(4), 324–330.

[53] van Maarschalkerweerd, M., & Husted, S., 2015. Recent developments in fast spectroscopy for plant mineral analysis. Frontiers in plant science 6.

[54] Peñuelas, J., Filella, I., Biel, C., Serrano, L., & Save, R., 1993. The reflectance at the 950–970 nm region as an indicator of plant water status. International journal of remote sensing 14(10), 1887–1905.

[55] Sandholt, I., Rasmussen, K., & Andersen, J., 2002. A simple interpretation of the surface temperature/vegetation index space for assessment of surface moisture status. Remote Sensing of environment 79(2), 213–224.

[56] Zawadzki, J., Przeździecki, K., & Miatkowski, Z., 2016. Determining the area of influence of depression cone in the vicinity of lignite mine by means of triangle method and LANDSAT TM/ETM+ satellite images. Journal of environmental management 166, 605–614.

[57] Gao, B., 1996. NDWI—A normalized difference water index for remote sensing of vegetation liquid water from space. Remote Sensing of Environment, 58(3), 257–266. https://doi.org/10.1016/s0034-4257(96)00067-3

[58] Ceccato, P., Flasse, S., Tarantola, S., Jacquemoud, S., & Grégoire, J. M., 2001. Detecting vegetation leaf water content using reflectance in the optical domain. Remote sensing of environment, 77(1), 22–33. https://doi.org/10.1016/s0034-4257(01)00191-2

[59] Jackson, T. J., Chen, D., Cosh, M., Li, F., Anderson, M., Walthall, C., ... & Hunt, E. R., 2004. Vegetation water content mapping using Landsat data derived normalized difference water index for corn and soybeans. Remote Sensing of Environment, 92(4), 475–482. https://doi.org/10.1016/j.rse.2003.10.021

[60] Wang, L., & Qu, J. J., 2007. NMDI: A normalized multi-band drought index for monitoring soil and vegetation moisture with satellite remote sensing. Geophysical Research Letters, 34(20). https://doi.org/10.1029/2007gl031021

[61] Penuelas, J., Filella, I., Biel, C., Serrano, L., & Save, R., 1993. The reflectance at the 950–970 nm region as an indicator of plant water status. International journal of remote sensing, 14(10), 1887–1905. https://doi.org/10.1080/01431169308954010

[62] Jackson, R. D., Slater, P. N., & Pinter, P. J., 1983. Discrimination of growth and water stress in wheat by various vegetation indices through clear and turbid atmospheres. Remote sensing of environment, 13(3), 187–208. https://doi.org/10.1016/0034-4257(83)90039-1

[63] Zhao, S., Wang, Q., Yao, Y., Du, S., Zhang, C., Li, J., & Zhao, J., 2016. Estimating and Validating Wheat Leaf Water Content with Three MODIS Spectral Indexes: A Case Study in Ningxia Plain, China. Journal of Agricultural Science and Technology, 18(2), 387–398.

[64] Choudhury, B. J., Ahmed, N. U., Idso, S. B., Reginato, R. J., & Daughtry, C. S., 1994. Relations between evaporation coefficients and vegetation indices studied by model simulations. Remote sensing of environment, 50(1), 1–17. https://doi.org/10.1016/0034-4257(94)90090-6

[65] Hunsaker, D. J., Pinter Jr, P. J., Barnes, E. M., & Kimball, B. A., 2003. Estimating cotton evapotranspiration crop coefficients with a multispectral vegetation index. Irrigation Science, 22(2), 95–104. https://doi.org/10.1007/s00271-003-0074-6

[66] Duchemin, B., Hadria, R., Erraki, S., Boulet, G., Maisongrande, P., Chehbouni, A., ... & Khabba, S., 2006. Monitoring wheat phenology and irrigation in Central Morocco: On the use of relationships between evapotranspiration, crops coefficients, leaf area index and remotely-sensed vegetation indices. Agricultural Water Management, 79(1), 1–27. https://doi.org/10.1016/j.agwat.2005.02.013

[67] González-Dugo, M. P., & Mateos, L., 2008. Spectral vegetation indices for benchmarking water productivity of irrigated cotton and sugarbeet crops. Agricultural water management, 95(1), 48–58. https://doi.org/10.1016/j.agwat.2007.09.001

[68] Glenn, E. P., Neale, C. M., Hunsaker, D. J., & Nagler, P. L., 2011. Vegetation index-based crop coefficients to estimate evapotranspiration by remote sensing in agricultural and natural ecosystems. Hydrological Processes, 25(26), 4050–4062. https://doi.org/10.1002/hyp.8392

[69] Moran, M. S., Clarke, T. R., Inoue, Y., & Vidal, A., 1994. Estimating crop water deficit using the relation between surface-air temperature and spectral vegetation index. Remote sensing of environment 49(3), 246–263.

[70] Suarez, L., Zarco-Tejada, P. J., Sepulcre-Canto, G., Perez-Priego, O., Miller, J. R., Jiménez-Munoz, J. C., & Sobrino, J., 2008. Assessing canopy PRI for water stress detection with diurnal airborne imagery. Remote Sensing of Environment, 112(2), 560–575. https://doi.org/10.1016/j.rse.2007.05.009

[71] Thenot, F., Méthy, M., & Winkel, T., 2002. The photochemical reflectance

index (PRI) as a water-stress index. International Journal of Remote Sensing, 23(23), 5135–5139. https://doi.org/10.1080/01431160210163100

[72] Sarlikioti, V., Driever, S. M., & Marcelis, L. F. M., 2010. Photochemical reflectance index as a mean of monitoring early water stress. Annals of Applied Biology, 157(1), 81–89. https://doi.org/10.1111/j.1744-7348.2010.00411.x

[73] Broge, N. H., & Leblanc, E., 2001. Comparing prediction power and stability of broadband and hyperspectral vegetation indices for estimation of green leaf area index and canopy chlorophyll density. Remote sensing of environment 76(2), 156–172.

[74] Georgios, P., Diofantos, H. G., Kyriakos, T., & Leonidas, T., 2010. Spectral vegetation indices from field spectroscopy intended for evapotranspiration purposes for spring potatoes in Cyprus. In Proc. of SPIE (Vol. 7824, pp. 782410-1).

[75] Huete, A. R., 1988. A soil-adjusted vegetation index (SAVI). Remote sensing of environment 25(3), 295–309.

[76] USDA-Natural Resources Conservation Service, 2014. Soil Survey Sta_. Keys to Soil Taxonomy, 12th ed.; USDA-Natural Resources Conservation Service: Washington, DC, USA.

[77] Driessen, P., Deckers, J., Spaargaren, O. and Nachtergaele, F. 2001. Lecture Notes on the Major Soils of the World. FAO World Soil Resources, Report 94, Rome

[78] Eswaran, H. and Cook, T., 1988. Classification and Management-related Properties of Vertisols IN Management of Vertisols in Sub Saharan Africa. Proceeding of a Conference held at ILCA, Addis Ababa, Ethiopia.

[79] Hophmayer-Tokich, S., 2010. The evolution of national wastewater management regimes—The case of Israel. Water, 2(3), pp. 439-460.

[80] Wigginton D., 2013 WATERpak. A guide for irrigation management in cotton and grain farming systems., 3rd ed. (Cotton Research and Development Corporation: Narrabri, NSW).

[81] Brouwer, C., & Heibloem, M., 1986. Irrigation water management: irrigation water needs. Training manual 3.

[82] NaanDanJain Irrigation Ltd. Tomato cultivation in open fields and greenhouses. NAANDANJAIN Ltd. 08/2012

[83] NaanDanJain Irrigation Ltd. Cotton. NAANDANJAIN Ltd. 03/2014

[84] Jackson, R. D., Clarke, T. R. and Moran, M. S., 1992. Bidirectional calibration results for 11 Spectralon and 16 BaSO4 reference reflectance panels. Remote sensing of environment vol. 40, no. 3, pp. 231–239

[85] Peñuelas, J., Pinol, J., Ogaya, R., & Filella, I., 1997. Estimation of plant water concentration by the reflectance water index WI (R900/R970). International Journal of Remote Sensing 18(13), 2869–2875.

[86] Beyer, F., Jarmer, T., Siegmann, B., & Fischer, P., 2015. Improved crop classification using multitemporal RapidEye data. In Analysis of Multitemporal Remote Sensing Images (Multi-Temp), 8th International Workshop on the (pp. 1-4). IEEE.

[87] Weichelt, H., Rosso, P., Marx, A., Reigber, S., Douglass, K., & Heynen, M. 2013. The rapideye red edge band. BlackBridge, Tech. Rep.

[88] Krauß, T., d'Angelo, P., Schneider, M., & Gstaiger, V., 2013. The fully automatic optical processing system CATENA at DLR. In ISPRS Hannover workshop (Vol. 1, pp. 177-181).

[89] Ge, Y., Bai, G., Stoerger, V., & Schnable, J. C., 2016. Temporal dynamics of maize plant growth, water use, and leaf water content using automated high throughput RGB and hyperspectral imaging. Computers and Electronics in Agriculture, 127, 625–632.

[90] Lichtenthaler, H. K., 1987. Chlorophyll fluorescence signatures of leaves during the autumnal chlorophyll breakdown. Journal of Plant Physiology, 131(1–2), 101–110.

[91] Zawadzki, J., Cieszewski, C. J., Zasada, M., & Lowe, R. C., 2005. Applying geostatistics for investigations of forest ecosystems using remote sensing imagery. Silva Fennica 39(4), 599.

[92] Park, H. M., 2008. Univariate analysis and normality test using SAS, Stata, and SPSS. Working Paper. The University InformationTechnology Services (UITS) Center for Statistical and Mathematical Computing, Indiana University.

[93] Tukey, J., 1977. W. Exploratory data analysis. Reading PA: Addison-Wesley..

[94] Hoaglin D. C., Iglewicz B., Tukey J. W., 1986. Performance of some resistant rules for outlier labeling. Journal of the American Statistical Association, 81 (396), 991–999.

[95] Daughtry, C. S. T., Walthall, C. L., Kim, M. S., De Colstoun, E. B., & McMurtrey, J. E., 2000. Estimating corn leaf chlorophyll concentration from leaf and canopy reflectance. Remote sensing of Environment 74(2), 229–239.

[96] Ballabio, D., 2015. A MATLAB toolbox for Principal Component Analysis and unsupervised exploration of data structure. Chemometrics and Intelligent Laboratory Systems, 149, 1–9. https://doi.org/10.1016/j.chemolab.2015.10.003

[97] Wold, S., Esbensen, K., & Geladi, P., 1987. Principal component analysis. Chemometrics and intelligent laboratory systems, 2(1–3), 37–52. https://doi.org/10.1016/0169-7439(87)80084-

[98] Abdi, H., & Williams, L. J., 2010. Principal component analysis. Wiley interdisciplinary reviews: computational statistics, 2(4), 433–459. https://doi.org/10.1002/wics.101

[99] Delgado-Vargas, F., Jiménez, A. R., & Paredes-López, O., 2000. Natural pigments: carotenoids, anthocyanins, and betalains—characteristics, biosynthesis, processing, and stability. Critical reviews in food science and nutrition, 40(3), 173–289. https://doi.org/10.1080/10408690091189257

[100] Gamon, J. A., & Surfus, J. S., 1999. Assessing leaf pigment content and activity with a reflectometer. New Phytologist, 143(1), 105–117. https://doi.org/10.1046/j.1469-8137.1999.00424.x

[101] Chalker-Scott, L., 1999. Environmental significance of anthocyanins in plant stress responses. Photochemistry and photobiology, 70 (1), 1–9. https://doi.org/10.1562/0031-8655(1999)070<0001:esoaip>2.3.co;2

[102] Cho, M. A., Sobhan, I., Skidmore, A. K., & De Leeuw, J., 2008. Discriminating species using hyperspectral indices at leaf and canopy scales. The International Archives of the Spatial Information Sciences, 369-376. https://doi.org/10.1016/j.jag.2007.02.001

[103] Nicotra, A. B., Hofmann, M., Siebke, K., & Ball, M. C., 2003. Spatial patterning of pigmentation in evergreen leaves in response to freezing stress. Plant, Cell & Environment, 26(11), 1893–1904. https://doi.org/10.1046/j.1365-3040.2003.01106.x

[104] Hovi, A., Forsström, P., Mõttus, M., & Rautiainen, M., 2017. Evaluation

of Accuracy and Practical Applicability of Methods for Measuring Leaf Reflectance and Transmittance Spectra. Remote Sensing, 10(1), 25. https://doi.org/10.3390/rs10010025

[105] Köksal, E. S., 2011. Hyperspectral reflectance data processing through cluster and principal component analysis for estimating irrigation and yield related indicators. Agricultural water management, 98(8), 1317–1328. https://doi.org/10.1016/j.agwat.2011.03.014

[106] Yang, J., Gong, W., Shi, S., Du, L., Sun, J., & Song, S. L., 2016. Estimation of nitrogen content based on fluorescence spectrum and principal component analysis in paddy rice. Plant Soil Environ, 62(4), 178–183. https://doi.org/10.17221/802/2015-pse

[107] Del Fiore, A., Reverberi, M., Ricelli, A., Pinzari, F., Serranti, S., Fabbri, A. A., ... & Fanelli, C., 2010. Early detection of toxigenic fungi on maize by hyperspectral imaging analysis. International journal of food microbiology, 144(1), 64–71. https://doi.org/10.1016/j.ijfoodmicro.2010.08.001

[108] Rhodes D, Nadolska-Orczyk A. 2001. Plant Stress Physiology. John Wiley & Sons, New York. DOI: 10.1038/npg.els.0001297

[109] Lichtenthaler, H.K. and Buschmann, C., 2001. Chlorophylls and Carotenoids: Measurement and Characterization by UV-VIS Spectroscopy. In: Current Protocols in Food Analytical Chemistry (CPFA) by Wrolstad, R.E., Acree, T.E., An, H., Decker, E.A., Penner, M.H., Reid, D.S., Schwartz, S.J., Shoemaker, C.F. and Sporns, P., John Wiley and Sons, New York, F4.3.1-F4.3.8

[110] Küpper, H., Seibert, S., & Parameswaran, A., 2007. Fast, sensitive, and inexpensive alternative to analytical pigment HPLC: quantification of chlorophylls and carotenoids in crude extracts by fitting with gauss peak spectra. Analytical chemistry 79(20), 7611–7627.

[111] Wu, Q., Song, C., Song, J., Wang, J., Chen, S., & Yu, B., 2018. Impacts of Leaf Age on Canopy Spectral Signature Variation in Evergreen Chinese Fir Forests. Remote Sensing, 10(2), 262. https://doi.org/10.3390/rs10020262

[112] Jay, S., Gorretta, N., Morel, J., Maupas, F., Bendoula, R., Rabatel, G., ... & Baret, F., 2017. Estimating leaf chlorophyll content in sugar beet canopies using millimeter-to centimeter-scale reflectance imagery. Remote Sensing of Environment, 198, 173–186. https://doi.org/10.1016/j.rse.2017.06.008

[113] Amirruddin, A. D., Muharam, F. M., & Mazlan, N., 2017. Assessing leaf scale measurement for nitrogen content of oil palm: performance of discriminant analysis and Support Vector Machine classifiers. International Journal of Remote Sensing, 38(23), 7260–7280. https://doi.org/10.1080/01431161.2017.1372862

[114] Hmimina, G., Merlier, E., Dufrêne, E., & Soudani, K., 2015. Deconvolution of pigment and physiologically related photochemical reflectance index variability at the canopy scale over an entire growing season. Plant, cell & environment, 38(8), 1578–1590. https://doi.org/10.1111/pce.12509

Chapter 8

Predictive Models for Reforestation and Agricultural Reclamation: A Clearfield County, Pennsylvania Case Study

Zhi Yue and Jon Bryan Burley

Abstract

Natural resource scientists, concerned citizens, and government officials are interested in reconstructing disturbed environments for reforestation and agricultural productivity. We examined Clearfield County in Pennsylvania, USA, to develop a predictive model to reconstruct the landscape for seven agronomic crops (corn, corn silage, oats, alfalfa hay, red clover, bluegrass, and soybeans) and thirteen woody plants (white cedar, lilac, highbush cranberry, Amur maple, gray dogwood, peashrub, white spruce, white pine, red maple, red pine, jack pine, nannyberry, and white ash). A significant predictive model ($p \leq 0.001$) was generated explaining 96.94% of the variance, with percent clay, bulk density, hydraulic conductivity, available water capacity, pH, percent organic matter, percent rock fragments, slope, topographic position, and electrical conductivity explored as main effect terms, plus squared terms, and first order interaction terms. The model is not over-specified and each predictor is significant ($p \leq 0.05$). The modeling effort suggests that there are at least several clusters of vegetation preference dimensions based upon the terrain of the landscape. The model provides insight into how to reconstruct the disturbed environment for vegetation in the study area.

Keywords: neo-soils, forestry, agronomy, environmental design, pastureland, landscape architecture, reclamation, soil science, horticulture

1. Introduction and literature

Reclamation scientists and partitioners are interesting in restructuring disturbed soils (neo-sols) for maintaining vegetation productivity in a sustainable manner [1]. Along with this interest, they are concerned with constructing predictive models (equations) to quantitatively assess the inherent productivity of a soil column. The literature addressing this interest originated in the 1980's to study reclaiming large surface coal mines [1]. However, the quest was perplexing with many unanswered questions such as: did a different equation need to be developed for each and every plant material? How much of the soil column required measuring and did the soil column have a weighted contribution? And, which variables should be measured? This article is primarily about one researcher's quest and the colleagues he is affiliated with to address this issue.

By the late 1980s and early 1990s a methodology was developed that answered these questions [2, 3]. The framework for this methodology is illustrated in **Figure 1**. The approach attempts to predict the productivity of the soil column itself and is not a real time productivity model that assesses the current plant production based upon immediate weather conditions or greatly added nutrients beyond modest levels. Therefore, weather and soil additives are beyond the modeling effort.

Soil and vegetation productivity can be actually rather vague and variant in definition. This variation in definition may surprise some who believed they had a very firm idea what constituted soil, especially in the biological and agronomic sciences. The broadest view comes from sol engineers who divide the terrestrial surface into two categories: bedrock and soil. Bedrock are expansive stone-like structures that cannot be dislodged or moved and soil consists of particles that can be moved [4]. Thus, almost any inclusion can become a soil particle such as plastics, organisms, large boulders, and many other objects. This viewpoint can be quite different from the classically trained agronomic soil scientist's sensibilities concerning what is soil. The divergence in thinking occurs between one group who utilizes soil for vegetation (softscapes) and another group who primarily utilize soils for construction for buildings, walls, roads, and paving (hardscapes). However, in soil productivity studies, it is soil properties that are measurable, acting as a construct representing the soil profile.

A similar issue exists when defining soil productivity. It is a general idea with no firm definition. However, vegetation measures such as plant height per year or weight per area are constructs representing vegetation productivity.

A constructed model would be applicable to the study area where the soils and vegetation are sampled. The ideal study area would have all vegetation of interest grown across all soils in the study area and across normal, dry, and wet years. To initiate such a comprehensive study would take numerous field plots measuring plant growth for at least ten years [5]. This is an extensive modeling project, taking up to 1 million USA dollars to accomplish. Most research projects last only a few years and are funded at much less levels [5]. However, the United States Department of Agriculture, directed the Natural Resource Conservation Service to conduct such work county by county. Not all counties in the United States have been evaluated; yet, the American federal government maintained a long term vision to collect this data in an effort that is nearing 100 years old. The federal government was excellent at collecting the data and publishing the data, accessible to all for free. The data is available to investigators who have the statistical ability to analyze the gathered information. This American database led to the development of the methodology **Figure 2**) [2, 3].

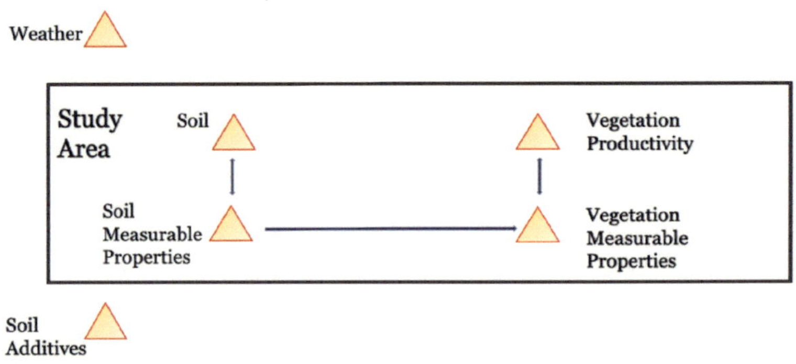

Figure 1.
Framework for creating neo-sol productivity equations.

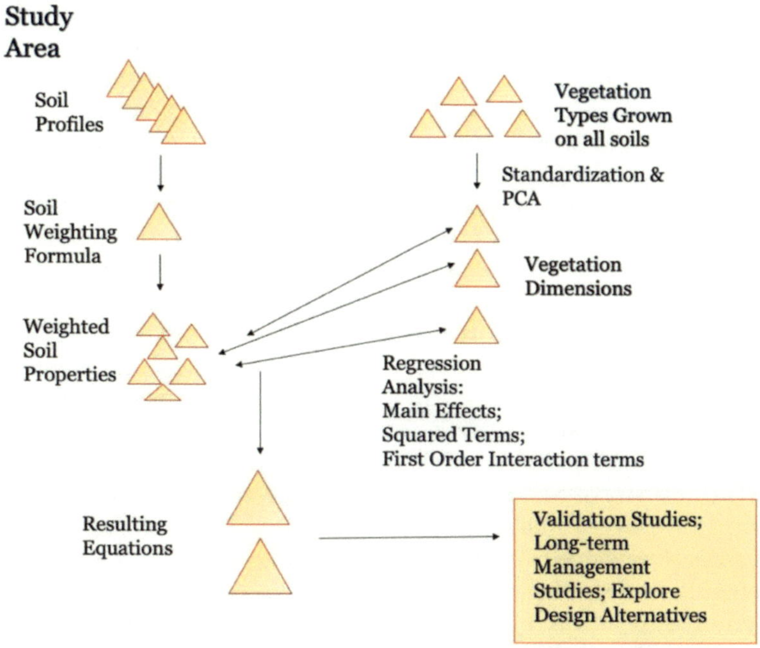

Figure 2.
Methodology to develop neo-sol productivity equations.

Recently published research articles by Wen and Burley and Corr et al. review many of the authorities and related modeling efforts to produce similar and related Equations [5, 6]. The focus of the literature review in this chapter will concentrate upon those studies that followed the methodology in **Figure 2**. The first reported equation was in a study by a team from the University of Manitoba of an equation for Clay County, Minnesota, published in 1989 [7]. This study suggested that many agronomic crops covary together concerning preferences for soil, meaning that an equation could be generated for a set of crops at one time, as opposed to having an equation for each individual crop. The team also produced equations for woody plants and for a combination of woody plants and crops [8, 9]. The team also discovered that sugarbeets (*Beta vulgaris* L.) could be productive in a different type of soil, with poor aeration, wet, and higher in electrical conductivity (more dissolved ions) [10]. The results from these investigations were presented in a 1992 soil productivity meeting hosted by the University of Illinois [11]. The data set from Clay County was combined with a data set from Cass, County, North Dakota, to present an equation from a larger study area and presented at a conference in Sudbury, Ontario, Canada [12].

At the time, it was somewhat unusual for a master's thesis to generate numerous scholarly articles (five journal articles and three conference articles), as many such landscape thesis results in few if any publications. However, the University of Manitoba encouraged such publications and activities. In addition, it was even more unusual that a landscape architect would generate that many articles. Zhi Yue a co-author of this book chapter wondered how Dr. Burley found a way to develop these equations, "When I was quite young (age 6), I lived in Edmonton, Alberta, and my parent's friends were American academics who worked at the University of Alberta in disciplines/professors such as anthropology, wildlife biology, and music. It was there that I met my first landscape architect (when I was 17), R. H. Knowles, who gave me a copy of his book [13]. So, it seemed natural to me and expected that scholarly efforts would result in publication. When I was 22, I wrote my first article as an undergraduate and had it published when I was 23 [14]. I later learned that this modest output was greater than all the landscape architecture output in 1978 from my eventual home academic institution. In other words, academic landscape architects did not publish much back then. But it seemed natural to me that a curious landscape architect dedicated to academic scholarship might be the one who eventually developed these neo-sol productivity equations. The equations could have been developed by agronomists, horticulturists, soil scientists, foresters, or environmental engineers. Yet I learned that many disciplines are deep but not broad in education like a landscape architect and did not ask the same practical and applied questions a landscape architect might ask. Plus, I had the fortune of working on a research-oriented Plan A Thesis (most landscape architecture master's students do a project as a Plan B Thesis), meaning that my committee members at the University desired that I take courses in statistics, as much as I could take. So, for my coursework, I took introductory statistics, non-parametric statistics, regression analysis, analysis of variance, philosophy of science, and linear algebra. Later, I took multivariate analysis and statistical autocorrelation. My University of Manitoba professors prepared me so well, that by the time I went to the University of Michigan for my Ph.D., the professors there in the School of Natural Resources waived any requirements for me to take a statistics course. Still, I took a course in epidemiological statistics, auto-correlation, and a course in risk analysis at my leisure. It was this statistical background that assisted me in the modeling efforts with skills and abilities often not present with others who were searching for a way to develop neo-sol equations." observed Dr. Burley.

The next equation to be published was a study of Polk County, Florida [15]. The study was initiated through Anthony Bauer, FASLA, a well-known landscape architect from Michigan State University, specializing in surface mine reclamation who gave a comprehensive exam question for Jon Bryan Burley in his quest for a Ph.D. The study revealed two different sets of vegetation preferences: a mesic preferring group of plants and another group preferring wet conditions. The previous studies in Minnesota and North Dakota had revealed primarily equations for mesic settings. It can be quite unusual that an exam question generated a paper, but this did not seem unusual to Jon Bryan Burley, as several papers were generated from assignments in graduate level courses in risk assessment, field studies, remote sensing, and anthropology [16–19].

After the Florida study, the next reported mesic preference neo-sol equations were reported about three counties in the North Dakota coal fields: Oliver, Mercer, and Dunn counties [20, 21]. A paper was published that illustrated the spatial application of the equation surface mine setting to maximize productivity and minimize costs [22]. Another paper examined the relationships between softscape soil and hardscape soils, identifying some soils that are suitable for both applications [23]. In addition, the effort addressed various American state laws concerning the deployment and use of neo-sol equations [24].

These efforts lead to the publication of a surface mine reclamation book, and two national American Society of Landscape Architects (ASLA) research awards [25]. By 2005, Dr. Burley became the American Society of Reclamation Sciences (ASRS) researcher of the year and contributed towards his induction as a 2010 Fellow in ASLA for his research contributions. Dr. Burley's goal was to see if he could develop a set of North American reclamation equations, even a global equation. But the research sputtered as there seemed little new interest in funding or construction such equation studies. "I was urged by my department chair to abandon my neo-sol productivity equations research and go where the money was, such as in healthy cities or climate change." noted Dr. Burley. "But I am rather stubborn. So much of the earth was being disturbed by human activities that the equations could be helpful in a wide variety of applications where the original soil profiled is disturbed — I did not want to have the reputation as an academic money ambulance chaser. And then something interesting happened. As an aging professor being successful in conducting research and publishing, I found international students and professors wanted to work with me and often they were interested in developing new equations as a means of learning how to do research."

The result of the renewed interest was started by a French team who worked with Dr. Burley and developed a mesic preference equation for Grand Traverse County in Michigan [26]. Then, Chinese scholars began working with Dr. Burley. There was a movement in P.R. of China for Chinese academics to learn research methods and publish. This resulted in the study and publication of a silica mining region in Chippewa County, Wisconsin and a kaolinite mine area in Georgia [27]. Coal mining in Montana, Wyoming, Colorado, and Texas was also explored [5, 28]. Corr et al. (Dustin Corr was an American graduate student of Dr. Burley) studied developing equations in the iron mining region of the Upper Peninsula of Michigan, deriving mesic and the first xeric set of equations, concluding the current set of equations that have been developed with this methodology [6].

Zhi Yue, a professor in landscape architecture from Nanjing Forestry University, Nanjing, P. R. of China was interested in applying this methodology to a study area in Clearfield County, Pennsylvania. This book chapter reports upon the results concerning the application of this methodology in this study area. The study represents a continuing effort to construct a set of equations and data sets to potentially derive a set of universal equations for the eastern 2/3rds of the United States.

2. Study area and methodology

2.1 Study area

Clearfield County is the study area, located in Pennsylvania (**Figure 3**). The county is composed of angular hills, farms, small towns, and forests (**Figure 4**). Coal mining and clay mining occur in the county [29]. The county's soil survey is one of the oldest in the United Sates being published in 1916, but updated in 1988 [30]. From southeast to northwest the terrain and soils change along ridges and river bottoms, resulting in a county that is physically quite diverse. Rose et al. published a paper concerning some of the environmental issues associated with coal mining in the county associated with acid mine drainage [31]. Brown and Parizek examined hydrological grown water flow for two mines in the county [32]. Skousen and Zipper describe a recent overview of coal mining the Appalachian region [33].

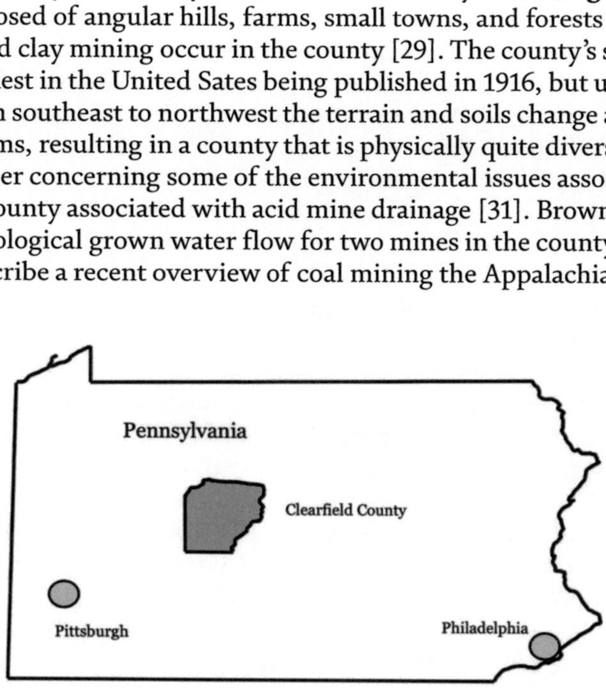

Figure 3.
Clearfield County, Pennsylvania.

Figure 4.
An image of the rolling hills in Clearfield County, Pennsylvania (copyright © 2020 Tim P. Danehy, used by permission, all rights reserved).

2.2 Methodology

The methodology has been reported in detail by several publications [2, 3, 5, 6]. For this study, crop and plant harvest and growth information are sorted by soil type. The variables employ different measurements by weight per acre or height per year. Each variable is standardized with a mean of "0" and a standard deviation of 1. This standardization prevents measurement scales with large numbers from dominating the results of scales expressed in smaller units. For example the weight measurement kg per hectare is a different type of measurement scale than the volume measurement of hectoliters per hectare. Standardizing allows apples and oranges (in this case corn and alfalfa hay) to be compared, as first proposed by Kendall [34]. Then the standardized variables are assessed with principal component analysis (PCA). The analysis examines the covariance of crops and woody plants across soil types, developing latent dimensions with vector coefficients. Each dimension is orthogonal (independent) to other dimensions. The maximum number of dimensions is equal to the total number of variables. Each dimension has an associated eigenvalue. The sum of the eigenvalues equals to the number of dimensions. The larger the eigenvalue, the greater the proportion of variance the dimension explains. Typically, eigenvalues greater than one are considered potential candidates for further analysis. Often the first few eigenvalues explain 70% or more of the variance in the data set [35]. The eigenvector coefficients facilitate the creation of a linear combination of values, when summed, represent the expected vegetation response to the soil, a single dependent variable per soil profile [2, 3, 5, 6].

The independent variables are composed by gathering the soil variables of interest for a depth of 1.22 meters. Each variable is weighted by depth at 30.5 cm increments. As the layer nearest the top contributes approximately 40% to plant growth, the second layer 30%, the third layer20%, and the fourth layer 10% [36]. By employing a weighting equation, one value per variable, per soil type can be computed. The effort by Doll and others settled the issue of where to measure soil variables and how to derive a single variable value for variables such as soil reaction and percent organic matter to describe the soil profile [36, 37].

The data source for the study has been published by Hallowich et al., [30]. The vegetation employed in the investigation include: corn (*Zea mays* L.), corn silage, oats (*Avena sativa* L. (1753)), alfalfa hay (*Meticago sativa* L.), red clover (*Trifolium pratense* L.), bluegrass (*Poa pratensis* L.), soybeans (*Glycine max* (L.) Merr.), Eastern white cedar (*Thuja occidentalis* L.), common lilac (*Syringa vulgaris* L.), highbush cranberry (*Viburnam trilobum* Marshall), Amur maple (*Acer ginnala* Maxim.), gray dogwood (*Cornus racemosa* Lam.), Siberian peashrub (*Caragana arborescens* Lam.), white spruce (*Picea glauca* (Moench) Voss), white pine (*Pinus strobus* L.), red maple (*Acer rubrum* L. 1753), red pine (*Pinus resinosa* Sol. ex Aiton), jack pine (*Pinus banksiana* Lam.), nannyberry (*Viburnum lentago* L.), and white ash (*Fraxinus American* L.). The soil properties employed in the study were: percent clay, bulk density, hydraulic conductivity, available water capacity, pH, percent organic matter, percent rock fragments, slope, topographic position, and electrical conductivity. Recently other soil profile variables have been proposed for potential inclusion [5, 6].

Regression analysis was performed employing main effects, squared terms, and first order interaction terms as independent variables from the soil profiles [38]. Doll et al. proposed a hypothetical multi-order interaction model, but supplied no evidence that such a model actually represented any true predictive power [36]. In addition, no investigator has demonstrated any theories or statistical models to suggest that independent variables beyond first order interactions are necessary or represent biological responses in soil profiles. Linear combinations derived from the vegetation dimensions formed the dependent variables [2, 3, 5, 6]. In this study, one of the linear combinations will be selected for equation development.

Past published equations have been somewhat complex containing many main effects, squared terms, and first order interaction terms as variables, often over 10 predictor variables. When selecting the best variable in step-wise regression, several criteria are employed. The first is that all of the regressors must be significant ($p \leq 0.5$) under Type III sums of squares, meaning in SAS the regressor's p-value is assessed as though it was the last predictor added to the regression model [39]. Second, an equation presenting the most the largest possible R-square is preferred, as it explains a larger proportion of the variance. Finally, an equation which does not violate Mallows' C statistic is preferred, as then the model is not over-specified, meaning the C_p value must be larger than the number of regressors, thereby avoiding multi-collinearity issues [40]. Once the best equation is selected, it is ready for interpretation and examination.

When interpreting the selected equation, the model may present significant variables that pose soil–plant relationships that have been poorly studied, especially when examining interaction terms. Main effect terms are often more widely studied and known, as illustrated by Buta et al. [41]. With the number of possible variables to include in a model concerning soil properties, in many respects soil science has examined many of the main effect, but, has yet to study many of the interaction properties [42]. Squared terms often indicate the limitations of any main effect or interaction term, counter-balancing the contribution of mail effect variables and suggesting a curvilinear relationship.

3. Results

The first five eigenvalues produced dimensions that have potential for equation development, as they are all greater than 1.0 (**Table 1**). The eigenvector coefficients are presented in **Table 2**. The results in **Table 2** suggest that there is no linear combination that is suitable for all the of the plants (no set of coefficients that are all positive) in the study and that the plants are divided into various preferences (each vector has positive and negative values).

Eigenvalues of the Covariance Matrix				
	Eigenvalue	**Difference**	**Proportion**	**Cumulative**
1	6.15159932	2.41453947	0.3076	0.3076
2	3.73705986	1.20377714	0.1869	0.4944
3	2.53328272	0.58384454	0.1267	0.6211
4	1.94943818	0.32725561	0.0975	0.7186
5	1.62218257	0.65412013	0.0811	0.7997
6	0.96806244	0.23982765	0.0484	0.8481
7	0.72823479	0.07943225	0.0364	0.8845
8	0.64880255	0.17616840	0.0324	0.9169
9	0.47263415	0.09123738	0.0236	0.9406
10	0.38139678	0.15364829	0.0191	0.9596
11	0.22774848	0.02736187	0.0114	0.9710
12	0.20038661	0.05686144	0.0100	0.9810
13	0.14352517	0.05544150	0.0072	0.9882
14	0.08808367	0.01406198	0.0044	0.9926
15	0.07402169	0.04205498	0.0037	0.9963
16	0.03196671	0.00428406	0.0016	0.9979
17	0.02768266	0.01735311	0.0014	0.9993
18	0.01032955	0.00676746	0.0005	0.9998
19	0.00356209	0.00356209	0.0002	1.0000
20	0.00000000		0.0000	1.0000

Table 1.
The eigenvalue principal component results.

	Prin1	**Prin2**	**Prin3**	**Prin4**	**Prin5**	**Prin6**
Corn	0.332479	0.145541	0.139158	0.077016	–.210171	–.264714
Corn Sil.	0.331979	0.147924	0.144917	0.099242	–.189641	–.276274
Oats	0.332213	0.186423	0.076989	0.045459	–.238382	–.152100
Alfalfa	0.309692	–.083102	0.122073	–.090182	0.099994	0.482045
Clover	0.291940	–.057589	0.152609	–.112666	0.059356	0.492065
Bluegrass	0.339909	0.138275	0.047272	–.011512	–.175828	–.086347
Soybeans	0.180228	0.054345	–.297047	0.071782	–.228810	0.350508
W. Cedar	0.250380	–.328201	–.232750	0.000319	0.032382	–.080211
Lilac	–.231681	0.350874	0.268621	–.001134	–.067886	0.090684
Cranberry	–.231340	0.350864	0.268989	–.004255	–.057389	0.082154
A. Maple	–.181483	0.262071	0.191192	0.077677	–.314753	0.282259
Dogwood	0.067687	0.063333	–.052826	0.579159	0.329169	0.141369
Peashrub	0.127188	0.278547	0.014126	0.120655	0.548169	–.090302
Spruce	0.204018	0.011463	0.247830	–.306748	0.106609	0.198899
W. Pine	0.206066	0.195899	–.057413	–.354656	0.134030	–.134368
R. Maple	–.089841	–.350998	0.312328	–.032084	0.041710	–.081588

	Prin1	Prin2	Prin3	Prin4	Prin5	Prin6
R. Pine	0.039054	−.277065	0.452915	0.173473	0.062951	−.032255
Jack Pine	0.008561	−.277714	0.443575	−.063398	0.010186	−.136655
Nanny	0.145735	0.062012	0.138996	0.502470	0.103156	0.018392
W. Ash	0.033906	0.257350	0.057782	−.304160	0.451231	−.110072

Table 2.
The eigenvector coefficients for the first six principal components.

Table 3 presents the best model in the regression analysis with the second principal component. The analysis results suggests the dimension is suitable for predicting plant growth for corn (*Zea mays* L.), corn silage, oats (*Avena sativa* L. (1753)), red bluegrass (*Poa pratensis* L.), soybeans (*Glycine max* (L.) Merr.), common lilac (*Syringa vulgaris* L.), highbush cranberry (*Viburnam trilobum* Marshall), Amur maple (*Acer ginnala* Maxim.), gray dogwood (*Cornus racemosa* Lam.), Siberian peashrub (*Caragana arborescens* Lam.), white spruce (*Picea glauca* (Moench) Voss), white pine (*Pinus strobus* L.), nannyberry (*Viburnum lentago* L.), and white ash (*Fraxinus American* L.).

The regression results contain terms that are main effects (one), squared terms (three), and first order interaction terms (seven). The model is not over specified

Analysis of Variance					
Source	**DF**	**Sum of Squares**	**Mean Square**	**F Value**	**Pr > F**
Model	12	137.66854	11.47238	68.73	<.0001
Error	26	4.33970	0.16691		
Corrected Total	38	142.00824			
Variable	**Parameter Estimate**	**Standard Error**	**Type II SS**	**F Value**	**Pr > F**
Intercept	27.50898	1.80555	38.74519	232.13	<.0001
TP	−14.60620	0.93164	41.02630	245.80	<.0001
SL^2	−0.00540	0.00145	2.30527	13.81	0.0010
CL^2	0.02569	0.00154	46.31271	277.47	<.0001
HC^2	0.17347	0.03325	4.54356	27.22	<.0001
AW^2	2268.76334	122.04697	57.67805	345.56	<.0001
TPBD	11.83549	0.74327	42.32138	253.56	<.0001
TPHC	−2.17939	0.14944	35.50112	212.69	<.0001
SLOM	0.03402	0.01332	1.08937	6.53	0.0168
FRAW	3.38680	0.37729	13.44987	80.58	<.0001
CLHC	0.51478	0.02704	60.50262	362.48	<.0001
CLPH	−0.25991	0.01607	43.68400	261.72	<.0001
BDAW	−374.54092	22.30985	47.04261	281.84	<.0001

Note: TP = Topographic Position; SL = % Slope; CL = % Clay; HC = Hydraulic Conductivity; BD = Bulk Density; OM = % Organic Matter; FR = % Rock Fragments; AW = Available Water Holding Capacity; PH = soil reaction.

Table 3.
Results from regression analysis of the best model from the second dimension.

as the C-plot score is 22.1294, suggesting that the results for this regression iteration would not be over specified until there were nearly 22 regressors in the model. The r-squared for the results in **Table 3** is 0.9694. In other words, the proposed equation predicts 96.94 percent of the variance in the second dimension.

4. Discussion

Unlike most of the previous studies that have been conducted where the previous results produced a universal mesic equation for all agronomic crops and woody plants, the results in **Table 2**, suggest that this was not possible. Often the first set of eigenvector coefficients would be all positive, indicating a universal covariance and soil preference amongst the vegetation types studied [9]. This was not true for Clearfield County. In addition, past results in Northern Michigan and in Florida suggested the vegetation studied was divided into two soil zones: in Michigan a mesic and xeric zone; in Florida a mesic and hydric zone. However, in Clearfield County, Pennsylvania, the vegetation may be responding to latent dimensions not as clearly identified and understood. In other words, the landscape of Clearfield County may be more complex and diversified. In comparison, a large three county study area in North Dakota presented a more uniform landscape than Clearfield County, a smaller area [21].

If a reclamation team was interest in reclaiming surface mine in Clearfield County, the results of the first dimension indicated that an ordination of the seven crops variables might lead to a universal crop equation, as all the eigenvector coefficients are positive for crops in the first dimension. But in such landscape, the choice of woody plants for adjacent reclaimed areas may be limited. For examples, the reclaimed soil may not be suitable for lilac, highbush cranberry, Amur maple, and red maple.

The equation derived from **Table 3** is presented in Eq. 1. This equation can be employed in Clearfield Country for a soil profile to predict plant growth. According to the statistical results, it will be wrong only one time in ten thousand applications. The value of such equations is that they provide an opportunity for the reclamationist to consider how to reconstruct the soil profile. The equations provide feedback.

Equations one suggests that topographic positions on the top of ridges should be avoided and that maximizing water holding capacity should be addressed. High topographic positions can have denser, clay soils. Well drained clay soils will be more productive. Increasing slopes eventually will reduce soil productivity; however steeper slopes should contain more organic matter. High clay content and high soil reaction reduces productivity. Abundant rock fragments can be beneficial as long as the available water holding capacity is high. For the most part, the equation is suggesting the management of water and aeration. These general principles are derived by interpreting the equation.

$$\begin{aligned} Y = 27.51 &+ (-14.606 * TP) + (-0.01 * SL^2) + (0.03 * CL^2) \\ &+ (2268.8 * AW^2) + (11.84 * TP * BD) + (-2.18 * TP * HC) \\ &+ (0.04 * SL * OM) + (3.39 * FR * AW) + (0.51 * CL * HC) \\ &+ (-0.26 * CL * PH) + (-374.54 * BD * AW) \end{aligned} \tag{1}$$

Where:
Y = Vegetation Productivity.
TP = Topographic Position.

SL = % Slope.
CL = % Clay.
HC = Hydraulic Conductivity.
BD = Bulk Density.
OM = % Organic Matter.
FR = % Rock Fragments.
AW = Available Water Holding Capacity.
PH = Soil Reaction.

When Eq. 1 is applied to predict soil productivity, highly productive soils have scores between the values of 2 to 3. For example, a soil similar in structure to the native Clymer Channery loam found in the county, which is a deep loamy soil residing upon 8% and 15% slopes, is a fairly productive soil, with a computed score of 2.24. The equation corroborates the expected high productivity of this soil. In contrasts, moderately productive soils had scores near zero. The Ernest silt loam is only a modestly producing soil. For soils with a similar soil profile, Eq. 1 predicts a score of −0.34. Finally, poorly producing soils had scores of −2 to −3. The Berks shaly silt loam in 15 to 25% slopes is a low productivity soil. When Eq. 1 is applied to soil profiles, similar to the Berks shaly silt loam, the calculated values is −2.75.

The value of the equation is to predict vegetation productivity prior to reclamation. Reclamationists can propose various reconstructed profiles with the soil resources available as illustrated by Burley in 1999 [22]. On a site being reclaimed, there may be a variety of soil profiles on various topographies and slopes. The sum of the total productivity per mine site area or disturbed environment can be computed and compared.

The resulting equation like the one presented in this study often present intriguing questions concerning the properties of soils. While some might believe that soils have already been overly studied, the truth is that many of the interacting properties from a soil have been only modestly investigated with primarily the main effects being explored. The various equations produced over the years, provide insight into which interactions merit further study. In addition, soil scientists have been exploring and assessing new and different soil properties [5, 6]. These properties could be folded into future modeling efforts.

A full exploration of the potential models suggested by **Table 2** would require a longer discourse and narrative than possible in a book chapter. For example, Corr et al. took 43 pages to describe the various combinations of equations they discovered in their study [6]. However, this book chapter does sufficiently cover the fundamentals concerning the literature, methodology, and the results from a new study area. This effort has been ongoing for over 30 years and has only explored the fringes of possibilities.

There are still many unanswered questions in the reconstruction of soil profiles, especially the long-term stability and productivity of any reconstructed profile. In addition, very few equations have been validated with studies growing crops and woody plants and comparing the results to past predictions [21].

5. Conclusion and future prospect

This investigation illustrates that it is possible to develop neo-sol productivity equations that are highly specific and rigorous. The science for this effort has been operational for at least 30 years. But the databases to conduct such work have often been collected and available for over 100 years, awaiting analysis. The databases are expensive and time consuming to build.; yet when constructed and analyzed, they may offer insight into reclaiming disturbed landscapes. The study of Clearfield County revealed that in a landscape complex of large hills/small mountains and large broad valleys, vegetation preferences may be diverse, divided into dimensions of preference. While the first equations were developed for reclaiming environments disturbed from surface mining, landscapes are disturbed by many more types of human activities. In the future, these equations may render service in guiding the reconstruction and management of the soil for vegetation across many forms od disturbance.

Conflict of interest

The authors declare no conflict of interest.

Dedication

The authors of this book chapter wish to acknowledge the scholarly contributions of the late Dr. Kimery Vories (1946–2019). Dr. Vories was a member of the American Society for Surface Mining and Reclamation, now the American Society of Reclamation Sciences [43]. He earned a master's degree at Western State University in Colorado and conducted Ph.D. work at the University of Amherst, Massachusetts and at Colorado State University. Back in the mid 1980s, Kimery was one of the scholars encouraging the initiation of the development of soil productivity equations for disturbed landscapes. His encouragement led to this line of research. Dr. Vories is known for his bat conservation activities with abandoned mines.

Author details

Zhi Yue[1] and Jon Bryan Burley[2]*

1 Nanjing Forestry University, Nanjing, Jiangsu, China

2 Michigan State University, E. Lansing, Michigan, USA

*Address all correspondence to: burleyj@msu.edu

References

[1] Burley, J.B., Thomsen, C. Landscape architecture: continuing investigations into creative site design for surface mining and post-mining land-use. In: Everything Up-to-Date. Sudbury, Ontario: Canadian Land Reclamation Association, Sudbury, 1987. p. 203-216.

[2] Burley JB., Thomsen C. Multivariate techniques to develop vegetation productivity models for neo-sols. In: Graves, D. H., editor. Proceedings, 1987 Symposium on Surface Mining, Hydrology, Sedimentology and Reclamation (Lexington, Kentucky). Washington, D.C.: Office of Scientific and Technical Information, United States Department of Energy, 1987. p. 153-161.

[3] Burley, J.B. Methodology for building soil-based vegetation productivity equations: a statistical approach. In: Daniels, W.L., Burger, J., Zipper, C. E., editors. Proceedings Thirteenth Annual Meeting American Society for Surface Mining and Reclamation: Successes and Failures: Applying Research Results to Insure Reclamation Success (May 18-23, 1996, Knoxville, Tennessee). Blacksburg, Virginia: Virginia Tech Research Division, Powell River Project, 1996. p. 789-798. DOI: https://doi.org/10.21000/JASMR96010789

[4] Hausmann, M.R. Engineering Principles of Ground Modification. New York, New York: McGraw-Hill, Inc., 1990.

[5] Wen, B., Burley., J. B. 2020. Soil-based vegetation productivity model for Coryell County, Texas. Sustainability, 2020, 12(5240);1-14. DOI: https://doi.org/10.3390/su12135240

[6] Corr, D.L, Burley, J. B., Cregg, B., Schutzki, R. Soil based vegetation productivity models for reclaiming northern Michigan landscapes: a monograph. Journal of the American Society for Mining and Reclamation, 2020, 9(4);1-43. DOI: http://dx.doi.org/10.21000/JASMR20040001

[7] Burley, J.B., Thomsen, C., Kenkel, N. Development of an agricultural productivity model to reclaim surface mines in Clay County, Minnesota. Environmental Management, 1989, 13(5);631-638. DOI: https://doi.org/10.1007/BF01874969

[8] Burley, J.B., Thomsen, C. H. 1990. Application of an agricultural soil productivity equation for reclaiming surface mines: Clay County, Minnesota. International Journal of Surface Mining and Reclamation, 1990, 4;139-144. DOI: https://doi.org/10.1080/09208119008944181

[9] Burley, J.B. 1991. Vegetation productivity equation for reclaiming surface mines in Clay County, Minnesota. International Journal of Surface Mining and Reclamation, 5:1-6. DOI: https://doi.org/10.1080/09208119108944279

[10] Burley, J.B. Sugarbeet productivity model for Clay County, Minnesota. Journal of Sugar Beet Research, 1990, 27(3 and 4);50-57.

[11] Burley, J.B. Vegetation productivity equations: an overview. In: Dunker, R.E., Barnhisel, R. I., R.G. Darmody, R. G., editors. In: Proceedings of the 1992 National Symposium on Prime Farmland Reclamation. Urbana, Illinois Department of Agronomy: University of Illinois, 1992. p. 259-265.

[12] Burley, J.B. A multi-county vegetation productivity equation for soil reclamation. In: Hynes, T.P., M.C. Blanchette, M. C., editors. In: Proceedings of Sudbury '95 - Mining and the Environment. Ottawa, Ontario: CANMET, 1995, Volume 3. p. 1113-1122.

[13] Knowles, R. H. Woody Ornamentals for the Prairie Provinces with Particular Reference to Alberta Conditions. Edmonton, Alberta: University of Alberta, 1967.

[14] Burley, J.B. Home landscaping in tune with environment, function. Minnesota Horticulturist, 1978, 106(5);142-143.

[15] Burley, J.B., Bauer, A. Neo-sol vegetation productivity equations for reclaiming disturbed landscapes: a central Florida example. In: Zamora, B.A., Conally, R. E., editors. The Challenge of Integrating Diverse Perspectives in Reclamation: Proceedings of the 10th Annual National Meeting of the American Society for Surface Mining and Reclamation. Spokane, Washington: ASSMR, 1993. p. 334-347. DOI: https://doi.org/10.2100/JASMR93010334

[16] Burley, J.B., Burley., C. J. A risk assessment of landscape hazards for building sites in the front range of Colorado. Landscape Research, 1996, 21(2);137-158. DOI: https://doi.org/10.1080/01426399608706482

[17] Burley, J.B. Visual and ecological environmental quality model for transportation planning and design. Transportation Research Record, 1997, 1549:54-60. DOI: https://doi.org/10.1177/0361198196154900107

[18] Burley, J.B. Vegetation survival strategies in building site development. Michigan Academician, 1993, 25;383-395.

[19] Burley, J.B. Gheo-Shih: an ancient Pre-Columbian plaza. Landscape Research, 1992, 17(2):74-78. DOI: https://doi.org/10.1080/01426399208706365

[20] Burley, J.B., Polakowski, K. J., Fowler., G. Constructing vegetation productivity equations by employing undisturbed soils data: an Oliver County, North Dakota case study. In: Daniels, W.L., Burger, J., C.E. Zipper, C. E., editors. Proceedings Thirteenth Annual Meeting American Society for Surface Mining and Reclamation: Successes and Failures: Applying Research Results to Insure Reclamation Success, May 18-23, 1996, Knoxville, Tennessee. Blacksburg, Virginia: Tech Research Division, Powell River Project, 1996. p. 393-401. https://www.osti.gov/biblio/588806-constructing-vegetation-productivity-equations-employing-undisturbed-soils-data-oliver-county-north-dakota-case-study

[21] Burley, J.B., Fowler, G. W., Polakowski, K., Brown, T. J. Soil based vegetation productivity model for the North Dakota coal mining region. International Journal of Surface Mining, Reclamation, and Environment, 2001, 15(4);213-234. DOI: https://doi.org/10.1076/ijsm.15.4.213.7415

[22] Burley, J.B. A spatial application of a vegetation productivity equation for neo-sol reconstruction. In: 1999 Proceedings Mining and Reclamation for the Next Millennium, August 13-19, 1999 American Society for Surface Mining and Reclamation 16th Annual National Meeting in conjunction with Western Region Ash Group 2nd Annual Forum. Scottsdale, Arizona: American Society for Surface Mining and Reclamation, 1999. p. 708-714. DOI: https://doi.org/10.2100/JASMR99010708

[23] Burley, J.B., D. Gray. 2001. Soil ordination: implications for post-mining disturbance land-uses. In: Vincent, R., Burger, J. A., Marino, G. G., Olyphant, G. A., Wessman, S. C., Darmody, R. G., Richmond, T. C., Bengson, S. A., and J.R. Nawrot, J. R., editors. Land Reclamation -- A Different Approach. June 3-7, American Society for Surface Mining and Reclamation. Albuquerque, New Mexico, American Society for Surface Mining and Reclamation, 2001.

p. 241-245. DOI: https://doi.org/10.21000/JASMR01010241

[24] Burley, J.B. Vegetation productivity equation compatibility with selected state environmental reclamation laws and regulations. In: International Land Reclamation and Mine Drainage Conference and the Third International Conference on the Abatement of Acidic Drainage: Volume 4 or 4: Abandoned Mined Lands and Topical Issues. Washington, D. C.: United States Department of the Interior, Bureau of Mines Special Publication SP 06D-94, 1994. p. 294-303. DOI: https://doi.org/10.2100/JASMR4040294.

[25] Burley, J.B. Environmental Design for Reclaiming Surface Mines. Lewiston, New York: Edwin Mellen Press, 2001.

[26] Le Cleac'h, G., Salles, M., Burley, J. B. Vegetation productivity model for Grand Traverse County, Michigan. 21st National Meeting of the American Society of Mining and Reclamation and The 25th West Virginia Surface Mine Drainage Task Force. Morgantown, West Virginia, April 18-24, 2004, Lexington, Kentucky: American Society for Mining and Reclamation, 2004. p. 1176-1191. DOI: https://doi:10.21000/JASMR04011176

[27] Bai, Y., Chang, Q., Guo, C., Burley, J.B, Partin, S. Neo-sol productivity models for disturbed lands in Wisconsin and Georgia, USA. International Journal of Energy and Environment, 2016, 10;52-60. https://www.naun.org/main/NAUN/energyenvironment/2016/a142011-065.pdf

[28] Burley, J.B., Wu, Z., He, S., Li, X. Soil-based vegetation productivity models for disturbed lands along the northern and central, western Great Plains, USA. Journal of Advanced Agricultural Technologies, 2020,7 (1);1-7. http://www.joaat.com/index.php?m=content&c=index&a=show&catid=65&id=361

[29] Western Pennsylvania Conservancy. Clearfield County Natural Heritage Inventory. Clearfield, Pennsylvania: Clearfield County Planning Office, 2004. https://www.naturalheritage.state.pa.us/CNAI_PDFs/Clearfield%20County%20NAI%202004.pdf

[30] Hallowich, J. S., Topalanchik, A. R., Jensen, E. L., Braker, W. L., Ellenberger, N. B. Soil Survey of Clearfield County, Pennsylvania. Washington, D. C.: US Department of Agriculture, 1988. https://www.nrcs.usda.gov/Internet/FSE_MANUSCRIPTS/pennsylvania/PA033/0/Clearfield.pdf

[31] Rose, A. W., Phelps, L. B., Parizek, R. R., Evans, D. R. Effectiveness of lime kiln flue dust in preventing acid mine drainage at the Kauffman Surface Coal Mine, Clearfield County, Pennsylvania. In: Proceedings American Society of Mining and Reclamation. Gillette, Wyoming: American Society of Mining and Reclamation, 1995. p. 159-171. DOI: https//doi:10.21000/JASMR95010159

[32] Brown, R L, Parizek, R R. Shallow Ground-water Flow Systems Beneath Strip and Deep Coal Mines at Two Sites, Clearfield County, Pennsylvania. Harrisburg, Pennsylvania: Department of Environmental Resources, SR-84, 1971. https://scholarsphere.psu.edu/resources/3b55baac-f7f3-418a-910c-f9a3435365a7

[33] Skousen, J., Zipper, C.E. Coal mining and reclamation in Appalachia. In: Zipper, C.E.. Skousen, J., editors. Appalachia's Coal Mined Landscapes. Cham, Switzerland, Springer, 2021. https://doi.org/10.1007/978-3-030-57780-3_3

[34] Kendall, M.G. The geographical distribution of crop productivity in England. Journal of the Royal Statistical Society,1939, 102;21-48, https://doi.org/10.2307/2980138.

[35] Johnson, R.A., Wichern, D.W. Applied Multivariate Statistical Analysis. Englewood Cliffs, New Jersey: Prentice Hall, 1988, second edition.

[36] Doll, E.C., Wollenhaupt, N. C. Use of soil parameters in the evaluation of reclamation success in North Dakota. ASSMR Second Annual Meeting. American Society for Surface Mining and Reclamation: Denver, Colorado, 1985. p. 91-94. https://doi.org/10.21000/JASMR85010094

[37] Doll, E. C., Merril, S. D., Halvorson, G. A. Soil replacement for reclamation of strip mined lands in North Dakota. Bulletin of the North Dakota Agricultural Experiment Station, 1984, 514.

[38] Bowerman, B. L., O'Connell. Linear Statistical Models: an Applied Approach. Boston, Massachusetts: PWS-Kent Publishing Company, 1990, second edition.

[39] SAS Institute Inc. SAS/STAT® 13.1 User's Guide. Cary, NC: SAS Institute Inc, 2013.

[40] Mallows, C. L. Some comments on *Cp*. Technometrics, 1973, 15(3);661-675.

[41] Buta, M., Blaga, G., Paulette, L., Pacurar, I., Rosca, S., Borsai, O., Grecu, F., Sinziana, P. E., Negrusier, C. Soil reclamation of abandonded mine lands by revegetation in northwestern part of Transylvania: a 40-year retrospective study. Sustainability, 2019, 11. http://doi.org/10.3390/su11123393

[42] Williams, D., Schuman, G. E., editors. Reclaiming Mine Soils and overburden in the Western United States. Ankeny, Iowa: Soil Conservation Society of America, 1987.

[43] Skousen, J. editor. Kimery Vories. Reclamation Matters, 2016;16-17.

Chapter 9

Dynamic-Catenal Phytosociology for Evaluating Vegetation

Sara del Río, Raquel Alonso-Redondo, Alejandro González-Pérez, Aitor Álvarez-Santacoloma, Giovanni Breogán Ferreiro Lera and Ángel Penas

Abstract

The conservation of nature is a problem that has concerned the scientific community for many years. Plants and plant communities play a main role in evaluation and land management studies, owing to their importance as natural and cultural resources. Several studies from the perspective of flora and vegetation have been carried out in the last fifty years (some of them directly related to Phytosociology). According to that, the Dynamic-Catenal Phytosociology must be considered as a very useful tool to evaluate the conservation status of vegetation and to establish suitable models for land management. The fundamental phytosociological concepts to take into account in the evaluation processes of the conservation status of vegetation are reviewed in this study.

Keywords: Dynamic-Catenal Phytosociology, evaluation, conservation, vegetation, methods, plant communities

1. Introduction

The conservation of nature is a problem that has concerned the scientific community for many years. The biological and landscape assessment of a territory is a major phase in land use planning studies and is carried out at an early stage to determine the environmental implications as well as to assess the suitability of the territory for its planned use [1]. Although several factors (such as physical, social, economic, biotic, etc.) are known to be involved in the conservation assessment, this study only analyses the role of vegetation due to its importance as a resource and both cultural and natural heritage.

Plants and plant communities play a main role in this type of evaluation in land management studies, owing to their importance as natural and cultural resources. Several authors [2–6] have indicated that plant communities result from the interplay of many environmental variables, representing practically the completely floristic diversity, as well as many of the ecological relationships between the organisms involved. For this reason, knowledge of the flora and vegetation of a territory is essential for studying and evaluating its conservation status [7].

A great number of environmental evaluation studies from the perspective of flora and vegetation had been carried out in the last fifty years (some of them directly related to Phytosociology) can be mentioned: [2, 5–48].

Since Phytosociology either classic Phytosociology or Dynamic-Catenal Phytosociology is the science that deals with the study of floristic composition of plant communities, their distribution and dynamism, besides the environmental factors that characterize them, this science must be considered as a useful tool to evaluate the conservation status of vegetation and to establish suitable models for land management. In this sense, it can be mentioned the comments pointed out by European researchers worried about nature conservation.

"La phytosociologie est. sans doute plus que toute autre science indispensable à l'étude et à la solution des délicats problèmes d'environnement (notamment d'évaluation et de valorisation des territoires) auxquels l'humanité est. confrontée" [22]. "Avoir une profonde connaissance phytosociologie d'un territoire permet de multiples applications de la phytosociologie pour la planification et la gestion des espaces naturels qui composent a territoire... La phytosociologie joue un rôle fondamental dans l'évaluation des valeurs biologiques du territoire..." [18].

It is noteworthy to say that in this type of studies that the biological evaluation of a territory will be more objective and easier to perform if formulations can be established to quantify the variety of environmental factors involved.

2. Fundamental phytosociological concepts to take into account in the evaluation processes of the conservation status of vegetation

Phytosociology is the part of Geobotany or Ecology that studies plant communities and their relationships with the environment. Not only classifies them and order the plant communities in a hierarchical system but also it is the science of syntaxons and association is the fundamental unit [49].

The terms Vegetal Sociology, Phytocenology and Plant Synecology have been used as synonyms. This ecological science studies biocenosis from a botanical perspective (phytocenosis), that is, it deals with plant communities, their relationships with the environment and the temporal processes that modify them.

With all this information and through an inductive and statistical method based on relevé it tries to create a universal hierarchical typology in which the association is the basic unit of the typological system.

Nowadays, the Dynamic-Catenal Phytosociology is recognized in addition to classic or Braun-Blanquet's phytosociology. The fundamental units of this science are the vegetation series or sigmetum, the permaseries or permasigmetum and the minoriseries or minorisigmetum in the Dynamic Phytosociology and the geoseries or geosigmetum, the geominoriseries or geominorisigmetum and the geopermaseries or geopermasigmetum in the Catenal Phytosociology.

The Dynamic-Catenal Phytosociology also called "Landscape Phytosociology" or "Global Phytosociology", tries to express through geobotanical and environmental sciences the biodiversity, structure and succession of the plant landscape, specifically of the natural, seminatural and anthropic terrestrial ecosystems [49].

It seems appropriate at this time to distinguish between the Dynamic Phytosociology and Catenal Phytosociology. The Dynamic Phytosociology deals with the vegetation series, with perennial and annual substitution stages and permaseries, without perennial non-nitrophilous substitution communities. The Catenal Phytosociology deals with vegetation geoseries, geominoriseries and geopermaseries.

Some concepts should be defined for a better understanding of the processes studied by the Dynamic-Catenal Phytosociology.

Let us firstly look at the concept of **succession**, which is the natural process of vegetation by which some plant communities are replaced over time by others within the same holotesela, originating different vegetative structures, which along the time can reach their final balance stage, which it is the head series or climax [49].

The succession process has been extensively studied in terrestrial ecosystems. Because of this, numerous well-known sequential models and vegetation series, both structurally and biocenotically as well as functional, are nowadays available allowing recognizing and designating the phytocenoses that go from the first colonizer stages to the climax.

But it is necessary to distinguish between the sequences that lead to the climax in a holotesela (progressive succession or progression) and those that by anthropozoogenic or natural actions remove from it (regressive succession or regression). It may also be useful to differentiate the notion of primary succession from the secondary one. The primary succession begins in environments not yet colonized and lacking of developed soils (rocky, recent fluglacial sediments, areas with totally eroded soils, etc.) while the secondary succession operates within the subserial stages of series of vegetation and pre-existing degraded soils [49].

Serial or subserial stages are all those stages that are not climax. The North American school of Clements distinguishes between the priserial associations or previous to the climax and subserial associations or after the destruction of the climax.

From a functional point of view, the initial communities only incorporate a small part of their energy availability as permanent biomass, while in the climacic forest stages the process is upturned, since energy production is mostly intended to growth, to respiration, and the maintenance of the system in its state of balance.

We have previously mentioned that the advances in Dynamic-Catenal Phytosociology have allowed us to establish well-known sequential models. Some examples are mentioned:

The characteristic climatophilous vegetation in a temperate oceanic territory (supratemperate humid bioclimatic belt) is a deciduous or marcescent forest. This vegetation can suffer alterations due to anthropic or natural (such us fire) phenomena that can cause changes in the structure and floristic composition, but also in the horizons of the soil on which it develops to a greater or lesser degree depending on the alteration intensity (**Figure 1**).

This means that deforested areas with loss of edaphic horizons are colonized almost immediately by communities of phanerophytes or nano-phanerophytes that constitute the pre-forests which are usually formed by softwood trees or brooms. These brooms are preferably constituted by *Cytisus* and *Genista* genera (**Figure 2**).

These communities sometimes share space with the communities of perennial hemicryptophytes that will make up the perennial grasslands in places with the greatest edaphic deterioration. These perennial grasslands will be replaced at the same time by chamephytic shrubland in areas where deeply destruction of the original forest soil has occurred. These soils are finally colonized in this successional process by communities of therophytes (**Figure 3**).

This example is perfectly transferable to other territories, in which depending on the climate and the soil, as well as the functionality of its vascular flora, we will find cases that are not the same but similar.

Thus, in tropical pluvial bioclimate usually an arboreal climax is replaced in a regressive successional process by up to three plant communities of phanerophytes and after that by two nanophanerophytic communities to be later substituted by a community dominated by hemicryptophytes. The last and more regressive stage is a community of chamephytic character. In fact, there would be eight serial stages in this vegetation series.

Figure 1.
Climax and serial stages of a climatophilous and acidophilous beech forest. Blechno spicant-Fagetum sylvaticae; Cytiso cantabrici- Genistetum polygaliphyllae; Pterosparto cantabrici-Ericetum aragonensis; Merendero pyrenaicae-Cynosuretum cristati. *Cofiñal (León, Spain).*

Figure 2.
*Forest (*Blechno spicant-Fagetum sylvaticae*) and nanophanerophytic shrubland (*Cytiso cantabrici-Genistetum polygaliphyllae*). Riaño (León, Spain).*

This model or sequence of serial stages is defined as **vegetation series** in the Dynamic-Catenal Phytosociology. It is also known **as sigmetum, sigmeto** (Spanishize) or **sinassociation.** It expresses the entire set of plant communities or steps that can be found in related tessellated spaces (superholotesela) as a result of the process of succession, which includes both the association representative of the mature and series head, like initial or subserial associations that can replace it.

Understood in this way, the vegetation series or sigmetum represents the basic unit or essential model of the Dynamic Phytosociology.

Figure 3.
Climax and serial stages of an edaphoxerophilous and basophilous beech forest. Epipactido helleborines-Fagetum sylvaticae, Lithodoro diffusae-Genistetum occidentalis *and* Helianthemo cantabrici-Brometum erecti. *Bernesga Valley (León, Spain).*

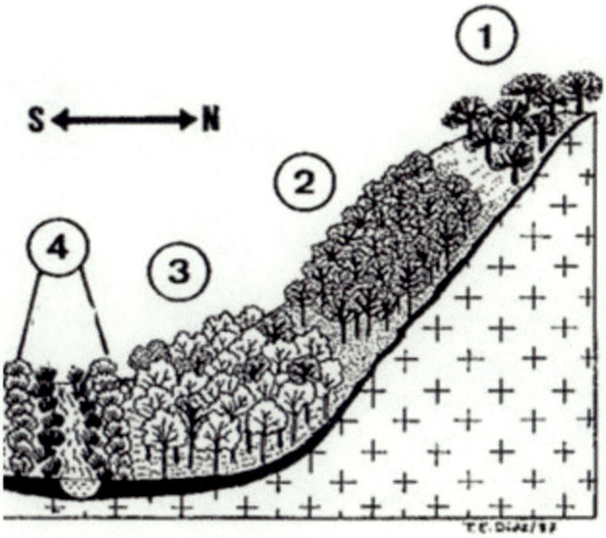

Figure 4.
1. Edaphoxerophilous vegetation series, 2. Climatophilous vegetation series, 3. Temporihygrophilous vegetation series, 4. Edaphohygrophilous vegetation series.

It is possible to differentiate between climatophilous, edaphoxerophilous, temporihygrophilous and edaphohygrophilous vegetation series (**Figure 4**).

The **climatophilous** or zonal vegetation series are those that are located in mature soils according to mesoclimate and only receive rainwater (**Figure 5**).

The **edaphoxerophilous** vegetation series are found in soils or in biotopes especially dry or xerophytic such as: lithosols, leptosols, arenosols, gipsisols, etc. These vegetation series grow in dunes, very windy places, steep slopes, ridges, slopes, etc. (**Figure 6**).

Figure 5.
Teucrio baetici-Quercetum suberis, *head series of the aljibic climatophilous and acidophilous cork oak forest.*

Figure 6.
Climax and chamephytic serial stage of an edaphoxerophilous vegetation series. Juniperetum sabino-thuriferae, Lithodoro diffusae-Genistetum scorpii. *Mirantes de Luna (León, Spain).*

The **temporihygrophilous**, **mesophytic** and **mesohygrophytic** vegetation series have extraordinary hydric contributions due to topographic reasons. These series develop in slightly flooded or very humid soils only part of the year. The soil horizons are well drained or aerated at least during the summer or the dry season.

Lastly, **edaphohygrophilous** vegetation series are those that occupy particularly humid soils and biotopes such as fluvisols, halools, histosols, etc. They are found in river beds, marshy areas, salt marshes, peat bogs... (**Figure** 7).

Subseries and vegetation faciations can be used as units of lower rank than the vegetation series. Those of higher rank, superseries, macroseries, megaseries and hyperseries can be used [45].

Figure 7.
Climax of an edaphohygrophilous vegetation series. Populo nigrae-Salicetum neotrichae. *Benavente (Zamora, Spain).*

There are other different models apart from the aforementioned vegetation series in the Dynamic-Catenal Phytosociology. **Minoriseries of vegetation** or **minorisigmetum** (Latinized expression) are plant communities with their corresponding perennial and annual substitution stages. These are found in the tessellated spaces and in their jurisdictional territories but that by exceptional environmental causes do not reach, in progressive succession, the mature stage of the habitual climatophilous or edaphophilous series head in the corresponding biogeographic and bioclimatic environment.

These plant communities have their own perennial substitution stages and can be frequent in certain zones and ecotonic areas such as: coastal cliffs with strong sea air, coastal dune systems, riverbeds, extensive peat bogs and marshes, estuaries, ridges and windy peaks, ecotonic altioreine territories with forests, etc.

Lazare [50] has recently proposed these units of Dynamic-Catenal Phytosociology with the names of "truncated series and geoseries: curtosigmetum and geocurtosigmetum", which we have preferred to translate by grammatical issues as: minoriseries and geominoriseries of vegetation (minorisigmetum and geominorisigmetum).

This is the case of the Asturian and Galician acidophilous laurels plant communities which represent the climax in highly protected areas from the very strong north-western winds on the cliffs. Their substitution stages and grasslands are dominated by *Armeria maritima.*

Also we can find nano-phanerophytic formations in the hills of the mountain passes where the wind is a determining factor for the development of the vegetation. These formations do not correspond to the bioclimatic and edaphic characteristics of the surrounding area, where the climax vegetation is a forest.

Both situations are an example of minoriseries or minorisigmetum.

Finally, we will also find **permaseries of vegetation** which are perennial, stable plant communities that configure permateselas or related permateselar complexes of exceptional biotopes such as: polar territories, hyper-deserts, peaks of high mountains, walls, cliffs, dunes, cliffs and coastal rocks beaten by sea, peat bogs, salt marshes, lagoon margins, etc. The mature or climax stage corresponds to a perennial vascular association (permassociation or permaseries), generally poorly

Figure 8.
*Geopermaseries of dune systems (*Ammophiletea australis*). Troia Peninsula (Portugal).*

Figure 9.
*Lacustrine permaseries (*Potamogetum natans *plant community). Villadangos del Páramo (León, Spain).*

stratified, lacking of perennial non-nitrophilous substitution communities. It means that, apart from the annual or ephemeral species and communities that may temporarily establish in the open or degraded spaces of such locations, it is only the perennial plants that participate in the mature community that can thrive to reorganize the same permanent plant community [49] (**Figures 8–11**).

The change of the environmental conditions of each permassociation or permaseries supposes its disappearance and its replacement by another permassociation or permaseries belonging to the same geopermaseries, without the possibility of returning to the previous situation through natural succession.

Figure 10.
*Glericolous permaseries (*Linario filicaulis-Crepidetum pygmaeae*). La Vueltona (Picos de Europa, Spain).*

Figure 11.
*Rupicolous permaseries (*Centrantho lecoqii-Saxifragetum canaliculatae*). Beberino (León, Spain).*

It is the case of a freshwater lagoon geopermaseries in which the persistence of water is not constant throughout the year. It undergoes partial or total desiccation during the summer because it loses the water table throughout the spring to recover it in autumn and having its maximum during winter (**Figure 12**).

The annual drying process previously mentioned makes the water-free spaces to be colonized by different vegetation circles (permaseries or permanent communities) along the year. It is possible to see up to seven different plant associations in some cases. Some of them are pre-spring nature; others have spring–summer character; others summer-autumnal and some annual, although these hygrophytes are all permanent. In other words, each one of them would be a permassociation (permaseries) included in the geopermaserie described.

We have to take into consideration what would happen if the lagoon was subjected to a natural process of silting. This would suppose an increasingly reduced space for the subsistence of the vegetation circles mentioned above; reaching in the future that some or all of them would disappear depending on whether the silting process is partial or total. In this last situation, the area occupied by the lagoon could be subjected to a new progressive successional process that might conclude in a climax of a series of edaphohygrophilous vegetation.

But the dynamic or successional processes not only are those that allow us to establish vegetation series or sigmetum, minoriseries or minorisigmetum or permaseries or permasigmetum. There are also those that have a progressive character towards the climax from communities structurally and floristically very distant from the mature stage corresponding to the ecological conditions (mainly climate and soil) of the territory where it is located.

These progressive processes towards the climax, according to the ecological theory, determine the recovery of the mature stage in equilibrium.

Nevertheless, it is not the most frequent process, existing ways of progression known as "pathways of urgency". They imply the overcoming in a single process of one or more of the pre-existing regressive successional stages that will preferably depend to a certain extent on the edaphic factors.

Furthermore, it is possible for a climax not to be the starting point of the regression process but being essential to go through it to reach the original mature stage. **Figure 13** shows the regressive and progressive dynamic models of vegetation.

This is the case happening when the regression process has been so strong. The plant community, which occupies the specific area where the deterioration takes place, happened to occupy a soil whose edaphoclimate does not correspond to the original one.

Giving an example:

Let us suppose a beech forest in southern exposure, at the limit of its distribution area between the Eurosiberian and Mediterranean region and therefore in a temperate oceanic bioclimate with a submediterranean variant, that undergoes a regressive

Figure 12.
Lagoon geopermaseries. Valdepolo León, Spain.

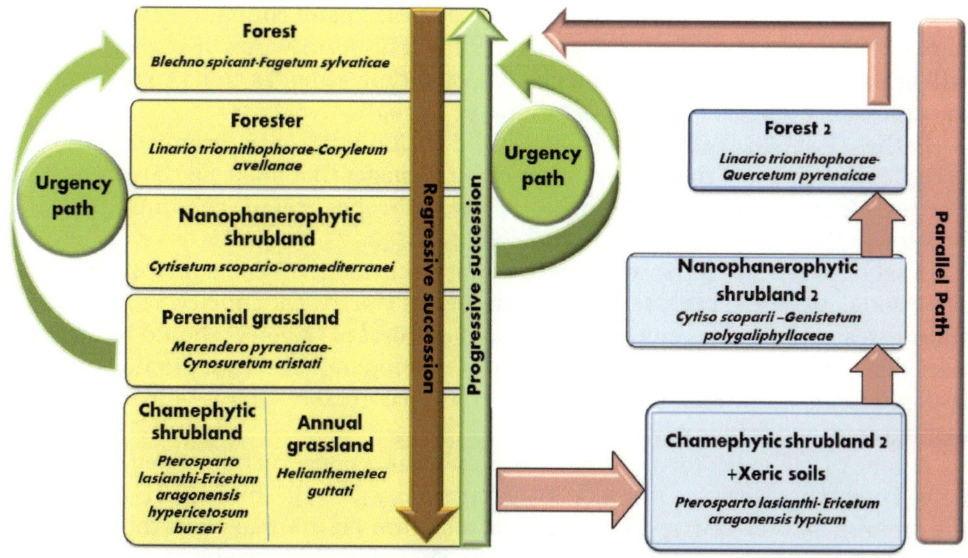

Figure 13.
Diagram showing regressive, progressive and parallel paths in vegetation dynamic.

process of wide magnitude, until reaching a serial stage or successional the chamephytic stage mainly due to soil erosion.

Under these conditions, the edaphoclimate of the soil on which this plant community grows will be clearly more xeric than in origin due to the solar radiation received. This will determine that the progressive process towards the original climax is modified, necessarily going through a marcescent forest, after reaching the original climax stage.

Picking up the **geopermaseries** concept, also called **geopermasigmetum**, it is the catenal expression of a set of contiguous permaseries or permasigmetum delimited by changing topographic or edaphic situations.

They are conditioned by extreme climatic situations (high mountains and polar areas), microtopographic or exceptional edaphic ones (walls, rocky areas, sea cliffs, salt marshes, etc.). It creates a large number of biotopes occupied by permanent perennial communities (contiguous permaseries), lacking non-nitrophilous perennial serial communities, which seem to have reached their equilibrium.

The most favorable places for the existence of geopermaseries or geopermasigmetum are cliffs, rock crevices, cliffs and littoral crags battered by seawater, peat bogs, snowdrifts, mobile dunes, shores of lagoons, water springs, etc. In addition, they are those locations that correspond to types of potential vegetation of extreme bioclimates of high mountains and Polar Regions.

The study of these groups of contiguous perennial permanent communities must be carried out within the framework of a single bioclimatic belt, in precise geomorphological limits and trying to follow the gradient of the ecological factor which determines the catena.

3. How to use these concepts for assessing the conservation status of vegetation in a specific territory?

As we have already mentioned, the greater the knowledge of the vegetation of the territory studied, the more precise will be the value of the conservation status of the vegetation.

We will always have to take into account that each plant community will be part of one of the units described above, that is, it will be permaseries of vegetation or it will be part of a minoriseries or a series of vegetation, either as climax or as a serial or successional stage, either in a regressive or a progressive sense.

In these last two cases, we must assess the position that each plant community occupies in relation to the total number of serial stages that make up the minoriseries or the vegetation series, including the anthropozoogenic communities existing in them.

A set of phytocenotic parameters (diversity, representativeness, maturity, naturalness and replaceability), territorial parameters (rarity, endemicity, relictism, finicole feature), besides environmental and cultural parameters should be added to evaluate individually the vegetation types existing in the territory under study. It would be also convenient to take into account parameters such as vulnerability, demography and fragility [7].

A very useful index for assessing the global status conservation of vegetation is the Potentially Distance Index (PDI) proposed by [7, 44]. This index is based on the premise that the conservation status of a territory is optimal when the whole of its vegetation is at the climax stage. Knowing the vegetation series and the distance of its serial stages from the climax, enable us to evaluate its conservation status independently of other territories. Therefore, PDI is founded on Dynamic-Catenal Phytosociology concepts. To apply this index, it is first necessary to establish the previous-actual vegetation cartography of the studied territory using syntaxa or syntaxon complexes according to phytosociological principles previously mentioned.

The PDI has the advantage of giving a universal application, which can provide us with an easy and valid comparison of the vegetation conservation status between different territories on the basis of their dynamic characteristics and their naturalness index. This index can offer objective and quantitative information to be applied in land management and nature conservation studies. Furthermore, it is a useful tool for restoration of potential natural vegetation in degraded areas.

Therefore the basis and interest of this index lies in the following aspects:

1. it is based on the Dynamic-Catenal Phytosociological concepts

2. the conservation status of a territory is optimal when the whole of its vegetation is at the climax stage (permaseries, climax of vegetation series and minoriseries)

3. it is objective and measurable

4. each plant community can be evaluated independently and numerically according to its inherent dynamic characteristics and its naturalness index

5. its application is universal

6. it is very useful for the restoration of potential natural vegetation in degraded areas

The formula for calculating the Potentiality Distance Index (PDI) is:

$$PDI = \sum_{i=1}^{n} \frac{\Omega i * DIi}{\Omega \, total} \tag{1}$$

Phytosociological classes	S	M	P	E	A
Adiantetea capilli-veneris Br.-Bl. *in* Br.-Bl., Roussine & Nègre 1952			•		
Greeonivio aureae-Aeonietea Santos 1976			•		
Alnetea glutinosae Br.-Bl. & Tx. *ex* Westhoff & al. 1946	•				
Euphorbio paraliae-Ammophiletea australis Géhu & Rivas-Martínez *in* Rivas-Martínez & al. 2011			•		
Anomodonto viticulosi-Polypodietea cambrici Rivas-Martínez 1975			•		
Artemisietea vulgaris Lohmeyer, Preising & Tüxen *ex* von Rochow 1951					•
Asplenietea trichomanis (Br.-Bl. *in* Meier & Br.-Bl. 1934) Oberd. 1977			•		
Betulo carpaticae-Alnetea viridis Rejmánek *ex* Bœuf, Theurillat, Willner, Mucina et Simler in Bœuf & al. 2014	•				
Bidentetea tripartiate Tüxen, Lohmeyer & Preising *ex* von Rochow 1951					•
Cakiletea maritimae Tx. & Preising *ex* Br.-Bl. & Tx. 1952					•
Calluno vulgaris-Ulicetea minoris Br.-Bl. & Tx. *ex* Klika & Hadac 1944			•	•	
Carici rupestris-Kobresietea myousuroidis Ohba 1974 *nom. mut.*			•		
Cisto-Lavanduletea stoechadis Br.-Bl. *in* Br.-Bl. & al. 1940				•	
Crithmo-Staticetea Br.-Bl. *in* Br.-Bl. & al. 1952			•		
Parietarietea Rivas-Martínez *in* Rivas Goday 1964			•		
Cytisetea scopario-striati Rivas-Martínez 1974		•	•	•	
Chamaecytiso-Pinetea canariensis Rivas Goday & Esteve *ex* Esteve 1969 *nom. mut.*	•				
Kobresio myosuroidis-Seslerietea caerulea Br.-Bl. 1948 *em.* Ohba. 1974 *nom. mut.*			•		
Epilobietea angustifolii Tx. & Preising *ex* von Rochow 1951					•
Festucetea indigestae Rivas Goday & Rivas-Martínez 1971			•	•	
Festuco hystricis-Ononidetea striatae Rivas-Martínez. & al. 2002			•	•	
Festuco valesiacae-Brometea erecti Br.-Bl. & Tx. *ex* Klika & Hadăc 1944				•	
Galio aparines-Urticetea maioris Passarge *ex* Kopecký 1969					•
Geranio purpurei-Cardaminetea hirsutae (Rivas-Martínez & al. 1999) Rivas-Martínez & al. 2002					•
Halodulo wrightii-Thalassietea testudinum Den Hartog *ex* Rivas-Martínez, Fernández-González & Loidi 1999			•		
Helianthemetea guttati (Br.-Bl. in Br.-Bl. & al. 1952) Rivas Goday & Rivas-Martínez 1963				•	
Isoeto-Nanojuncetea Br.-Bl. & Tx. *ex* Westhoff & al. 1946			•		

Phytosociological classes	**S**	**M**	**P**	**E**	**A**
Juncetea maritimi Br.-Bl. *in* Br.-Bl. & al. 1952			•		
Caricetea curvule Br.-Bl. 1948 *nom. conserv.*			•		
Junipero sabinae-Pinetea sylvestris Rivas-Martínez 1965 *nom. inv.*	•		•		
Kleinio neriifoliae-Euphorbietea canariensis (Rivas Goday & Esteve 1965) Santos 1976			•	•	
Koelerio glaucae-Corynephoretea canescentis Klika *in* Klika & v. Novák 1941			•	•	
Lauro azoricae-Juniperetea brevifoliae Rivas-Mart. & al. *in* Rivas-Mart. & al. 2002	•				
Lemnetea Tüxen *ex* O. Bolòs & Masclans 1955			•		•
Littorelletea uniflorae Br.-Bl. & Tx. *ex* Westhoff, Dijk & Passchier. 1946			•		
Loiseleurio procumbentis-Vaccinietea microphylii Eggler *ex* Schubert 1960	•		•	•	
Lygeo sparti-Stipetea tenacissimae Rivas-Martínez. 1978 *nom. conserv. propos.*				•	
Molinio caerulae-Arrhenatheretea elatioris Tx. 1937				•	
Montio fontane-Cardaminetea *amarae* Br.-Bl. & Tx. *ex* Br.-Bl. 1948			•		
Mulgedio-Aconitetea Hadăc &Klika *in* Klika & Hadăc 1944			•		
Nardetea strictae Rivas Goday *in* Rivas Goday & Rivas-Martínez 1963			•	•	
Nerio oleandri-Tamaricetea Br.-Bl. & O. Bolòs 1958		•	•		
Rhamno crenulatae-Olecetea cerasiformis Santos *ex* Rivas-Martínez 1987 *nom. inv.*		•		•	
Oryzetea sativae Miyawaki 1960					•
Oxycocco-Sphagnetea Br.-Bl. & Tx. *ex* Westhoff & al. 1946			•		
Pegano harmalae-Salsoletea vermiculatae Br.-Bl. & O. Bolòs 1958					•
Petrocoptido pyrenaicae-Sarcocapnetea enneaphyllae Rivas-Martínez, Cantó & Izco *in* Rivas-Martínez & al. 2002			•		
Phagnalo saxatilis-Rumicetea indurati (Rivas Goday & Esteve 1972) Rivas-Martínez, Izco & Costa 1973			•		
Phragmito-Magnocaricetea Klika *in* Klika & v. Novák 1941			•		•
Poetea bulbosae Rivas Goday & Rivas-Martínez *in* Rivas-Martínez 1978					•
Polycarpaeo niveae-Traganetea moquini Santos *ex* Rivas-Martínez & Wildpret 2002	•	•			
Polygono-Poetea annuae Rivas-Martínez 1975					•
Posidonietea Den Hartog 1976			•		
Potametea Klika *in* Klika & V. Novák 1941			•		
Pruno hixae-Lauretea novocanariensis Oberd. 1965 *corr.* Rivas-Martínez, T.E. Díaz, Fernández-González, Izco, Loidi, Lousa & Penas 2002	•				

Phytosociological classes	S	M	P	E	A
Quercetea ilicis Br.-Bl. *ex* A. & O. Bolòs 1950	•	•		•	
Querco roboris-Fagetea sylvaticae Br.-Bl. & Vlieger *in* Vliegar 1937	•	•		•	
Rhamno catharticii-Prunetea spinosae Rivas Goday & Borja *ex* Tüxen 1962		•		•	
Rosmarinetea officinalis Rivas-Martínez, T.E. Díaz, F. Prieto, Loidi & Penas 2002				•	
Ruppietea maritimae J. Tx. 1960			•		
Saginetea maritimae Westhoff & al. 1962			•		
Salicetea herbaceae Br.-Bl. 1948			•		
Salici purpureae-Populetea nigrae (Rivas-Martínez & Cantó *ex* Rivas-Martínez & al. 1991) Rivas-Martínez & Cantó 2002	•		•		
Sarcocornietea fruticosae Br.-Bl. & Tx. *ex* A. Bolo's 1950 *nom. mut.*			•		
Scheuchzerio palustris-Caricetea nigrae Tx. 1937 *nom. mut.*			•		
Sedo albi-Scleranthetea biennis Br.-Bl. 1955				•	
Spartinetea maritimae Tx. *in* Beeftink 1962			•		
Stellarietea mediae Tüxen, Lohmeyer & Preising *ex* von Rochow 1951					•
Stipo giganteae-Agrostietea castellanae Rivas-Martínez & al. 1999				•	
Thero-Suaedetea Rivas-Martínez 1972			•		
Thlaspietea rotundifolii Br.-Bl. 1948			•		
Trifolio medii-Geranietea sanguinei T. Müller 1962				•	•
Utricularietea intermedio-minoris Pietsch 1965			•		
Vaccinio-Piceetea abietis Br.-Bl. *in* Br.-Bl., Sissingh & Vliegar 1939	•		•		
Zosteretea marinae Pignatti 1954			•		

If (S), Successional Stages, Minoriseries (M), Permaseries (P), Vegetation Series (E) are included in them or if they have Anthropozoogenic character (A) are expressed with • [51–53]

Table 1.
Phytosociological classes of Spain and Portugal are shown in alphabetical order

Where i = i-th plant community, Ωi = surface area of i-th plant community, Ω total = total surface area occupied by all communities, DI_i = potentiality distance value of i-th community.

Its formula is as follows: $DI_i = 1 - [(3\,Pi - NIi)/3n)]$, Pi = position of i-th plant community in its respective vegetation series and minoriseries in relation to the series head, NIi = naturalness index of i-th community, n = number of serial stages in the vegetation series or minoriseries resulting from the succession process. In permaseries n = 1 because such plant communities are considered as series head.

We have:

NI = 1 for plant communities with low natural conservation and with strong human influence (distance to the optimum stage >50%);

NI = 2 for plant communities with relatively high natural conservation and with a low, but appreciable, human influence (distance to the optimum stage 30–50%);

NI = 3 for plant communities with high natural conservation without (or with very little) human influence (distance to the optimum stage <30%) [7, 44].

PDI are divided into four groups showing the distance to the series head of the studied area and therefore the global conservation status of vegetation. If the PDI value is ≤0.25 the vegetation is very distant from the climax and the conservation status is classified as "Poor". When it is >0.25 and ≤ 0.50 the vegetation is distant from the climax, and the conservation status is "Moderate". Vegetation is shown as moderately distant from climax when values are >0.50 and ≤ 0.75 and the conservation status is "Good". The conservation status is "very good" when the value is >0.75, and vegetation is not distant/too much distant to climax.

In **Table 1** are shown the relationship between vegetation series, minoriseries and permaseries with Spanish and Portuguese phytosociological classes.

4. Conclusions

The Dynamic-Catenal Phytosociology can be used to establish suitable models for land management. Fundamental concepts of this science can be applied to the Potentially Distance Index for assessing the global status conservation of vegetation. The index is universal, objective and measurable, for this reason it is a very useful tool for the restoration of potential natural vegetation in degraded areas.

Author details

Sara del Río[1]*, Raquel Alonso-Redondo[2], Alejandro González-Pérez[2], Aitor Álvarez-Santacoloma[2], Giovanni Breogán Ferreiro Lera[2] and Ángel Penas[1]

1 Department of Biodiversity and Environmental Management (Botany), Faculty of Biological and Environmental Sciences, Mountain Livestock Institute (CSIC-ULE), University of León, Spain

2 Department of Biodiversity and Environmental Management (Botany), Faculty of Biological and Environmental Sciences, University of León, Spain

*Address all correspondence to: sriog@unileon.es

References

[1] Ortega R, & Rodríguez I. Manual de Gestión del Medio Ambiente. Fundación Maphre; 1996; Madrid

[2] Arnáiz C. Utilización de criterios ecológicos para la valoración del territorio. Séminaire de phytosociologie appliquée. 1981; 3, 119-126.

[3] Blandin P. Bioindicateurs et diagnostic des systèmes écologiques. Bulletin Écologie; 1986; 17(4), 214-307.

[4] Asensi A. Aplicación de la fitosociología a la evaluación del territorio. Monografías Flora Vegetación Béticas; 1990; 4-5,91-100.

[5] Géhu J.M. Sur la fragilité structurale de la frange terrestre des bords de mer et son analyse par les méthodes de la phytosociologique moderne dans un souci conservatoire. Colloques Phytosociologiques; 1992; 19, 37-44.

[6] Loidi J. Phytosociology applied to nature conservation and land management. In: *Applied Ecology*; 1994; Y.Song, H. Dierschke & X. Wang (eds.), Proceed. 35th Symposium IAVS in Shanghai, East China Normal Univ. Press, 17-30.

[7] Penas A, Herrero L, & del Río S. Valuation methods in vegetation and its use in land management. Acta Botanica Gallica; 2010; 157(4): 735-748.

[8] Géhu J.M. Phytosociologie et conservation de la nature. Colloques Phytosociologiques; 1988; 15, 1-7.

[9] Nef L. Considérations sur le choix des critères d'évaluation écologique. Évaluation des écosystèmes, *Mons*; 1977; 5-9.

[10] Nef L. Problèmes concernant les critères et l'évaluation biologique de l'environnement. *Séminaire de phytosociologie appliquée*; 1981; 3, 1-6.

[11] Froment A. Méthodologie des premières études d'impact écologique réalisées en Belgique pour les projets d'infrastructure routière. *Séminaire de phytosociologie appliquée*; 1981; 3, 15-26.

[12] Báscones J.C, & Ursúa, C. Criterios para la selección de espacios naturales protegidos en Navarra. *Monografías Instituto Pirenaico de Ecología de Jaca (Homenaje P.Montserrat);* 1988; 4, 121-125.

[13] Dumont J.M. La conception, la réalisation et l'utilisation de la carte d'évaluation biologique de la Belgique. Colloques Phytosociologiques; 1988; 15, 117-126.

[14] Boullet V. Approches méthodo logiques et phytosociologiques de la protection et de la gestion des pelouses calcicoles planitiaires. Colloques Phytosociologiques; 1988; 15,417-430.

[15] Delpech R. & Philippe T. Possibilités d'application de la phytosociologie à la gestion écologique et conservatoire des communautés herbacées d'altitude de grand intérêt biologiques. Colloques Phytosociologiques; 1988; 15,487-507.

[16] Stein J. Un intermédiaire écologique adéquat en aménagement du territoire : la végétation naturelle potentielle. Colloques Phytosociologiques; *1988;* 15, 569-579.

[17] Onaníndia M, Benito I, & García I. Descripción y valoración de los ecosistemas vegetales en el entorno del embalse de Urkullu (Guipúzcoa). Lurralde, 1988; 11,123-136.

[18] Bou i Tomas J. La phytosociologie et le réseau des espaces d'intérêt naturel de la Catalogne. *Colloques Phytosociologiques*, 1988; 15, 695-711.

[19] Géhu J.M. Étude phytocoenotique analytique et globale de l'ensemble des vases et prés salés et saumâtres de la

façade atlantique française; 1979; Ministère de L'Environnement, Bailleul.

[20] Géhu J.M. Introduction au 3e seminaire de phytosociologie appliquée. *Séminaire de phytosociologie appliquée;* 1981; 3.

[21] Géhu J.M. L'analyse symphyto sociologique et géosymphyto sociologique de l'espace. Théorie et méthodologie. *Colloques Phytosociologiques*, 1991; 17, 11-46.

[22] Lucas A. Une échelle de cotation des milieux naturels. *Pen ar Bed n.s.,* 4, 72, 1-6.

[23] Géhu J.M. & J. Géhu-Franck, 1979.- Essai d'évaluation phyto coenotique de l'artificialisation des paysages. Colloques Phytosociologiques; 1973; 3, 95-120.

[24] Géhu J.M, & Géhu-Franck J. Essai d'évaluation biologique des milieux naturels. Exemples littoraux. *Séminaire de phytosociologie appliquée;* 1981; 3

[25] Géhu J.M, & Géhu-Franck J. Les landes de La Poterie près de Lamballe (22), un exemple remarquable de diversité phytocénotique relictuelle; arguments pour leur protection. Colloques Phytosociologiques; 1988; 15, 299-316.

[26] Géhu J.M. & Géhu-Franck J. Essai d'évaluation phytocoenotique de l'artificialisation des paysages. Colloques Phytosociologiques; 1991; 17, 479-515.

[27] Theurillat J.P, Delarze R & Werner Ph. Inventaire des terrains secs de Suisse : le cas du Valais. Colloques Phytosociologiques; 1988; 15, 273-280.

[28] Costa M., Pérez R. & Soriano P. La vegetación como elemento de diagnóstico y valoración en la conservación del litoral. Colloques Phytosociologiques; 1988; 15, 281-298.

[29] Martín M.V, & Asensi A. Evaluación biológica del Parque Natural de la Sierra de Grazalema. Colloques Phytosociologiques; 1988; 15, 677-693.

[30] Asensi A, Díez Garretas B, Ferre E, & Martín Osorio M.V. Aplicación de la metodología fitosociológica al análisis integrado del paisaje en el Parque Natural de la Sierra de la Nieve, Serranía de Ronda (Málaga, España). Colloques Phytosociologiques; 1991; 17, 159-175.

[31] Asensi A, Díez Garretas B, Ferre E, & Martín Osorio M.V. Geomorfología y vegetación en el análisis del paisaje. Colloques Phytosociologiques; 1991; 17, 483-496.

[32] Asensi A, Díez Garretas B, & Navarro P. Utilización de datos fitosociológicos en la evaluación y restauración del territorio Sierra de Gador-Campo Dalias (Almería, España). Colloques Phytosociologiques; 1997; 17, 933-950.

[33] Meaza G, & Ormaetxea O. Propuesta metodológica de valoración fitogeográfica de unidades de paisaje vegetal. *Cuadernos de Sección, Historia-Geografía;* 1993; 20, 21 pp.

[34] Sesma J, & Loidi J. Estudio de la vegetación de Monte Peña (Navarra) y su valoración naturalística. *Príncipe de Viana, suplemento de Ciencias;* 1993; 13(3), 127-168.

[35] Cano E, García-Fuentes A, Nieto J, & Torres J.A. Ensayo para la evaluación de asociaciones pertenecientes al Hábitats-(2000). *XIV Jornadas de Fitosociología*; 1994; Bilbao.

[36] Izco J. Biodiversidad fitocenótica: aproximación conceptual y métodos de valoración. *XIV Jornadas de Fitosociología*; 1994; Bilbao, 15.

[37] Díaz González T.E, Fernández Prieto J.A, Felicísimo Pérez A.M, Martínez Martínez J, & García

Rodríguez, A. *Valoración mediante un sistema de información geográfica de la cartografía de hábitats derivada de la Directiva 92/43/CEE existentes en Cantabria.* Consejería de Agricultura. Comunidad Autónoma de Cantabria; 1996; Santander

[38] Díaz González T.E, Fernández Prieto J.A, Felicísimo Pérez A.M, Martínez Martínez J, & García Rodríguez, A. *Tratamiento de los datos del inventario de hábitats derivado de la Directiva 92/43/CEE*. Consejería de Agricultura, Principado de Asturias; 1996; Oviedo.

[39] Díaz González T.E, Fernández Prieto J.A.Un nuevo método para la evaluación de la cubierta vegetal de un territorio. Colloques Phytosociologiques; 1997; 27, 727-738.

[40] García-Baquero G, & Valle Gutiérrez C.J. Ensayo de valoración naturalística en el centro-oeste ibérico. Salamanca (hoja 13-19 E. 1:50.000). Studia Botanica; 1998; 17, 9-22.

[41] Cadiñanos J.A, & Meaza G. *Bases para una biogeografía aplicada: criterios y sistemas de valoración de la vegetación.* Ed. Geomorfa. 1998; 144 pp. Logroño.

[42] Giménez E, & Gómez F. Valoración naturalística del paraje natural de Punta Entinas-Sabinar. *XVII Jornadas de Fitosociología*, 1999; Jaén.

[43] Díaz González T.E, & García-Rodríguez A. Aplicaciones de la Fitosociología en la evaluación de la cubierta vegetal de un territorio: Monte la Lamiella, Valdés (Asturias). *En: Vegetación y Cambios Climáticos.* 2001

[44] Penas A, del Río S, & Herrero L. A new methodology for the quantitative evaluation of the conservation status of vegetation: the potentiality distance index (PDI). Fitosociologia, 2005; 42(2), 23-51.

[45] Duval J. Indices agro-écologiques. *Séminaire de phytosociologie appliquée*; 1980; 3, 47-49.

[46] Duval J. Adjonction d'un critère "M" (valeur écosystémique d'un milieu) à l'échelle de cotation des milieux naturels de Lucas. *Séminaire de phytosociologie appliquée*; 1980; 3, 76-93.

[47] Batista T, Cavaco J, Mendes P, Vila-Viçosa C, Penas Á, del Río S, Cabezas J, Fernández-Pozo L, Arancibia N, & Pinto-Gomes C. Conservation status of vegetation in the North and Central area of Pardiela river basin (Évora, Portugal), Acta Botanica Gallica, 2012; 159(2): 177-185, DOI: 10.1080/12538078.2012.696932.

[48] Long P.R, Nogué S, Benz D, Willi K.J. Diversity a method to remotely model and maptistibution of natural landscapes in Europe with the greatest recreational amenity value (cultural services). *Frontiers of Biogeography;* 2021; 13 (1): 1-13. DOI. 10.21425/F5FBG47737

[49] Rivas-Martínez S, & al. Mapa de series, geoseries y geopermaseries de vegetación de España. [Memoria del mapa de vegetación potencial de España]. Parte I. Itinera Geobotanica; 2007; 17: 5-436.

[50] Lazare J. J. 2009 Phytosociologie dynamico-caténale et gestion de la biodiversité. Acta Botanica Gallica; 156(1): 49-61.

[51] Rivas-Martínez S, Fernández-González F, Loidi J, Lousã M, & Penas A. Syntaxonomical checklist of vascular plant communities of Spain and Portugal to association level. Itinera Geobotanica; 2001; 14: 7-341.

[52] Rivas-Martínez S, Díaz González T.E, Fernández-González F, Izco J, Loidi J, Lousã M, & Penas A. Vascular plant communities of Spain and Portugal. Addenda to the

syntaxoniomical checklist of 2001. *Itinera Geobotanica*; 2002; 15 (1-2): 5-922.

[53] Rivas-Martínez S, et al. Mapa de series, geoseries y geopermaseries de vegetación de España. [Memoria del mapa de vegetación potencial de España], Parte II. Itinera Geobotanica; 2011; 18 (1-2): 5-800.

Chapter 10

Germination and Seedling Growth of *Entandrophragma bussei* Harms ex Engl. from Wild Populations

Samora M. Andrew, Siwa A. Kombo and Shabani A.O. Chamshama

Abstract

Entandrophragma bussei Harms ex Engl. (wooden banana) is an important indigenous multipurpose tree species endemic to Tanzania. The species has a long history of human use but recent increased utilization pressure, deforestation and high mortality rate of seedlings threaten the survival of natural populations in the wilderness. Therefore, to facilitate domestication, two experimental studies were conducted to evaluate variations in seed germination and seedling growth of three wild populations at the Directorate of Tree Seed Production Laboratories in Morogoro, Tanzania. Germination percentage, mean germination rate, final germination rate and germination index varied significantly among the populations. In terms of seedling growth there was a significant difference in number of leaves among the populations at 3 months of age. The number of course roots and seedling shoot fresh weight varied significantly among the studied populations at 10 months of age. Ruaha population had the highest survival (56%) followed by Kigwe (41%) and Tarangire being the last (36%). The two experiments have clearly demonstrated the existence of considerable variation in germination and seedling growth traits in *E. bussei*. These traits may prove to be important tools for selection of suitable seed sources for domestication and tree improvement programmes.

Keywords: Height, root collar diameter, seed, germination traits, seedling traits, Tanzania

1. Introduction

Although often not pronounced, arid and semi-arid areas are critical areas for biodiversity conservation and as sources of livelihoods to many communities. Of the total area of Tanzania, arid and semi-arid cover more than 74% of total land equivalent to 88.6 million hectares comprising about 74% of plant species found in East Africa [1]. *Entandrophragma bussei* Harms ex Engl. (wooden banana) is a high value multipurpose tree species found in arid and semi-arid areas of Tanzania. It occurs as an emergent species from deciduous *Commiphora* thicket, often associated with *Cordyla densiflora* and *Adansonia digitata* but can also occur in deciduous woodland and bushland, at an elevation range of 785–1220 m [2, 3]. Nearly all parts of the tree are used for certain purposes and the species contributes to rural and urban incomes. The species is found in the Meliaceae family and shares many of the characteristics with genuine mahogany and thus can be used as an alternative. The tree produces heavy timber with good finish and reddish-brown colour and therefore used as source of construction and handicrafts materials [4]. The tree is also used locally as a source of dye and tannin in Tanzania [5]. Research has shown that extracts of the tree parts (roots, leaves and barks) contain secondary metabolites and other lead compounds used as source of medicine to cure diarrhoea, anaemia, worms, hypertension, asthma, urinary infection, trypanosomiasis, chest and abdominal problems, and general ailments [6–10]. Most importantly, the species has been prioritized to be among the top 10 multipurpose tree species to be domesticated by the World Agroforestry Centre in Tanzania [11].

Despite the importance, *E. bussei* is increasingly threatened from over exploitation and deforestation [5]. Next to over exploitation and deforestation, high mortality rate of seedlings due to fire poses another threat to the survival of *E. bussei* populations in the wilderness. Under natural conditions, seeds of *Entandrophragma* germinate abundantly but seedling growth requires light shade environment at initial stages but after sometime they should be exposed gradually to lighter environment [2]. Germination and seedling growth information is available for other *Entandrophragma* species [12, 13], but there is inadequate information for *E. bussei*.

To design effective domestication and tree improvement programmes for high value important species with wide distribution range like *E. bussei*, knowledge on silviculture is necessary. This knowledge will not only help researchers and planners in designing the programmes but other stakeholders including plantation managers and forest and extension officers would use the knowledge to encourage the wide use of the species in afforestation/reforestation projects. This study was therefore designed with the overall objective to evaluate variation in germination and seedling growth from seeds collected in three different agroecological zones of Tanzania. It is hoped that this study will facilitate development of efficient conservation and tree improvement strategies both within and outside the natural habitats of the species. A previous study done by Andrew et al. (*in press*) was the first work to report diversity in fruit and seed morphology of *E. bussei* in Tanzania. The present study supplements the previous one by focusing on variation in seed germination and seedling growth traits of three populations of *E. bussei* found in Tanzania. Differences in germination characteristics depending on climatic zone are commonly observed for widely distributed plant species [14].

2. Materials and methods

2.1 Study area

This study was carried out between February 2020 and January 2021 at the Directorate of Tree Seed Production (DTSP) Laboratories in Morogoro, Tanzania. The DTSP is one of the Directorates of Tanzania Forest Services Agency (TFS) with the mandate to produce, procure and market high quality tree seeds and other propagating materials in Tanzania.

Morogoro region is located at 6.8278° S and 37.6591° E and experience sub-humid climate with average annual temperature of 25°C, annual rainfall of about 935 mm, relative humidity of about 75% and altitude of around 550 m.a.s.l. Seeds of *E. bussei* used in the germination and seedling growth studies were obtained from three sites (**Figure 1**) found in three agroecological zones i.e. arid, semi-arid and southern highlands, respectively (**Table 1**).

2.2 Fruits collection and processing

Fruits were collected from a total of 15 parent plus trees (5 from each site) in the three regions in August 2019 and stored under room temperature at the DTSP

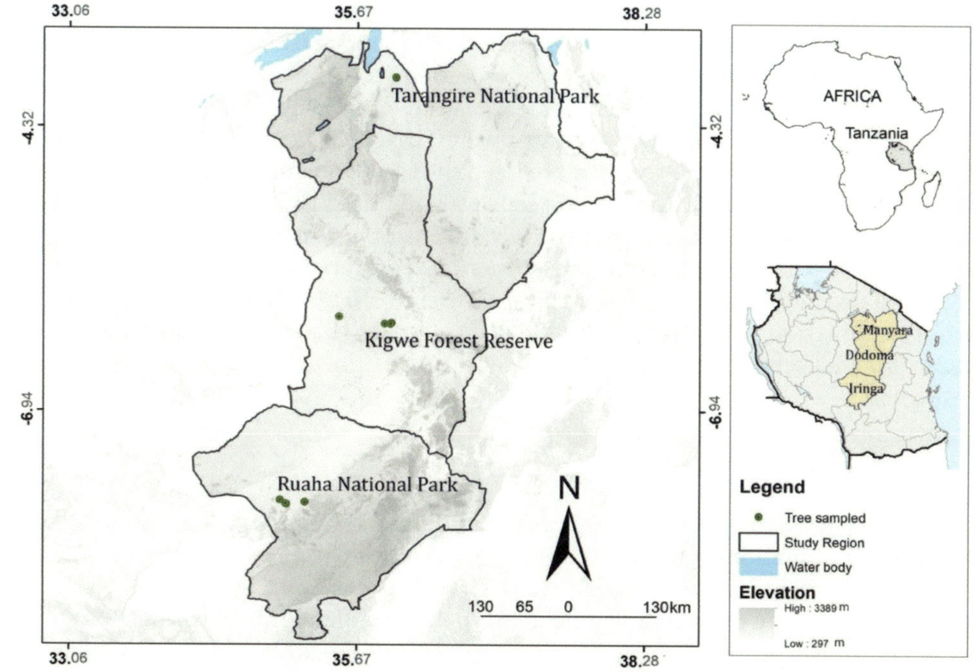

Figure 1.
Map of Tanzania showing location of Entandrophragma bussei *natural populations used in the study.*

Population	Administrative Region	Agroecological zone	Location	Elevation (m)	Rainfall (mm)
Ruaha	Iringa	Southern Highlands	7°75'S, 34°98'E	945	1100
Kigwe	Dodoma	Semi-arid	6°08'S, 35°51'E	1039	650
Tarangire	Manyara	Arid	3°87'S, 36°01'E	1195	550

Table 1.
Site data for the three wild populations of Entandrophragma bussei *used in germination and seedling growth studies in Tanzania.*

until February 2020. The parent trees had heights and diameter at breast heights (DBH) ranging from 15 to 20 m and 54–104 cm, respectively. From each plus tree, 20 ripe fruits without any damage or malformation were collected making 100 fruits per population and 300 fruits for the three populations. The collected fruits were packed and labelled appropriately and transported to DTSP for further processing. Fruits were left to dry under the house shade for 14 days to allow natural opening of the capsules. Seeds were extracted by shaking the capsules using hands and cleaned by hands to remove debris. The extracted seeds were finally stored under shade until time of use for seed germination and seedling growth experiments.

2.3 Germination experiment

Seed germination study was conducted in the germination laboratory at DTSP where temperature ranged from 10–25°C. The experiment was laid out in a randomized arrangement with three populations replicated four times. During the experiment, twelve rectangular germination trays (24 x 18 x 11 cm) that contained sand (that had been washed to remove silt and organic matter) were used for each population. In each tray, 25 cleaned seeds were sown (after the removal of the wings) to a uniform depth of 1 cm making a total of 300 seeds per population and 900 seeds per experiment. The sand was water-irrigated manually twice per day (in the morning and in the evening) to keep the sand continuously moist without becoming waterlogged. Germinated seeds were counted first on the 12th day after seeds were sown and the emergence of a visible protrusion of cotyledons above the substrate surface. During the experiment, the number of dead seeds was also recorded on the 34th day. The seeds count was done for 34 days after which no more germination was observed. At the end of the 34-day observation period, ungerminated seeds were removed and condition of the embryos was physically inspected.

2.4 Seedling growth experiment

Seedling growth experiment was also established in a randomized arrangement with three populations replicated three times to assess the development of *E. bussei* seedlings under nursery conditions. One seed was sown in open end black polythene tube of 6 cm diameter and 11.5 cm depth, filled with woodland soil mixed with sawdust in 3:1 ratio. In this experiment, 75 seedlings for each population were raised. Watering was done using watering can once a day to maintain the ideal soil moisture condition. Assessment of seedling growth traits (i.e. shoot height, root collar diameter and number of leaves per seedling) were conducted on the 30th, 45th, 60th, 90th and 105th days after sowing between April and July 2020. Seedling shoot height (cm) and root collar diameter (mm) were measured using a standard ruler and micro-calliper, respectively. After the 105th day, seedlings were maintained until when they were 303 days (10 months) old when shoot height, root collar diameter, tap root length, total number of course roots per seedling, fresh and dry weights of all course roots and shoots, were assessed from twelve (12) randomly selected individuals of each population. Seedlings had shed the leaves at the time of final assessment thus the leaves were not evaluated at the age of 303 days. To determine root and shoot parameters, the seedlings were removed from the polythene tubes and soil washed off carefully using water in the trays to avoid losing roots. Drying of shoot/stem samples was done in the oven at 95°C to constant weight. Fresh and dry weights (in grams) were then measured using a digital balance.

2.5 Data analysis

Germination period (GP) was determined as number of days from first observed germination (FOG) to where no more germination was observed (NMG) i.e. GP = NMG – FOG [15]. Germination percentage (GC) was determined as the ratio of the total germinated seeds (TGS) to the total of the seeds sown (TSS) i.e. GC = (TGS/TSS) x 100. Germination value (GV) was computed as GV = ($\sum$DGs/N) GP/10 where GV = Germination value, GP = Germination percentage at the end of the test, DGs = Daily germination speed, obtained by dividing the cumulative germination percentage by the number of days since sowing, N = the number of daily counts, starting from the date of first germination and 10 = Constant [2, 16].
Mean germination rate (MGR) was calculated as MGR = $\sum$ F/ $\sum$ FX where F = Number of germinated seeds on a particular day and X = Number of days taken for seeds to germinate. Final germination (FG) was computed as FG = GS/ Dt where

GS = Number of seeds germinated when there is no more germination and Dt = Total number of days taken for particular seeds to germinate. Germination index (GI) was obtained from GI = $\sum$ Gt/Dt where Gt = is the number of germinated seeds on day t and Dt = is the time corresponding to Gt in days. We examined assumptions of parametric test for germination and growth traits using standard diagnostic plots in package ggplot2 and Shapiro–Wilk's Test. Growth variables that did not meet the assumptions were log transformed and germination percentage was arcsine transformed to reduce skewness in frequency distribution and to improve homoscedasticity. Comparison in germination and growth traits between populations was done using One-way ANOVA and means were compared by using Tukey's Honestly Significant Difference (Tukey's HSD) post hoc test. Pearson Product Moment Correlation (r) was used to evaluate the relationship between different germination and seedling growth traits. Standard diagnostic plots were used to check appropriateness of the models to our data sets. All data analyses were done in R free software version 4.0.3 [17].

3. Results

3.1 Seed germination

Germination in the laboratory started within 12 days for all the populations (**Figure 2**). Gradual increase in seed germination was experienced until day 24 for both Kigwe and Tarangire populations after which germination levelled off (**Figure 2**). However, Ruaha population had the gradual increase in germination up to day 32 after which the germination fell off (**Figure 2**). There was no new seedling that emerged after day 34. Physical examination of the seeds at the end of 34 days germination period revealed that all the un-germinated seeds were rotten. Germination period ranged from 10 to 12 days but was not statistically different

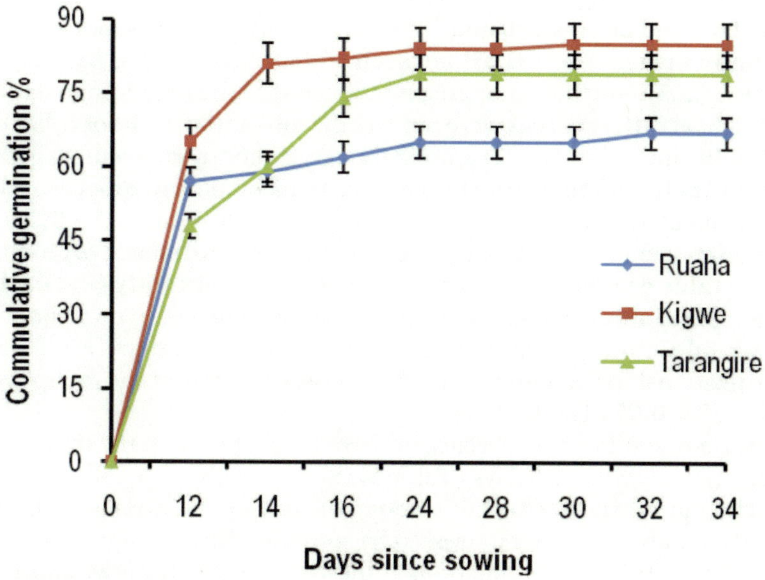

Figure 2.
Germination of seeds of three wild populations of Entandrophragma bussei *in Tanzania. Vertical bars are Standard Errors.*

($P > 0.05$) among the three populations. Ruaha population had the germination period of 12 days followed by Kigwe (11 days) and the least was Tarangire population (10 days).

There were significant differences ($P < 0.001$) in germination percentages amongst the three *E. bussei* populations with Kigwe and Tarangire populations having higher germination percentages than Ruaha (**Table 2**). There were no significant differences (in all cases $P > 0.05$) in germination period and germination value among the three study populations. The mean germination rate differed significantly ($P < 0.05$) among the three populations, with Kigwe population having the highest value followed by Tarangire and Ruaha being the last (**Table 2**). Kigwe and Tarangire populations had significantly higher ($P < 0.05$) final germination rate than Ruaha population (**Table 2**). There was a significant difference in germination index among populations ($P < 0.05$), with Kigwe having higher value than Ruaha and Tarangire (**Table 2**).

There were significant correlations among some of the germination traits (**Table 3**). Germination value had a negative significant correlation with germination index ($r = -0.41$, $P < 0.05$), mean germination rate ($r = -0.73$, $P < 0.001$) and germination percentage ($r = -0.73$, $P < 0.001$) (**Table 3**).

3.2 Seedling growth

The first measurement of seedling height, root collar diameter and number of leaves was taken at the 30th day since germination. Seedlings from Kigwe and Tarangire populations exhibited more or less the same trend of gradual increase in shoot height (**Figure 3A**). Seedlings of all the three populations had similar trend of gradual increase in root collar diameter over time (**Figure 3B**). There was a similar steady increase in number of leaves bore by seedlings from Tarangire and Ruaha populations (**Figure 3C**).

Regardless of the population, there were significant differences ($P < 0.001$) in seedling growth traits among assessment age groups i.e. 30, 45, 60, 90 and 105 days (**Table 4**). There were significant differences ($P < 0.001$) in shoot height between 30 and 90 days, and between 30 and 105 days ($P < 0.001$). The shoot height was also significant different ($P < 0.05$) between 45 and 105 days, and between 60 and 105 days ($P < 0.05$). Further comparison revealed that the remaining relationships in shoot height at different ages were not significant (**Table 4**). Root collar diameter differed significantly ($P < 0.001$) among the age groups in the seedling growth experiment (**Table 4**). There were no significant differences in root collar diameter between 45 and 60 days and between 60 and 90 days (in all cases $P > 0.05$). The rest of the relationships in root collar diameter between different age groups were significant (**Table 4**). The number of leaves differed significantly ($P < 0.01$) among assessment ages (**Table 4**). There were significant differences in number of leaves between 30 and 105 days ($P < 0.01$), and between 30 and 90 days ($P < 0.05$). The rest of the relationships in number of leaves between different age groups were not significant at $P < 0.05$ (**Table 4**).

There was no significant difference in shoot height between the three studied populations ($P > 0.05$) at 105 days (**Table 5**). Similarly, the root collar diameter did not differ significantly ($P > 0.05$) among the three populations at 105 days. However, the number of leaves possessed by seedlings differed significantly ($P < 0.01$) among the three populations at the same age, with Kigwe and Tarangire populations having higher number of leaves than Ruaha population (**Table 5**) as also attested by **Figure 3C**.

Positive and significant correlations were observed among the evaluated seedling traits at the age of 303 days (**Table 6**). The lowest significant correlation was

Population	Germination percentage (means ± SD)	Germination period (means ± SD)	Germination value (means ± SD)	Mean germination rate (means ± SD)	Final germination rate (means ± SD)	Germination index (means ± SD)
Ruaha	63.4 ± 1.3b	12 ± 1.02a	1.03 ± 0.04a	15.8 ± 3.06b	0.06 ± 0.20b	0.75 ± 0.31a
Kigwe	81.4 ± 2.4a	11 ± 1.03a	1.01 ± 0.02a	20.3 ± 2.60a	0.13 ± 0.37a	0.96 ± 0.34b
Tarangire	72.1 ± 4.2c	10 ± 0.62a	1.03 ± 0.03a	18.0 ± 4.00c	0.12 ± 0.32a	0.83 ± 0.26a

Means followed by a common letter(s) in the same column are not significantly different at P < 0.05, Tukey's HSD Test; Means are followed by the Standard Deviation (SD) of the mean.

Table 2.
Variation in seed germination traits of three wild populations of Entandrophragma bussei *in Tanzania.*

Germination trait	Germination period	Germination value	Germination index	Mean Germination Rate	Final Germination Rate
Germination value	−0.16				
Germination index	0.02	−0.41*			
Mean germination rate	0.03	−0.73***	0.16		
Final germination rate	−0.05	−0.14	0.17	0.22	
Germination percent	0.03	−0.73***	0.16	0.25	0.22

*Significant at 0.05 probability level.
***Significant at 0.001 probability level.

Table 3.
Pearson correlation coefficients for seed germination traits of three wild populations of Entandrophragma bussei *in Tanzania.*

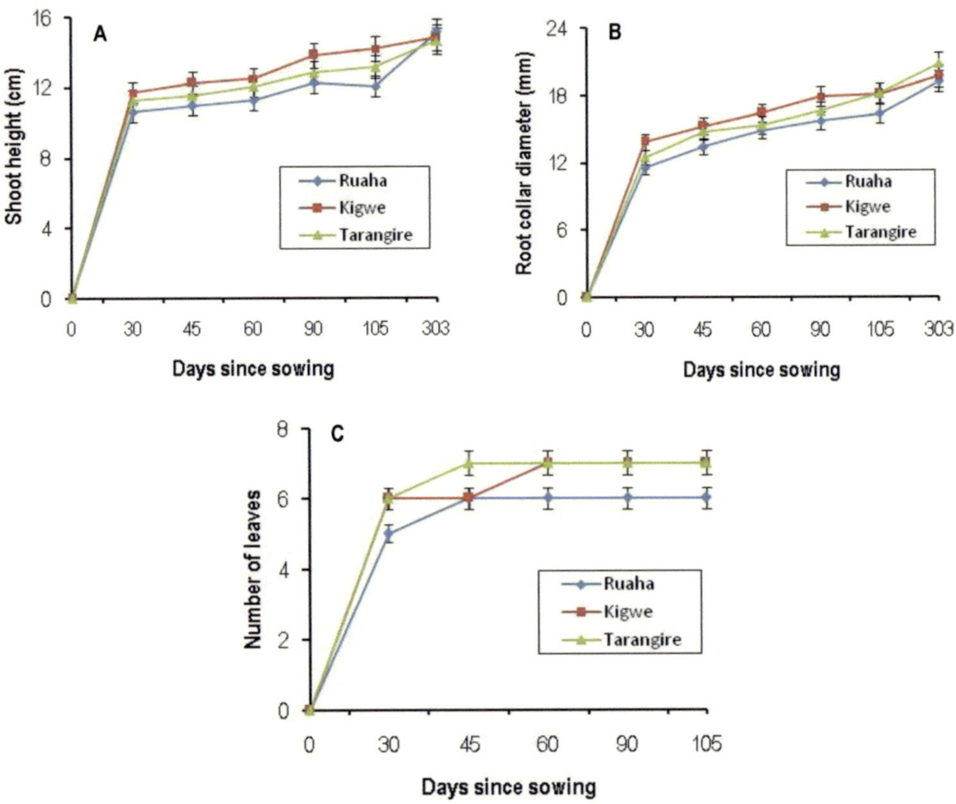

Figure 3.
Nursery growth trends of three wild populations of Entandrophragma bussei *in Tanzania. Vertical bars are Standard Errors.*

observed between shoot dry weight and root collar diameter (r = 0.35, P < 0.05) while the highest was between root fresh weight and root dry weight (r = 97, P < 0.001) (**Table 6**).

Time period	Shoot height (cm) (means ± SD)	Root collar diameter (mm) (means ± SD)	Number of leaves (means ± SD)
Day 30	11.1 ± 2.8a	12.4 ± 2.4b	5.8 ± 1.6abd
Day 45	11.4 ± 2.8ab	14.3 ± 2.7a	6.3 ± 1.8bef
Day 60	11.8 ± 2.8ab	15.4 ± 2.8a	6.5 ± 1.9bdeg
Day 90	12.8 ± 3.0c	16.5 ± 3.0 ac	6.5 ± 1.9cfg
Day 105	13.2 ± 2.9c	17.7 ± 2.9d	6.6 ± 2.1bc

Means followed by a common letter(s) in the same column are not significantly different at P < 0.05, Tukey's HSD Test; Means are followed by the Standard Deviation (SD) of the mean.

Table 4.
Evolution of seedling growth traits of three wild populations of Entandrophragma bussei *in Tanzania.*

Population	Shoot height (cm) (means ± SD)	Root collar diameter (mm) (means ± SD)	Number of leaves (means ± SD)
Ruaha	12.03 ± 3.0a	16.35 ± 3.5a	5.8 ± 1.7b
Kigwe	14.20 ± 3.2a	18.13 ± 2.8a	7.2 ± 1.4a
Tarangire	13.13 ± 2.4a	18.17 ± 2.6a	7.3 ± 1.8a

Means followed by a common letter(s) in the same column are not significantly different at P < 0.05, Tukey's HSD Test; Means are followed by the Standard Deviation (SD) of the mean.

Table 5.
Variations in seedling growth traits of three wild populations of Entandrophragma bussei *at 105 days in Tanzania.*

Trait (Unit)	Shoot height	Root collar diameter	Number of course roots	Tap root length	Shoot fresh weight	Shoot dry weight	Root fresh weight	Root dry weight
Shoot height (cm)								
Root collar diameter (mm)	0.23							
Number of course roots	0.29	0.16						
Tap root length (cm)	0.06	0.23	0.02					
Shoot fresh weight (g)	0.27	0.52**	0.52**	0.29				
Shoot dry weight (g)	0.38*	0.35*	0.31	0.28	0.68***			
Root fresh weight (g)	0.32	0.55**	0.30	0.54**	0.63***	0.59***		
Root dry weight (g)	0.30	0.49**	0.24	0.50**	0.53**	0.57***	0.97***	
Root to shoot ratio	0.15	0.17	−0.16	0.27	0.19	−0.12	0.53**	0.58***

Significant at 0.05 probability level.
**Significant at 0.01 probability level.*
***Significant at 0.001 probability level.*

Table 6.
Pearson correlation coefficients for the seedling growth traits of three wild populations of Entandrophragma bussei *at 303 days in Tanzania.*

Growth traits	Ruaha (means ± SD)	Kigwe (means ± SD)	Tarangire (means ± SD)
Shoot height (cm)	15.1 ± 2.2a	14.8 ± 1.9a	14.6 ± 1.7a
Root collar diameter (mm)	19.2 ± 3.0a	19.7 ± 3.8a	20.9 ± 3.5a
Number of course roots	12 ± 3.4b	10 ± 4.1a	13 ± 6.2b
Tap root length (cm)	13.6 ± 3a	13 ± 3.6a	12.3 ± 3.4a
Shoot fresh weight (g)	26.9 ± 6.6b	22.9 ± 9.7a	28.8 ± 5.4b
Shoot dry weight (g)	3.5 ± 1.0a	3.5 ± 2.2a	3.6 ± 1.1a
Root fresh weight (g)	16.4 ± 8.2a	16.8 ± 7.2a	15.4 ± 8.8a
Root dry weight (g)	1.7 ± 1.2a	1.8 ± 1.0a	1.5 ± 1.2a
Root to shoot ratio	0.44 ± 0.2a	0.6 ± 0.3a	0.38 ± 0.2a

Means followed by a common letter(s) in the same column are not significantly different at $P < 0.05$, Tukey's HSD Test; Means are followed by the Standard Deviation (SD) of the mean.

Table 7.
Variation in seedling growth traits of three wild populations of Entandrophragma bussei *at 303 days in Tanzania.*

Most of the evaluated seedling growth traits did not differ significantly ($P > 0.05$) between the three populations on harvesting at 303 days (**Table 7**). However, the number of course roots differed significantly ($P < 0.05$) between the three populations with Tarangire (13) and Ruaha (12) having the higher number of roots per seedling than Kigwe (10). Similarly, Tarangire (28.8 g) and Ruaha (26.9 g) populations had significantly ($P < 0.05$) higher shoot fresh weight than Kigwe (22.9 g) (**Table 4**). In terms of survival in the nursery, Ruaha had the highest seedling survival (56%) followed by Kigwe (41%) and Tarangire (36%) being the last.

4. Discussion

4.1 Variation in germination

This study has demonstrated the differences in germination traits in three populations of *E. bussei* found in three agroecological zones of Tanzania. Germination of less than 100% may indicate dormancy or non-viability of the un-germinated seeds. The rapid germination observed in this species may indicate that this species establishes itself in the environment as quickly as possible to take advantage of the favourable conditions and to give it competitive advantage. All of the three populations germinated well within 34 days which is ideal for nursery production.

The study species *E. bussei* is a native species which is recommended for restoration and conservation strategies since it is adapted to local habitat conditions. So germination studies facilitate better understanding of optimal conditions for germination and influence of seed populations. The observed variation in germination among populations emphasizes the need for proper seed source selection to minimize waiting time and cost of nursery operations [18]. Commercial nurseries are often established to produce large number of high quality planting stocks in the possible shortest time. This study has shown that all the *E. bussei* populations can be produced under nursery conditions to supply strong materials for planting out in agroforestry and plantation forestry. Our study has shown that *E. bussei* can germinate well within reasonable time of five weeks.

Germination is an important factor in assessing the quality of any seed and determines early seedling performance and end products standard. In this study, seed cumulative germination percentage ranged from 63.4% to 81.4% indicating that the three populations possess quality seeds that can be used during domestication and tree improvement processes. The observed variation in germination percentage among the provenances is not attributed to differences in altitudes, environmental factors (day length, temperature, light quality, water availability and altitude), and climatic conditions of particular population, since this study was undertaken in one geographic location [19]. The results overall indicate that, maternal factors associated with individual seeds from each population could explain the observed variations [20]. It has been reported that position of seed in the fruit or tree and the age of the mother plant influence seed germination ability [21]. However in this study, such effects were not studied.

In this experiment no treatment was undertaken for germination of *E. bussei* but the overall germinations for all populations were good as opposed to other species in the genus *Entandrophragma*. For example, it has been reported that *E. cylindricum* needs pre-sowing to attain higher germination while *E. angolense* requires certain substrate type for germination in West Africa [22]. In order to support establishment of a nursery for *E. bussei* domestication, genotypes with superior germination traits are favoured [23]. Kigwe provenance had significantly higher values for most of the germination traits (i.e. germination percentage, mean germination rate and germination index) as compared to the other two populations. These observed results could be attributed to by the fact that Kigwe source had heavier seeds than the other two sources (Andrew et al., *in press*). Studying variation in *Uapaca kirkiana*, Mwase et al. [24] pointed out that seeds with heavy weight were producing higher cumulative germination percentages. High nutrients reserved in seeds from Kigwe have a chance of germinating at 81.4% within 16 days (**Figure 3**). The selection of seeds from this population would be appropriate during nursery development in tree domestication, agroforestry and plantation forestry.

4.2 Variation in seedling growth

Seedling traits including shoot height, root collar diameter, number of leaves, tap root length, total number of course roots per seedling, fresh and dry weights of all course roots and shoots variations are important determinants of seedlings to be planted in the field. This study has demonstrated that there are degrees of variation among various seedling growth traits suggesting that selection of any trait for improvement would be effective [18]. Number of leaves differed significantly among the three populations, with Kigwe and Tarangire having higher number of leaves than Ruaha, at 105 days (**Table 5**). Ruaha and Tarangire had higher number of course roots and shoot fresh weight than Kigwe population at 303 days (**Table 7**). Parker et al. [25] and Assogbadjo et al. [26] reported a positive influence of large seed size and seed reserve on the establishment and early growth of seedlings. So, higher seed width, weight and length might have contributed to the observed variation of *E. bussei* populations in this study. It is known that roots support plants growth by absorbing nutrients and water so presence of higher number of course roots contributed to the higher survival of Ruaha population. It has been reported that traits displayed by seedlings in the nursery are influenced by mostly genetic rather than phenotypic origin. This experiment was set at the DTSP nursery in Morogoro for all the three provenances. It is obvious that the observed results have been influenced mostly by seed genetic makeup (which were not tested in this study though) from each individual population rather than variation of environmental factors between the populations as reported by Freigoun et al. [20] on *Balanite aegyptiaca*. During the study period of ten months, some seedling growth traits were not significantly different but showed a positive and strong correlation among them (**Table 6**). Despite such observations, it is important to consider raising the seedlings in the nursery until plantable size when seedlings are at least 25–30 cm high and with root collar diameter of 3–4 mm to allow seedling maturation and biomass increase [27].

5. Conclusion

E. bussei is threatened by recent intensified utilization pressure, deforestation and high mortality rate of seedlings and hence in situ and ex situ conservation measures are urgently needed to restore and preserve genetic diversity. On site measures can include wide planting, extension services and enforcement of laws governing forestry-agriculture interphase. Domestication of *E. bussei* from the wild should consider collection of seeds that possess swift germination and seedling growth under managed nurseries. Establishment and/or incorporation of the tree into farms or agroforestry systems would help to reduce the pressure on the remaining natural populations. In this study, Kigwe had the best germination traits with high germination percentage, mean germination rate and germination index. However, a different trend was observed during the nursery experiment where Tarangire and Ruaha performed well in number of course roots and shoot fresh weight. Ruaha had the highest survival rate at the age of 10 months. The difference observed for the two experiments could have resulted from variations in genetic factors. Results of this study pave a way for further studies to confirm the best performing population(s) for in situ and ex situ conservation and tree improvement. It is therefore recommended that seedlings in the nursery are monitored until when they reach plantable size (i.e. height of 25–30 cm and root collar diameter of 3–4 mm).

Acknowledgements

G. Ngatena, M. Said, R. Khalfani, P. Mmanda and J. Mtika assisted during fieldwork. Tanzania Forest Services Agency (TFS) supported data collection.

Conflict of interest

None.

Author details

Samora M. Andrew[1]*, Siwa A. Kombo[1,2] and Shabani A.O. Chamshama[1]

1 Department of Ecosystems and Conservation, Sokoine University of Agriculture, Morogoro, Tanzania

2 Directorate of Tree Seed Production, Tanzania Forest Services Agency, Morogoro, Tanzania

*Address all correspondence to: smacrice@sua.ac.tz

References

[1] Kideghesho J, Rija A.A., Mwamende K.A. and Selemani I.S. (2013). Emerging issues and challenges in conservation of biodiversity in the rangelands of Tanzania. Nature Conservation 6:1-29

[2] Msanga, H. P. (1998). Seed Germination of Indigenous Trees in Tanzania, Including Note on Seed Processing, Storage and Plant Uses. University of British Columbia Press, Vancouver. 292 pp.

[3] Styles, B.T. and White, F. (1991). Meliaceae Flora of Tropical East Africa. Meliaceae, 67 pp.

[4] Hall, J. S. (2008). Seed and seedling survival of African mahogany (*Entandrophragma* spp.) in the Central African Republic: Implications for forest management. Forest Ecology and Management 255: 292-299

[5] Makonda, F.B.S. and Batiho, T.A. (2018). Benchmarking Current Status of *Entandrophragma bussei* an Endemic Species of Tanzania. Final KiLi meeting Moshi, Tanzania 2nd – 4th October 2018, 6 pp.

[6] Dery, B. B. (1999). Indigenous Knowledge of Medicinal Trees and Setting Priorities for Their Domestication in Shinyanga Region, Tanzania. World Agroforestry Centre, Nairobi, Kenya.

[7] Monela, G. C., Chamshama, S. A. O., Mwaipopo, R. and Gamassa, D. M. (2005). A study on the Social, Economic and Environmental Impacts of Forest Landscape Restoration in Shinyanga Region, Tanzania. Ministry Natural Resource and Tourisms, Dar es Salaam, Tanzania 223pp.

[8] Mouthé-Happi, G., Tchaleu Ngadjui, B., Green, I. R. and Fogué Kouam, S. (2018). Phytochemistry and pharmacology of the genus Entandrophragma over the 50 years from 1967 to 2018: A golden' overview. Journal of Pharmacy and Pharmacology 70(11): 1431 – 1460.

[9] Nibret, E., Ashour, M. L., Rubanza, C. D. and Wink, M. (2010). Screening of some Tanzanian medicinal plants for their trypanocidal and cytotoxic activities. Phytotherapy Research 24(6): 945 – 947.

[10] Rao, M.R., Palada, M.C. and Becker, B.N. (2004). Medicinal and aromatic plants in agroforestry systems. Agroforestry Systems 61:107-122.

[11] Dery, B. B. and Otsyina, R. (2000). The 10 priority medicinal trees of Shinyanga, Tanzania. *Agroforestry Today*12(1):5 – 8.

[12] Adewunmi, A. O., Bolanle–Ojo, O. T. and Oladejo, A. O. (2014). Seedling growth performance of *Entandrophragma angolense* (Welw.) CDC as influenced by different light intensities. *Journal of Agriculture, Forestry and the Social Sciences* 12(2): 197 – 206.

[13] Hall, J. S., Medjibe, V., Berlyn, G. P. and Ashton, P. M. S. (2003). Seedling growth of three co-occurring *Entandrophragma* species (Meliaceae) under simulated light environments: implications for forest management in central Africa. Forest Ecology and Management 179 (3): 135 – 144.

[14] Hamasha HR, Hensen I (2009). Seed germination of four *Jordanian Stipa* spp.: differences in temperature regimes and seed provenances. Plant Species Biology 24:127-132.

[15] Asiedu, J. B. K., Van Der Puije, G. C., Taah, K. J. and Dovlo, V. (2012). Effect of some presowing treatments on germination of *Bauhinia rufescens* seeds.

International Journal of Agricultural Research 7(4): 195 – 204.

[16] Djavanshir, K. and Pourbeik, H. (1976). Germination value: a new formula. Silvae Genetic 25: 79-83.

[17] R Development Core Team (2020). R: a language and environment for statistical computing. R Foundation for 343 Statistical Computing, Vienna.

[18] Ngulube, M. R., Hall, J. B. and Maghembe, J. A. (1997). Fruit, seed and seedling variation in Uapaca kirkiana from natural provenances in Malawi. Forest Ecology and Management 98(3): 209 – 219.

[19] Vazquez-yanes, A. C., Rincón, E., Toledo, J. R. and Barradas, V. L. (1990). Light beneath the litter in a tropical forest: Effect on seed germination. Ecology 71(5):1952 –1958.

[20] Freigoun, S. A., Raddad, E. Y. and Elagib, T. Y. (2019). Provenance variation in seed morphological traits and early seedling growth of *Balanites aegyptiaca* Del. Sudan Journal of Agricultural Research 27(2): 93 – 108.

[21] Gutterman Y (2000). Maternal effects on seeds during development. In: Fenner M (ed) Seeds: the ecology of regeneration in plant communities, 2nd edn. CABI Publishing, Wallingford, pp 59-84.

[22] Ambebe, T. F. and Achankeng, N. (2019). Influence of pre-sowing treatment and substrate on germination and seedling vigour of *Acacia angustissima* (Mill.) Kuntze, *Entandrophragma angolense* (Welw) CDC and *Newtonia camerunensis* Villiers. *Journal of Agriculture and* Forestry Sciences 7(2): 25 – 35.

[23] Mkwezalamba, I., Munthali, C. R. and Missanjo, E. (2015). Phenotypic Variation in Fruit Morphology among Provenances of *Sclerocarya birrea* (A. Rich.) Hochst. International Journal of Forestry Research 2015: 1 – 8.

[24] Mwase, W. F., Bjørnstad, Å., Ntupanyama, Y. M., Kwapata, M. B. and Bokosi, J. M. (2006). Phenotypic variation in fruit, seed and seedling traits of nine *Uapaca kirkiana* provenances found in Malawi. Southern African Forestry Journal 208(1): 15 – 21.

[25] Parker W.C., Noland T.L. and Morneault A.E. (2006). The effects of seed mass on germination, seedling emergence, and early seedling growth of eastern white pine (*Pinus strobus* L.). New Forests 32:33-49

[26] Assogbadjo, A. E., Glèlè Kakaï, R., Edon, S., Kyndt, T., & Sinsin, B. (2010). Natural variation in fruit characteristics, seed germination and seedling growth of *Adansonia digitata L.* in Benin. New Forests 41(1), 113-125.

[27] United Republic of Tanzania (URT) (2021). Technical Order No. 1 of 2021. Forestry and Beekeeping Division, Ministry of Natural Resources and Tourism.

Chapter 11

Spatial Dynamics of Forest Cover and Land Use Changes in the Western Himalayas of Pakistan

Amjad ur Rahman, Esra Gürbüz, Semih Ekercin and Shujaul Mulk Khan

Abstract

The current study deals with the mapping and evaluation of forest and land use cover changes in the western Himalayas, Pakistan. These forest types include i) Moist temperate forests ii) Mixed coniferous forests and iii) Sub-tropical broad leaved forests. Moist temperate forest mostly consists of evergreen conifers with some of oaks and deciduous trees. Subtropical pine forest are mostly dominated by *Pinus roxburghii*. These forest type are mostly mixed by *Pinus roxburghii* and other coniferous species like *Pinus wallichiana* at the upper ranges in Dewal, Angoori, Nambal, Aucha and Khanitak etc. The broad-leaved subtropical forests are recorded on the hills and in the lower slopes of Himalaya near Islamabad and Rawalpindi. The high quantity of vegetation index were observed in winter season as compared to summer. The Landsat satellite images of years 1988, 1998, 2008 and 2018 were classified into land-cover units. Vegetation land decreased in the total area whereas the bare land class increased in the total. Water class further reduced and the built-up class increased up in the Murree area, Pakistan.

Keywords: Murree mountains, Remote sensing, Landsat images, NDVI

1. Introduction

In mountain ecosystems, vegetation serves the very first trophic level. Vegetation is the plant composition of any given area which possesses characteristic physiognomy including various taxonomic groups and present in a particular microclimatic space [1, 2]. Plant communities are the characteristic assemblage of plant species which is determined by the interaction of vegetation with other biotic and abiotic component and can easily be differentiated from each other (RIFFAT NASEEM [3]; RIFFAT N [4]). A plant community is a group of plants that have collective relationships with each other and their immediate environment [5]. The climate, topography and soil affect the characteristics of each plant community. The course or form of the plant community or types of vegetation is also shaped by biotic factors, especially human influences [6]. It forms a reasonably uniform layer that is distinct from neighboring patches of various types of vegetation. The nature and development of plant communities represents the conditions in which they are developed [7]. Various aspects of vegetation studies also contribute to the conservation and management of plant diversity. These studies also evaluate of the ecological impact and uses of vegetation and analyses of potential future changes [8].

The unique species aggregation of an area reflects the effects of environment on vegetation. Vegetation complex fluctuates in correspondence with the environmental fluctuations, which might be a seasonal or long term in nature [9]. Vegetation of a region strongly depend on climatic, soil and variation in disturbance levels that itself affects other factors as well [10]. Vegetation thus provides valuable information about the health of an ecosystem. The concept of vegetation can historically be demonstrated as a means of organizing plant assemblies at various spatial scales. The composition of plants has changed, mostly over time, and human activities have become increasingly concerned with the esthetic and socio-economic values of natural resources [11, 12]. The information can be used to manage an ecosystem, habitat and productivity of the area. Different environmental variables have different effects on the vegetation but all the environmental variables have a cumulative dynamic effect on plant species composition of an area [13]. Phytogeographical and phytosociological research all over the globe try to classify vegetation into plant communities based on composition, development and co-occurrence of species [14] which is important in ecological research to explore areas for the first time [15]. It deals with the species composition of plant communities, their evolution and the relationships between the species present. Gradient analyses are its complementary tools to understand functioning and description of vegetation [16, 17].

2. Vegetation dynamics

Vegetation dynamics represent the net effects of several variables, including climate, biotic interactions, abiotic environment and the level and history of disturbances. There is an emerging trend in ecological research to know how these variables interact and influence the coexistence and productivity of species over a time and space in an ecosystem. Vegetation Dynamics are complex phenomena in many ways and need to be accessed via varied angles. Such dynamics are functions of the disturbance regimes in a particular spatiotemporal range. The higher biological diversity of specific sort of vegetation over the other is sometimes supported by the natural disorders. Forest management strategies thus need the knowledge of natural disturbances within a given region [18].

Several studies have shown that natural resources in northern Pakistan, especially forest resources, are constantly decreasing because of growing human population, increase in demand of fuel and timber wood, and expansion of land for crops. Careless in collection of medicinal plants, over grazing, and mismanagement are other causes of reduction in forests. Furthermore, the actual state of forest cover in the country is controversial and been assessed quite longer time ago [19–23].

2.1 Himalayan perspective of the vegetation dynamics

Himalaya is derived from the Sanskrit word which means "abode of snow" comprising a wide-ranging consistent arch about 2600 kilometers along northern border of subcontinent from the Indus river of Pakistan [24]. The Himalaya contains the highest mountains in the world with the highest ecological amplitude [25]. The Himalaya include the most inexperienced habitat on earth that cherish varying biodiversity of forest types due to critical climatic changes, topographical and soil composition from the foothills to alpine mountaintops. The Himalayan flora is diverse and varies in the southeast from tropical evergreen forest species to thorn steppe and alpine species in the northwestern regions [26]. The lesser elevation range (901-1501 m) of Himalayas were occupy by subtropical broad leaved and mixed pine forests, leading by *Pinus roxburghii*, *Dodoneaviscosa, Olea cuspidata*, *Pinus walichiana*, *Punicafloridaand Acacia nilotica* species etc. [27]. The moist temperate forests and cool moist temperate forests were prevailed above the lesser elevation range with *Abies pindrow*, *Pinus wallichiana*, *Cedrus deodara*, *Asculus indica* and *Quercus dilatata* [28].

Pakistani Himalaya is gifted with richness of plant biodiversity. The north western Himalayan zone is one of the 18 hotspots of the biosphere. Enormous geological, geographical and climatic variations in altitude, topography, temperature, precipitation, soil condition bring subsequent diversity in forestry, horticulture and wildlife of the region. These mountain ranges consist of a series of chains that run roughly parallel to each other for long distances and cover areas including a chain of valleys, and glaciers. Few high altitudinal regions of the Himalayan forests are comparatively protected due to their remoteness and less population densities. The lower subtropical and moist temperate forests are the most severe victims of anthropogenic stress resulting in massive forest losses. Three-quarters of the western Himalayan forest cover is reported to have been disappeared in the last century [29]. Prabhakar et al. [30] projected 60% forest deterioration in the states of Garhwal and Uttarkhand in Indian Himalayan. A rapid forest cover decline from 57–23% was recorded in Nepal from 1950 to 1980 with all of Nepal's subtropical forests either severely degraded or completely lost [31]. Due to heavy deforestation in Pakistan merely 4.8% of land remains covered with forests with an enduring deforestation annual rate of more than 3% [32]. Large scale logging activities in Yunnan of Chinese Himalayan province caused 20% forest cover deterioration from 1960s to 1990s [33]. Pakistan vanished 25% of forest cover in only 15 years during 1990 and 2005 [34]. Forest loss of a total of 23% in western Himalayas and 8% in Eastern Himalaya has been assessed by using GIS and remote sensing methods in last three decades [35]. The condition in Bhutan is quite different due to strict implementation of forest conservation plans, where about 60% of the country area remains forest covered, even though restricted exploitation remains [36].

2.2 Vegetation and climate change

The dynamics of vegetation are considered as a significant indicator for the regulation of the terrestrial equilibrium of carbon and climate change. This challenge is significant for the climate change assessment. Even though correlations between vegetation dynamics, temperature and precipitation have been widely studied, the correlated issues are linked to the relations between vegetation dynamics and other climatic variables. Monitoring the long term change of vegetation growth and exploring its relations with climate change is relevant for the global change study [37, 38].

Vegetation dynamics are very vulnerable to climate change in particular [39–41]. A prevalent research area has been the use of remotely sensed data to dynamically analyze the interannual variations of long-term sequence vegetation [42, 43]. Several researchers have used various vegetation indices and models to analyze and assess trends in vegetation change [44–46].

3. Quantitative ecological assessment of vegetation

Visual estimations have been used in vegetation assessment despite of more recent development of reliable numerical measurement techniques for the quantification of vegetation attributes. The evaluation of data by counting, measuring or even other ways of direct measuring is more comparable to approximate eye approximation. According to several ecologists, this does not provide a systematic way of evaluating vegetation parameters. Various methods have been suggested for this purpose in order to optimize the data collection on fields, supported via mathematical and statistical procedures in order to bring an accurate representation of vegetation [47].

The use of multivariate methodologies in the investigation of vegetation data has many advantages. Ecologists get support from computer based multivariate systematic statistic softwares to define structure in data sets and to assess effects on complete group of species. Nowadays computer technology evolving is fast and cheap [48]. Various softwares designed for vegetation sciences can be used for comprehensive models, interpretation and approaches of descriptive statistics of plant communities. Vegetation science approaches include sampling, classification of vegetation, gradients analysis and investigation of association between species distribution and their atmosphere.

Recently several ecologists have been working to determine the underlying mechanism of vegetation composition in the entire vegetation complexes. The use of multivariate statistical methods such as Canonical Correspondence Analysis (CCA), Detrended Correspondence Analysis (DCA), Cluster Analysis and other statistical techniques have advanced ecological techniques [49, 50]. Cluster analysis is a technique of classification used to characterize and combine ecological communities into associations or clusters. DCA utilizes an Eigen vector of indirect gradient method focusing on investigation of plants distribution [51–53]. As only plant plants data are needed for DCA study, it presents the results without interference. CCA, on the other hand, is a direct gradient, analytical method in which environmental factors regulate the distribution of plants. CCA is being used to establish association between plants and environmental factors [54–57]. Regression analysis is combined with either reciprocal averaging or correspondence analysis by the CCA method [58].

Vegetation assessment has been giving strong bases for improvement of ecological science for several decades. Plants at individual or at community levels in response to its environment can be measured by means of quantitative and qualitative ecology (Phytosociology). Phytosociology can also be used to explore plant community services over both quantifiable and qualitative methods from plants to community level, as it offers understanding of species diversity and significance values [59, 60]. It is a developed field that explains the diversity of the plant communities and relationship with the environment [61]. The distribution of individuals in a community of the same or dissimilar species is a function of micro environmental changes, biotic interactions, and time. Therefore, understanding of vegetation can be helpful in assessing plant assembly of species in community in a specific manner [12, 62, 63]. In ecology, natural resource management and ecosystem protection, knowledge of plant species is a crucial requirement. This understanding is essential for the evaluation of rare plant species and the development of management policies to protect and minimize habitat fragmentation [16, 64].

The investigation of plant diversity is a key concern to ecologists as it offers the foundation for global conservations policy [65]. Phytosociological procedures permit environmentalists to compute plant diversity, abundance and richness of plant species in ecosystems. These methods not only assist to comprehend almost conservation but moreover measure as indicators of specific habitat forms. In addition, important value indices (IVI) can be calculated from datasets that not merely provide an understanding of the heterogeneity of floral phenomena, but can also be used to provide an indication of plant conservation needs [66, 67]. Furthermore frequency, fidelity and constancy investigation helps to recognize the threatened plants and those habitats requiring protection [68, 69].

The use of several indicators and indices for better understanding in relation to anthropological activities are recommended [70, 71]. Single indicator cannot specify all aspects of biodiversity. On the basis of broad vegetation explanation and statistical investigation, indicator plants were recognized. Khan et al. [72] recognized indicator species based on the Indicator Species Analysis (ISA) in western Himalayas. At least one significant indicator was nominated from each of herb, shrub and tree layers in every community. Further vegetation studies along ecological gradients in mountains ecosystem have just matched the indices of diversity among communities. All species were treated correspondingly without seeing their ecological location and their importance in those specific ecosystems [73–75]. Plant species with greater fidelity rates were reflected to need the supreme conservation significance. Those type of species having limited distribution and perhaps patchy habitats at maximum danger [69, 76, 77].

4. Biotic and abiotic interaction and vegetation

Biotic interaction in the vegetation is the most significant factor for many plants affecting their surrounding plant species in environment. In ecology one of the most important debate focuses on the issue of the mechanisms by which plants interact with one another. Interactions between plants - plants vary from positive (facilitation) to negative (competition) impact on neighboring plants [78]. Plants which germinate on the floor of the forest are climbers. Climbers grow by winding around, anchoring or adhering to other plants to achieve great stature for at least part of their life, or when the forest closes up around them [79].

The analysis of plant species – abiotic environmental variables is considered as an important subject in the ecological and environmental sciences. This type of interrelationship between environmental variables and forests is essential to any assessment [80, 81]. Floristic composition and its relation to environmental variables has become a recent subject of research. Multiple studies have also shown that environmental variables are mostly correlated with the vegetation patterns and distribution, comprising local topographic variables (elevation, slope, aspect), soil factors (physical and chemical properties) and anthropogenic factors [81–86]. Among abiotic variables, soil factors are the most important features affecting plant diversity and abundance in an area [87]. In general, soil factors comprising of total nitrogen, organic carbon, and clay etc. primarily regulate the distribution of vegetation patterns. Main factors affecting plant species abundance, growth are the soil nutrients and physic-chemical soil properties [86, 88–90]. Each species needs a particular nutrient contents and chemical composition to develop. The composition of these variables defines the fundamental habitat of an organism, described as the variety of conditions and resources under which individuals of a species can survive. Physiological tolerances to abiotic variables decide the boundaries of a basic habitat. Thus, abiotic variables have been found to be the strong determining features of certain plant species' development. Therefore, the distribution of plants within their environment is determined by the combination of abiotic variables, some of which are more significant than the others. On a local scale, what particular abiotic variables determine the fundamental habitat of a species? And what variables lead to increased success in one area in some habitats, and not in others? [91, 92]. Various approaches are used to determine such a complex relationship.

4.1 Geo-informatics and vegetation evaluation

Data collection to produce logical information about the dynamics of an ecosystem had been expensive and time consuming process. Consequently, our knowledge of globally important ecosystems like Western Himalayas, especially those which are found in the developing countries like Pakistan have been insufficient. However, with the invention and applications of satellite remote sensing techniques, these areas are getting international attention with detailed studies towards monitoring of biodiversity and ecosystem conservation [93]. There is a need for speedy and innovative technologies for ecosystem management, inventories and valuation of biodiversity, environmental monitoring and species habitat suitability investigation. These technologies should be based on physical factors of the ecosystem and socio-economic situations. Habitat mapping provides knowledge about quality and quantity of vegetation cover, the physical set up and anthropological interactions. Technological development in the area of remote sensing and GIS holds the promise to collect and integrate different levels of information [94].

Remote sensing data provide evidence with respect to location and extent of available areas and its spatial dissemination for execution of several problems. In the recent time relations between ecology and remote sensing have been considerably increased because of developments in imaging spectroscopy [95, 96]. GIS is associated with a powerful reference base or geographical locations including maps of vegetation, topography, soil, bird migration, hydrology and distribution of other wildlife. Locating various features related with attributes could allow various data sets to be combined and compared. It may also be analyzed in a particular data-base to create new relationship between environmental properties and the diverse biota. GIS, therefore, is an effective and powerful means of tool to communicate a wide range of data within the shortest possible time scale [97]. GPS is a ground based satellite and radio navigation system that facilitates the user to fix the accurate positions on the surface of earth. Therefore GPS and remote sensing have given rise to the beginning of more accurate and geographically referenced data sets for improved analysis [98, 99].

Knowledge about the distribution and status of species are important for wildlife research and conservation policies. Remote sensing and GIS are progressively used in monitoring flora and fauna habitats. In order to discover potential habitats for species such as the hamadryas baboons (Papiohamadras) in Eritrea, for example knowledge about the distribution of the main habitat features such as food sources, water supply, sharp cliffs and elevation of the area were digitized from topographical maps and remote data sets. It is demonstrated that locations with a mixture of these features are deliberated to be potential habitat for the species concerned [100]. Therefore, remote sensing and GIS are broadly used to discover potential habitats, digitize the information and then mapping the appropriate habitats. There has been a quick rise in the usage of remotely sensed evidence for biodiversity assessment, land management, wildlife ecology, aquatic ecology as well as observing the effects of greenhouse gases and additional environmental problems [93].

Satellite imagery is a valuable basis for land use land cover information and urban land cover has been recognized and diagramed by remote sensed data with a reasonable spatial resolution [101, 102]. In the current years there has been a growing understanding of the impacts of geographical features in ecosystems. In specific important factors like spatial and scale configurations have become progressively significant in a huge range of ecological studies [103]. Remote Sensing currently provides ecologists and other scientists with regular information on the earth and its atmosphere at the regional to global scale. GIS provides resources to collect evaluate and visualizes spatial data containing those resulting from remote sensing altogether with related innovations in computational specialist tools and facilities [104, 105].

Spectral resolution of the Remote Sensing method has high potential for monitoring land use land cover behavior, natural resources environment and hazards of land degradation in forest areas in Pakistan. Remote sensing and GIS can subsidize to observing land use land cover in comprehensive means. Remote sensing has regularly been used to develop land covers evidence. Variations in land use and land cover are main variables affecting environmental system. Land use land cover types fluctuate significantly in their bio-geochemical cycling and hence information of their distribution is imperative in many ecological modeling studies. Land cover changes have major effects on matters fluctuating from climate change to biodiversity management. Given that the remotely sensed response is principally a function of land cover type there has been substantial interest in utilizing remotely sensed data sets as a source of evidence on land use land cover [106]. GIS and RS role is very significant to evaluate the spatial and temporal forest and urban land use classes. Zafar et al. [107] assessed land use variations by using Remote sensing datasets for zonation organization of Margalla Hills National Park, Islamabad Pakistan on the base of diverse environmental variables. Wardlow et al. [108] utilized time series remotely sensed data for judgment of crops and related land cover types. Wheat harvest was projected based on the analysis and interpretation of the images. Ashraf et al. [109] investigated satellite imageries datasets of drought and post-drought (2001 to 2006) phases in order to evaluate variations in vegetation cover and land uses over hybrid (digital and visual) explanation method.

Progressively organizations involved in forest conservation and management are utilizing this expertise to capture and analyze spatial occurrences. In conservation biology the emphasis has currently moved from individual species to entire ecosystems. GIS and Remote sensing methods could be utilized for inventorying, assessing and monitoring terrestrial biodiversity at local landscape and community ecosystem ranks [110]. Gap examination is a GIS based technique to identify breaches in the safeguard network [111]. In a gap investigation of Western Ghats, India, Ramesh et al. [112] establish that numerous regions of high biodiversity were omitted from the highest stages of protection. Recent developments in GIS and Remote sensing technologies have made it promising to quantify forest biodiversity from satellite imageries.

5. Vegetation of Pakistan and study area

The vegetation types of Pakistan can broadly be divided into; a) Tropical:- Tropical dry deciduous forests, Littoral and Swap forests, Tropical thorn forests, b) Sub-tropical:- Sub-tropical pine forests and Sub-tropical broad-leaved evergreen forests, c) Temperate:- Himalayan moist temperate forests and Dry temperate forests, and d) Alpine:- Alpine scrub and Sub-alpine forests [113].

Administratively study area is divided in six parts; 4 sub-divisions (Sambli, Ghora Gali, Sehr Bagla, and Lower Topa) and 2 ranges (Ban and Municipal range) shown in **Table 1**. Murree forest division is approximately 44% of the Murree Tehsil which contains of almost 19,135–20127 hectares of forested land (State owned).

Murree is the most famous hill station in Pakistan. Murree located at 33–34° north latitudes and 73.30° east longitudes and lies 50 kilometers northeast from Islamabad, Pakistani capital at an easy elevation ranges changing from 500 to 2300 m in the Himalayan foothills.

Murree is a mountainous region founding portion of the outer Himalayas in western side Pakistan. It comprises of four progressively growing foothills (**Figure 1**).

Subdivision/Range	Area (ha)
Sambli Subdivision	5,369.98
Ghora gali Subdivision	4,606.61
Sehr Regla Subdivision	3,182.12
Lower Topa Subdivision	2,439.12
Ban Range	2,368.53
Municipal Range	2,160.99

Table 1.
Murree area subdivisions/ranges.

Figure 1.
Murree brewery remains in the Murree.

On the highest among these is the Murree city itself situated at altitude of 2290-2300 m. Other peaks contain Patriata, Gharial and Kuldana. It is delimited in the east by River Jhelum, Haripur and Abbottabad districts of Khyber Pakhtunkhwa (KPK) to the West and North, in southwest Islamabad Capital Territory and in the South KotliSattian town of Rawalpindi district. Murree municipality was built in line with the European municipalities with Churches in the center and main The Mall road running along with commercial areas and organizational offices around this. The Mall road was and is still the center of charm. Non-Europeans were not permissible to entrance to Mall road till the independence in 1947.

6. Murree Forest division (MFD)

Murree Forest Division is part of Ecoregion of Western Himalayas which is familiar as among the Ecoregions of the World (referred to as G200/Global 200) based on biodiversity and ecological significance. Murree Forest Division is a famous hill station and very well-known tourist hotspot in Western Himalaya of Pakistan. It is situated along Islamabad - Kohala highway, 30 kilometers northeast of Islamabad. Murree is one of the large tehsil of District Rawalpindi, Punjab. Geographically MFD is lies and centered at 33°52′26.34″ north and 73°23′42.21″ east (**Figure 2**). MFD is located in diverse ecological zones from Himalayan moist

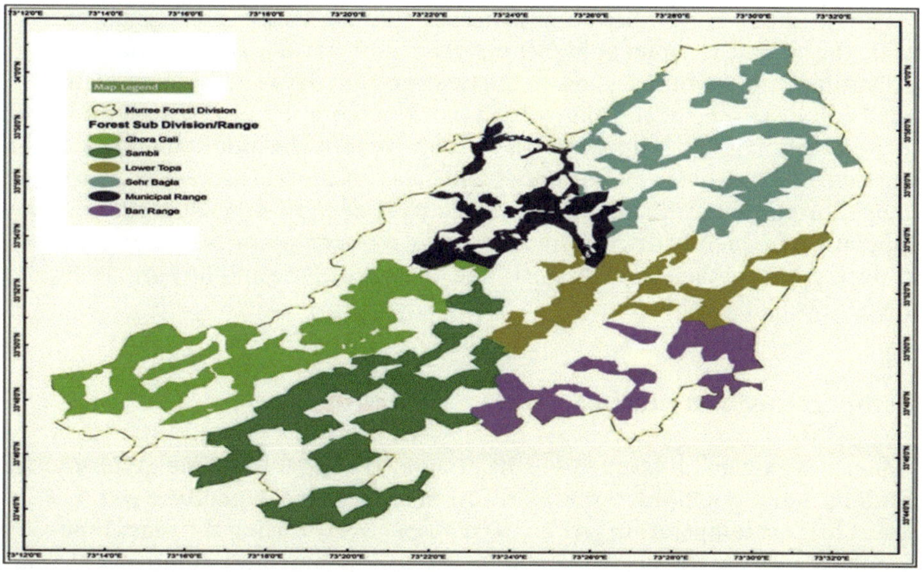

Figure 2.
Administrative division of Murree.

temperate to Broad-leaved deciduous forest at lower elevations [114]. İts elevation ranges 500-2380masl (1700-7800 ft). Murree hills came in to existence by the collision of Indian plate and Eurasian platen by a rapid raise in early Ecocene era [115]. The area gives secenic view and is important in having compact forest at higher elevations which typically includes Cedrus, blue and chir pine forests.

7. Geology, geography and climate

The Himalayan range which is 2500 km long was formed during Eocene period by the collision of the Eurasian plate with Indian plate the along a junction region about 20 million year back constructing quickly elevating zones [34, 116]. Murree region comprises of brittle rocks with hard gray reddish sandstones interbedded with soft red calcareous shales and alluvial deposits belongings to Shiwalik and Sirmar series of sub-Himalayas structure [116–119]. This region fall in tertiary and quaternary sediments dominate the zone with extensive rock formation of Shiwalik nature [120, 121]. Sedimentary rocks of the Murree area are highly mutilated due to active geological faults and tectonic pressures. These rocks have the uppermost affinity to landslide hazards [116, 122]. The Murree city is built in the European style that is the reason that has Church in middle and marketable regions alongside the main Mall road. At the time of foundation it comprised of only five major regions but with the passage of period there were several territorial variations in Murree area. At present Murree is divided into fifteen union councils (UCs) and cantonment zones. Some of the UCs contain Murree are Angoori, Ghoragali, Charhan, Dewal, Phagwari, Mosyari, Potha Sharif, Nambal, Tret and Sehrbagla etc. Another geographic significance of the Murree is that it links the Punjab plains to the Kohala, Azad Kashmir, and Abbottabad, KPK province.

Climatically the Murree hills are in divided into diverse regions from higher temperate zone to subtropical lowlands. Such a varied climate in a small geographical area to a diverse topography is triggered by the differences in elevation, depth of the snow accumulation amount of snowfall in winter and changing vegetation etc. The weather has four distinct seasons i.e. summer, winter, spring, and autumn. Usually the climate is cooler at higher elevation and warmer at lowlands with a short autumn and spring seasons. In the western Himalayas higher mountains located at the opening of the hills act as a block to the summer monsoon and bound its dissemination into the upper north western parts of the mountains [123, 124]. Winter normally start in the December and gets considerable snowfall. Murree and its neighboring parts are shielded with thick layer of snow in most of the winter. Temperature frequently drifts round about the freezing point. Summer duration lasts during May to the end of August. The months of June and July are the topmost tourists' season in the Murree.

8. Temperature and rainfall

Murree region exhibits extensive differences in temperature due to substantial altitudinal and topographic variations. The mean lowest temperature was 4–9°C whereas highest temperature was 27–30°C respectively during the years 1988–2018 (Metrological Department Islamabad, Pakistan). The warmest month of the year was June, 2018 with an average maximum temperature of 30°C. Spring season in Murree area lasts from March to the middle of May. Maximum temperature in this period fluctuates between 12 and 20°C and minimum 4–9°C respectively. Monsoon winds are the leading source of rainfall. The Murree hills receives the highest quantity of precipitation in Pakistan with an average of 1,640 mm- 1,904 mm and nearly 89 mean rainy days per year [117, 125, 126]. On the other hand several parts of the region receive fluctuating quantities of precipitation. The majority of the rainfall is received from July to August during the monsoon.

9. Murree bio-physical environment

Ecological research on habitat forms has not been performed comprehensively in some of the mountainous regions of Pakistan, particularly in the Himalayan regions. For the first time [113] designated the forest types of Pakistan using the wide-ranging groups: alpine scrub, subalpine forests, Himalayan dry temperate, Himalayan moist temperate, subtropical pine forests, tropical thorn forests, dry sub-tropical forests and marshlands etc. [28] well-defined key habitat forms as; cold desert, alpine vegetation, dry temperate, subalpine forests, moist temperate forests, subtropical forests, subtropical semi evergreen forests, tropical dry deciduous forests, tropical thorn forests and tropical swamplands. All these vegetation forms excluding the swamplands are characterized in the northwestern parts of the Pakistan [28, 113]. The Murree Mountains are placed on the foothills of the Western Himalaya, Pakistan and therefore forms a portion of the internationally acknowledged Western Himalayan floristic province (G200) of western Asiatic Irano-Turanian sub region. Its floristic, climatic, geological, geomorphological and geographical setting makes it among the unique biodiversity hotspots. This offers a particular phyto-geographical importance to the Murree hills and its flora.

10. Urban and rural division

Most of the urban populations live in the areas of two cantonments and Murree City (**Figure 3**). Permanent urban residents are few and most of the urban parts have private corporations, rest houses of government and summer resorts of elite

Figure 3.
Urbanization expansions in the Murree City.

class. Other important commercial institutes are the General Post Office, tailors and millinery and general merchants. Murree Brewery was established in 1860 at GhoraGali to satisfy to the drinking desires of the British in the area. About 88% of the rural population lives in small villages spread over all the top of Murree hills. The village residents have easy convenience to the local primary and secondary schools, clinics and bazaars. But water supply to the families of villages has constantly been a problem. People are migrating mostly for new job opportunities, the lack of other supplies like tap water, gas, roads and to meet the higher education needs of their children. Major portion of these people migrate near the low rental settlements of Rawalpindi and Islamabad.

11. Education and literacy

As per 1998 census Murree was reported with 69% literacy rate in the age of ten. Murree region is among the well-educated parts in Pakistan and simply exceeds main cities in this respect. Ausia area having education rate of 82.7% in the populations of just 4450 residents is among the maximum literate areas in the country. There is undoubtedly no other rural village of same extent with such extraordinary literacy rate anyplace in Pakistan. In the beginning primary schools were established in the Ausia, Murree, Karor and Tret. There has been one degree college (for girls and boys each) a present in the Murree area. At Phagwari region, one additional girl's college was established. Moreover there are 112 boys and 109 girls primary schools, two boys higher secondary schools at Tanda and Ausia, 16 boys secondary and 6 girls high schools whereas 12 boys middle and fifteen girls middle schools.

Murree too is renowned for its elite academic institutions that have attracted student from all over the region. Lawrence College was founded as Memorial Asylum (Lawrence) around 1860 at GhoraGali for kids of retired or serving British armed forces far from tropical environment of subcontinent. It was portion of four such schools chain which established through British India. Lawrence College is covering a space of more than 150 acres at an elevation of 1950 m and provides education from grade 1 to O and A level and is very famous among elite class in Pakistan (**Figure 4**).

Figure 4.
Lawrence College, GhoraGali, Murree.

Figure 5.
Front look of convent of Jesus and marry building.

Other colleges established in the course of the British for the children of British colonist but nowadays serving the Pakistani aristocracy includes Convent of Jesus and Marry and Saint Dynes (**Figure 5**). Saint Dynese has in the recent times closed its lodging to in order to accommodate the necessities of native population. After independence further schools established contain Army Public School, Cadet College Murree at Pindi Point and Cadet College Lower Topa. It would be discriminating not to discuss Murree Christian School which was unfortunately exposed to terrorist attack during 2002. Murree Christian School situated at Garial near JikaGali assist educational necessities of kids of missionaries who serve in Pakistan. It receives admission from grade 4–12 and is open for kids from Christian family even if they work in other occupations. It is head office of the Murree-Town (sub-division) of Rawalpindi district, Punjab of Pakistani.

12. Livelihood

Livelihood in most of the remote areas in the Murree Mountains is one of the challenges for survival. Typically, people have more than one kind of profession in order to keep sustainable livelihoods. Commonly each family keeps livestock grazing to meet their dairy and poultry requirements and to earn a living out of it as well. The quantities and types of livestock differ from a few to hundreds. Majority of the households keeps cows, buffalos and goats etc. The live stocks provide the livelihoods to the indigenous people for the reason that the rangelands of the Murree region are full of palatable and nutritious grass species [127, 128]. Grasses are harvested and stored which are supposed to be used as dry-fodder which is fed to cattles during the winter seasons. Now majority of the people are switching towards the adjacent cities of Islamabad and Rawalpindi for other professions.

As Murree area is mostly rain fed and hence the agricultural economy contingent to rainfall and to a particular level on water providing by mountains springs and streams. The area in Murree region is cultivated up to round 2000 m asl with fruits and cereals frequently on stepped slopes even though there are also large parts which are uncultivated and have thin soil with slight vegetation. In few parts of the Murree old ways of agriculture are still experienced where the fields are plowed with bullocks. The commonly grown crops in the Murree area are wheat, Millet, maize, Barley, Mustard, Sunflower, Turnip, Pulses, Tomato, Pumpkins, Radish, Cucumber, Lady finger and Potato etc. Fruits trees like Apple, pear, Citrus, Plum, Guava, Apricot, walnut and Peach are grown in the area [129].

Murree's livelihood also depends greatly on tourism during the January to mid-October of tourists' season. Monthly from 20,000 to 25,500 tourists visits these foothills ranges. Similarly each year more than one million tourist visits Murree area and the number is rising by 5% each year throughout the times of political serene [116]. Domestic tourism contributed Rs. 89 billion in 2018, contributed up to 30% of the total domestic travels and tourism expenditures.

13. Importance of the present study

Vegetation dynamics in the foothills of Western Himalayas have been rarely studied for ecological evaluation and vegetation dynamics. Vegetation diversity in these mountains is under huge anthropogenic pressures in the form of over grazing, poor collection approaches, flawed storing of medicinal plants, uneducated and ignorant people and unmaintainable Government policies [130]. The consistent abandoned harvesting of essential medicinal plants along with augmented habitat degradation and human interventions in their distribution zones has overwhelming effect on normal populations. Numerous essential plants having small ecological place and being constantly exploited are vulnerable to their presence [27]. In this situation basic phytosociological knowledge about vegetation dynamics and distribution is instantly mandatory to develop and launch a conservation plan.

Phytosociological studies disclose that many of the struggles in such disciplines performed individually; focusing on only one approach. The struggles done are not restricted to ecological either quantitative vegetation characteristics such as frequency, density, and cover or only floristic inventories of the vegetation but also utilized geo-informatics (GIS & RS) tools. Therefore in the current research an integrated struggle was made to study, investigate and argue both important sides of vegetation i.e. Ecological multivariate analysis and their association with environmental and anthropogenic variables and GIS & Remote sensing tools for long term historical forest dynamics assessment by land use land cover. This research work was planned to collect baseline information about phytosociology; species composition and distribution pattern, stand population structure, frequency, density, abundance and other essential species as well as community physiognomies in the western Himalayan region.

In the present study, an attempt has been made to examine and evaluate association of vegetation in relation to important environmental factors using multivariate analyses procedures. Very limited quantitative studies have been conducted so far in forests regions of western Himalaya of Pakistan to explain different population structure and forest types. Similarly very few studies have been carried out by using GIS and Remote sensing tools. Therefore this research was intended to describe quantitative description and population structure of diverse Himalayan forests in Murree Mountains, Pakistan. Another important aspect was to analyze the individual characteristics and mapping the forest vegetation into diverse forest types. Data acquired over a combination of these methods make available basic information for conservation planners and biodiversity managers to assess ecosystem services delivered by mountain ecosystem and to articulate sustainable management policies.

14. Forest types and vegetation diversity

MFD consists of three types of forests based on indicator species and environmental factors. These forest types include i) Moist temperate forests ii) Mixed coniferous forests and iii) Sub-tropical broad leaved forests (**Figure 6**).

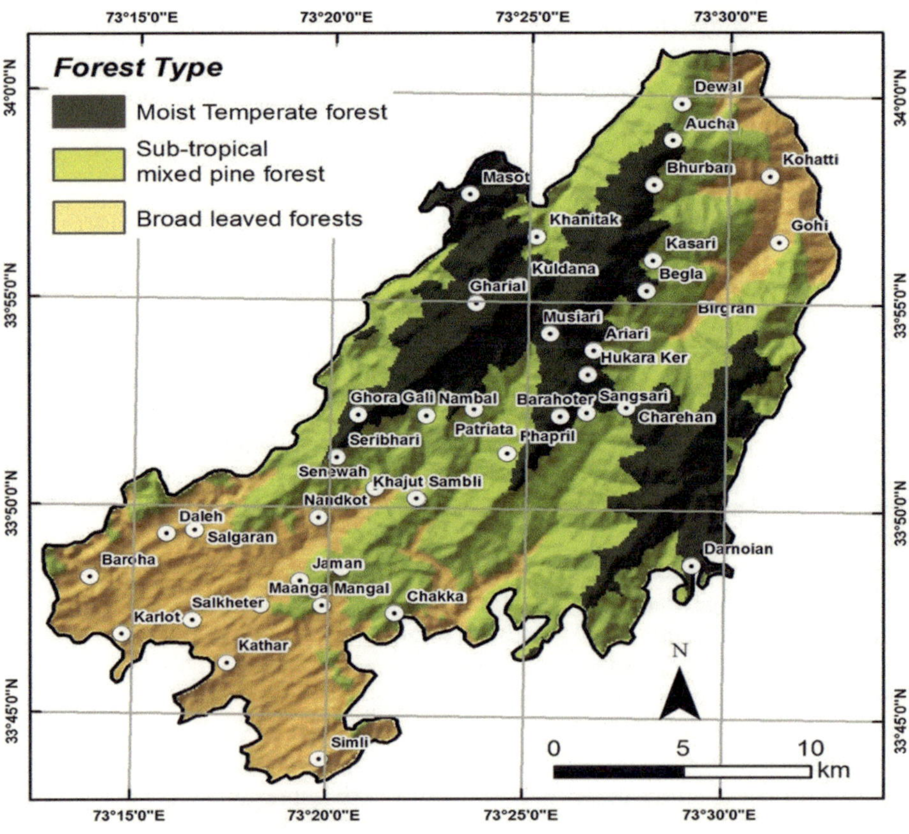

Figure 6.
Forest types in the Murree, Western Himalayas, Pakistan.

Figure 7.
Moist temperate forests in Pakistani Western Himalaya.

14.1 Western Himalayan moist temperate forest

The Himalayan moist temperate forest mostly consists of evergreen conifers with some of oaks and deciduous trees. Moist Temperate Forest occurs in the altitudinal zone ranging between 1639 m to 2078 m (**Figure 7**) and they are extended into dry temperate regions having more snowfall in winter. These forests are found in Kuldana, Gharial, Bhurban, Musiari, Masot, Ariari, Hukara Ker, Seribhari, Ghora Gali, Charehan, Darnoian, Patriata, Sangsari and Senewah.

These forests are separated into an upper and lower zones, in each of which definite species of conifers as well as in few oaks dominates. In the upper areas *Abies pindrow* and *Quercus semecarpifolia* are the dominant while in the lower zones, Deodar *(Cedrus deodara)*, *Abies pindrow*, *Picea smithiana* and *Pinus wallichiana* are the key conifer plants in order of increasing elevation, with *Quercus incana* at lower zones or altitudes and *Quercus dilatata* above 2000 m. The temperate deciduous associated tree species include *Aesculus indica*, *Prunus* cornuta, *Quercus incana, Q. dilatata* and few *Juglansregia* and *Ulmus wallichiana* etc are fairly general in these locations. The undergrowth shrubs are hardly dense and comprises of both evergreen as well as deciduous species like *Rubus ellipticus*, *Viburnum cotinifolium*, *Cotoneaster nummularia*, *Berberis calliobotrys*, *Sarcococca saligna*, *Indigofera heterantha*, *Rubus fruticosus, Rosa moschata* and *Rosa macrophylla* etc.

14.2 Subtropical mixed coniferous Forest

The subtropical pine forests found in between Himalayan moist temperate and sub-tropical broad-leaved forest. The altitudinal ranges of these forests are 1034 m to 1573 m from sea level in the Western Himalayan part and the south-west summer monsoon range (**Figure 8**). Subtropical pine forests are found in majority of the Murree hills which includes; Begla, Bara Hoter, Kasari, Phapril, Khajut, Jaman, Chakka, Birgran and Nandkot. Subtropical pine forest are mostly dominated by *Pinus roxburghii*. These forest type are mostly mixed by *Pinus roxburghii* and other coniferous species like *Pinus wallichiana* at the upper ranges like Dewal, Angoori, Sambli, Nambal, Aucha and Khanitak.

Figure 8.
Author in front of mixed coniferous subtropical forests.

The other associated species of *Pinus roxburghii* are *Quercus glauca*, *Quercus incana*, *Pistaciachinensis* ssp. *integerrima*, *Xylosmalongifolium*, and broad-leaved tree species like *Cornus macrophylla* and *Celtis australis* etc. are also found there. The commonly undergrowth shrub layer consists of *Myrsine africana*, *Carissa spinarum*, *Berberis lycium*, *Rubus ellipticus*, *Dodonaea viscosa*, *Mallotusphilippinenis*,

Figure 9.
Subtropical broad leaved forest at Daleh, Murree Hills, Western Himalaya.

Ziziphus jujuba, *Daphne papyracea*, *Daphne mucronata* and *Zanthoxylum armatum* etc. Common ground or Herbaceous flora are *Themeda anathera* and *Heteropogoncontortus*.

14.3 Subtropical broad-leaved forests

The broad-leaved subtropical forests are recorded on the hills and in the lower slopes of Himalaya near Islamabad and Rawalpindi. Subtropical broadleaved forest extended from 520 to 968 m elevation ranges adjacent to the subtropical pine forest in the upper ranges. These type of forest are relatively observed in drier and hot climate conditions with some xerophtic thorny plant species. These forests are found in the areas of Gohi, Simli, Karlot, Baroha, Salgaran, Maanga, Salkheter, Kohatti, Kathar, Daleh and Mangal etc. The subtropical broad leaved forest vegetation is mainly comprised of *Acacia modesta*, *Dodonaea viscosa*, *Woodfordiafruiticosa*, *Ziziphus jujuba*, *Berberis lycium*, *Justicia adhatoda*, *Mallotus philippensis*, *Punica granatum* and *Carissa spinarum*. Noteworthy species that are fastly being removed are *Acacia nilotica*, *Pistacia integerrima*, and *Olea ferruginea*. These forests subsequently hosts undersized shrubs which are oftenly intermittent by herbs and grasses (**Figure 9**).

15. Temporal and spatial vegetation dynamics

Vegetation in the western Himalayas experienced significant temporal & spatial changes in as shown in **Figures 10** and **11**. NDVI maps indicated that NDVI values for the Murree region in the Pakistani Himalayas varied regularly from −0.45 to 0.74 in the summer. The fluctuation of NDVI values were − 0.30 and 0.80 in the winter. It is apparent that in winter there was increased in vegetation cover due to more rainfall. The NDVI values above zero to one indicated that the forest vegetation increased in their maximum quantity.

The classification results for 1988, 1998, 2008 and 2018 are summarized in **Table 2**. Different classes along with their respective percentage on the basis of these results show the land use land cover practices observed in Murree area during 1988, 1998, 2008 and 2018. The results showed that main decline with respect

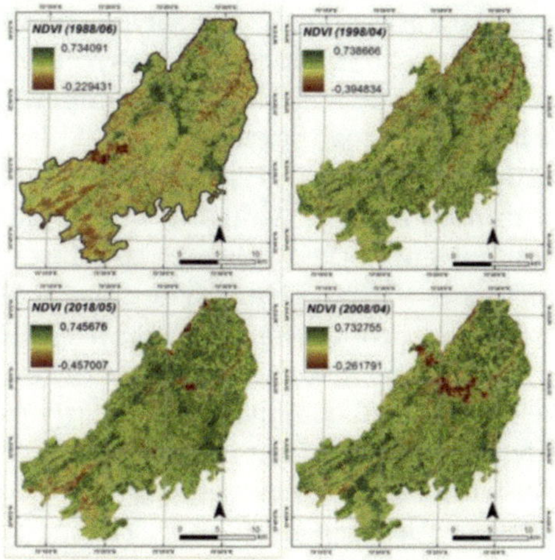

Figure 10.
Vegetation index in summer.

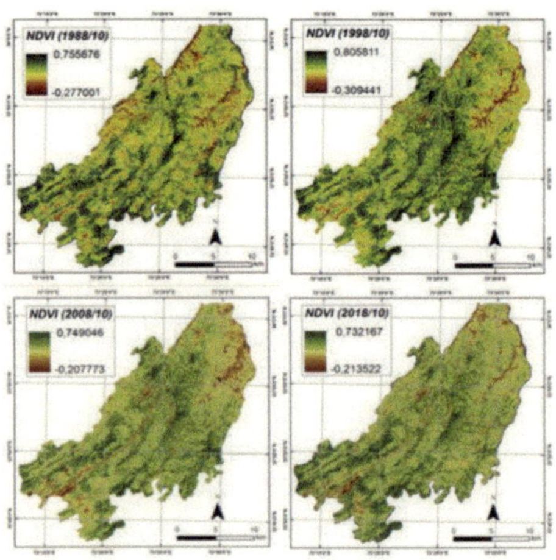

Figure 11.
Vegetation index in winter.

Year Land use	1988 (ha)	%	1998	%	%	2008%	2018	%
Forest cover	30819	71%	30772	70.9%	31720	73%	28904	66.5%
Build up	596.53	1.3%	2893	6.6%	2384.6	5.4%	4113.5	9.5%
Bare land	4586	11%	3175.6	7.3%	6794	15.6%	7123.5	16.5%
Water	452	1.04%	569	1.3%	183.65	0.42%	142.5	0.32%

Table 2.
Land use land cover statistics for four decades (1988–2018) in the Murree.

to area in Murree was observed in forest vegetation and conversely, the area of bare land, built up/Settlements area and water classes were increased. Vegetation land decreased from 71% to 66.5% of the total area whereas the bare land confronted an increase in the total share from 10.5% to 16.5%. Water class, which was minimum area covering class in 1988, reduced further area under its cover from 1.04% to 0.32% and the built up area was 1.3% of the total area which increased up to 9.5%in the Murree hills, Pakistan (**Figure 12**).

Land use land cover classification results supported the above shown facts that forest cover decreased over the past 30 years by 4.5% from 1988 to 2018. Forest vegetation in the Murree area chiefly includes moist temperate forest, mixed coniferous forest and Subtropical broad-leaved forests. This land cover class was also substituted by built up and bare land class. In addition to cutting of fuel wood by the indigenous communities, deforestation and widespread cattle grazing have malformed the vegetation present in the study area to forest patches, small bushes and several areas have left merely barren lands. The chief accelerators of forest deterioration in the area were the anthropogenic activities like due to high market value; illegitimate forest wood cutting and also the severe use of forest wood for fulfilling domestic necessities like heating and cooking and similarly for timber production. In addition to these ineffectual management of forests also played an important role in forest decrease. Causes were building new recreational areas, housing schemes and farmhouses that have been developed in the area in the past 30 years. Along with these expansions, there is rise in the construction of new roads and highways to access these areas. The area covered by water class has also observed decrease from 1988 to 2018. The land use land cover changes witnessed in all classes affected

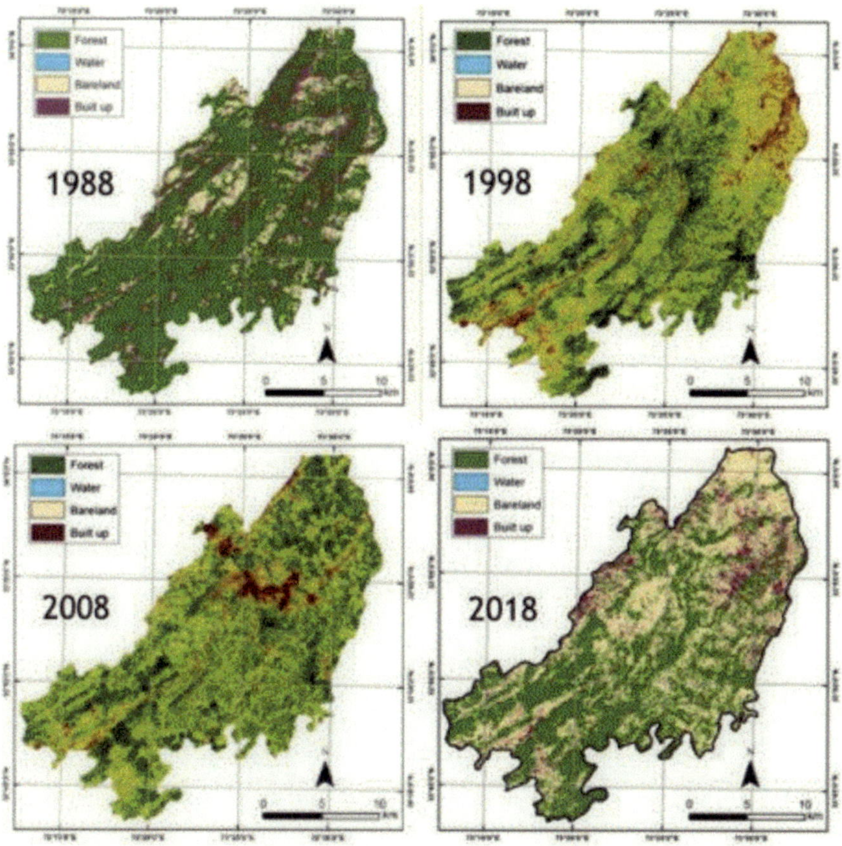

Figure 12.
Land cover classification of the study area (1988–2018).

the water class during three decades. One reason for the decrease was an increased rate of surface runoff due to deficiency of plant roots to withhold the water. Greater than before deforestation also added to the increase in surface runoff and is accountable for down flow of sediments and nutrients.

Author details

Amjad ur Rahman[1,2]*, Esra Gürbüz[3], Semih Ekercin[4] and Shujaul Mulk Khan[1]

1 Department of Plant Science, Quaid-i-Azam University, Islamabad, Pakistan

2 Department of Botany, University of Swabi, Pakistan

3 Department of Geomatics, Faculty of Engineering, Aksaray University, Turkey

4 Department of Geomatics, Faculty of Engineering, Necmettin Erbakan University, Turkey

*Address all correspondence to: rahman@uoswabi.edu.pk

References

[1] Hargreaves, P. (2008). Phytosociology in Brazil. J Plant Sci Biotechnol, *2*, 12-20.

[2] Hussain, F., Saljoqi, A., Shah, A., & Ilahi, I. (1992). Phytosociology of the vanishing sub-tropical vegetation of Swat with special reference to Docut Hills II: Spring aspect [Pakistan]. Sarhad Journal of Agriculture (Pakistan).

[3] Malik, R. N., & Husain, S. Z. (2006). Land-cover mapping: A remote sensing approach. Pakistan Journal of Botany, *38*(3), 559.

[4] Malik, R. N., & Husain, S. Z. (2008). Linking remote sensing and ecological vegetation communities: A multivariate approach. Pak. J. Bot, *40*(1), 337-349.

[5] Oommen, M. A., & Shanker, K. (2005). Elevational species richness patterns emerge from multiple local mechanisms in Himalayan woody plants. Ecology, *86*(11), 3039-3047.

[6] Grubb, P. (1987). *Global trends in species-richness in terrestrial vegetation: a view from the northern hemisphere.* Paper presented at the Symposium of the British Ecological Society.

[7] Malik, Z., Ahmad, S., & Hussain, F. (1994). Present status of subtropical chir pine vegetation of Samani Hills Azad Jammu and Kashmir. Sci. KPK. Pak, *7*, 51-58.

[8] Kent, M. (2011). *Vegetation description and data analysis: a practical approach*: John Wiley & Sons.

[9] Mandal, G., & Joshi, S. P. (2014). Analysis of vegetation dynamics and phytodiversity from three dry deciduous forests of Doon Valley, Western Himalaya, India. Journal of Asia-Pacific Biodiversity, *7*(3), 292-304.

[10] Hussain, F., Ilyas, M., & Takatsuki, S. (1997). Plant communities of Girbanr hills, Swat district, northwestern Pakistan. *Ecological Review= Seitaigaku Kenyku, 23*(4), 247.

[11] Gillison, A. (2006). A field manual for rapid vegetation classification and survey for general purposes. *International Forestry*, 3-10.

[12] Mueller-Dombois, D., & Ellenberg, H. (1974). *Aims and methods of vegetation ecology*: Wiley.

[13] Billings, W. D. (1952). The environmental complex in relation to plant growth and distribution. The Quarterly Review of Biology, *27*(3), 251-265.

[14] Ilyas, M., Qureshi, R., Akhtar, N., Munir, M., & Haq, Z. (2015). Vegetation analysis of Kabal valley, district Swat, Pakistan using multivariate approach. Pakistan J Bot, *47*, 77-86.

[15] Zhao, J., Ouyang, Z., Zheng, H., Zhou, W., Wang, X., Xu, W., & Ni, Y. (2010). Plant species composition in green spaces within the built-up areas of Beijing, China. Plant ecology, *209*(2), 189-204.

[16] Ewald, J. (2003). A critique for phytosociology. Journal of Vegetation Science, *14*(2), 291-296.

[17] Lee, Y. J., & Shukla, S. D. (2005). Pro-and anti-apoptotic roles of c-Jun N-terminal kinase (JNK) in ethanol and acetaldehyde exposed rat hepatocytes. European journal of pharmacology, *508*(1-3), 31-45.

[18] King, S. L., Antrobus, T. J., & Billups, S. (2000). Vegetation dynamics. *In: Gen. Tech. Rep. SRS-38. Asheville, NC: US Department of Agriculture, Forest Service, Southern Research Station. p. 29-31.*

[19] Butt, A., Shabbir, R., Ahmad, S. S., & Aziz, N. (2015). Land use change

mapping and analysis using Remote Sensing and GIS: A case study of Simly watershed, Islamabad, Pakistan. The Egyptian Journal of Remote Sensing and Space Science, *18*(2), 251-259.

[20] Jan, S., Hamayun, M., Khan, S. A., Ahmad, N., Ahmad, I., & Wali, S. (2015). Plant diversity of Hindu Kush mountain region of Utror and Gabral, Northern Pakistan. Pakistan Journal of Weed Science Research, *21*(2).

[21] Shehzad, K., Malik, Z., & Qureshi, R. (1999). Phytosociological survey of Samahni valley, Bhimber, Azad Kashmir (Pakistan). *Pakistan Journal of Forestry (Pakistan)*.

[22] Siddiqui, K., Mohammad, I., & Ayaz, M. (1999). Forest ecosystem climate change impact assessment and adaptation strategies for Pakistan. Climate Research, *12*(2-3), 195-203.

[23] Siddiqui, M. F., Shaukat, S. S., Ahmed, M., Khan, N., & Khan, I. A. (2013). Age and growth rates of dominant conifers from moist temperate areas of Himalayan and Hindukush region of Pakistan. Pak. J. Bot, *45*(4), 1135-1147.

[24] Krishnaswami, S., Trivedi, J., Sarin, M., Ramesh, R., & Sharma, K. (1992). Strontium isotopes and rubidium in the Ganga-Brahmaputra river system: Weathering in the Himalaya, fluxes to the Bay of Bengal and contributions to the evolution of oceanic87Sr/86Sr. Earth and Planetary Science Letters, *109*(1-2), 243-253.

[25] Dobremez, J.-F., & Jest, C. (1976). *Manaslu: Hommes et milieux des vallées du Népal central*: FeniXX.

[26] Behera, M. D., & Kushwaha, S. P. S. (2006). An analysis of altitudinal behavior of tree species in Subansiri district, Eastern Himalaya *Plant Conservation and Biodiversity* (pp. 277-291): Springer.

[27] Shinwari, Z. K., & Gilani, S. S. (2003). Sustainable harvest of medicinal plants at Bulashbar Nullah, Astore (northern Pakistan). Journal of Ethnopharmacology, *84*(2-3), 289-298.

[28] Beg, A. (1975). *Wildlife habitats of Pakistan*: Pakistan Forest Institute.

[29] Joshi, P., Singh, S., Agarwal, S., & Roy, P. (2001). Forest cover assessment in western Himalayas, Himachal Pradesh using IRS 1C/1D WiFS data. Current Science, 941-947.

[30] Prabhakar, R., Somanathan, E., & Mehta, B. S. (2006). How degraded are Himalayan forests? *Current science*, 61-67.

[31] Myers, N. (1986). Environmental repercussions of deforestation in the Himalayas. Journal of World Forest Resource Management, *2*(1), 63-72.

[32] Cronin, R. P., & Pandya, A. (2009). *Exploiting natural resources: Growth, instability, and conflict in the Middle East and Asia*: Henry L. Stimson Center.

[33] Li, D., Grueter, C. C., Ren, B., Long, Y., Li, M., Peng, Z., & Wei, F. (2008). Ranging of Rhinopithecus bieti in the Samage Forest, China. II. Use of land cover types and altitudes. *International Journal of Primatology, 29*(5), 1147.

[34] Abbasi, A., Khan, M., Ishfaq, M., & Mool, P. (2002). Slope failure and landslide mechanism in Murree area, North Pakistan. Geol Bull Univ Peshawar, *35*, 125-137.

[35] Pandit, M., Sodhi, N. S., Koh, L. P., Bhaskar, A., & Brook, B. W. (2007). Unreported yet massive deforestation driving loss of endemic biodiversity in Indian Himalaya. Biodiversity and Conservation, *16*(1), 153-163.

[36] Dhital, D. (2002). *Overview of forest policy reviews in Bhutan*. Paper presented at the Forest Policies and Forest Policy

Reviews. Proceedings of the Forest Policy Workshop.

[37] Mancino, G., Nolè, A., Ripullone, F., & Ferrara, A. (2014). Landsat TM imagery and NDVI differencing to detect vegetation change: assessing natural forest expansion in Basilicata, southern Italy. Iforest-Biogeosciences and Forestry, *7*(2), 75.

[38] Shao, Z., & Zhang, L. (2016). Estimating forest aboveground biomass by combining optical and SAR data: a case study in Genhe, Inner Mongolia, China. Sensors, *16*(6), 834.

[39] Shao, Z., Fu, H., Li, D., Altan, O., & Cheng, T. (2019). Remote sensing monitoring of multi-scale watersheds impermeability for urban hydrological evaluation. Remote Sensing of Environment, *232*, 111338.

[40] Zhang, S.-q., Zhang, S.-k., & Zhang, J.-y. (2000). A study on wetland classification model of remote sensing in the Sangjiang Plain. Chinese Geographical Science, *10*(1), 68-73.

[41] Zhang, Y., Chen, H. Y., & Taylor, A. R. (2017). Positive species diversity and above-ground biomass relationships are ubiquitous across forest strata despite interference from overstorey trees. Functional Ecology, *31*(2), 419-426.

[42] Ernst-Brock, C., Turner, L., Tausch, R. J., & Leger, E. A. (2019). Long-term vegetation responses to pinyon-juniper woodland reduction treatments in Nevada, USA. Journal of environmental management, *242*, 315-326.

[43] Teh, S. Y., Turtora, M., DeAngelis, D. L., Jiang, J., Pearlstine, L., Smith, T. J., & Koh, H. L. (2015). Application of a coupled vegetation competition and groundwater simulation model to study effects of sea level rise and storm surges on coastal vegetation. Journal of Marine Science and Engineering, *3*(4), 1149-1177.

[44] Antilla, L. (2005). Climate of scepticism: US newspaper coverage of the science of climate change. Global environmental change, *15*(4), 338-352.

[45] Cui, L., Wang, L., Qu, S., Singh, R. P., Lai, Z., & Yao, R. (2019). Spatiotemporal extremes of temperature and precipitation during 1960-2015 in the Yangtze River Basin (China) and impacts on vegetation dynamics. Theoretical and Applied Climatology, *136*(1-2), 675-692.

[46] Elhakeem, A., & Elshorbagy, W. (2015). Hydrodynamic evaluation of long term impacts of climate change and coastal effluents in the Arabian Gulf. Marine Pollution Bulletin, *101*(2), 667-685.

[47] Curtis, J. T., & Mcintosh, R. P. (1950). The interrelations of certain analytic and synthetic phytosociological characters. Ecology, *31*(3), 434-455.

[48] Hjorth, J. U. (1993). *Computer intensive statistical methods: Validation, model selection, and bootstrap*: CRC Press.

[49] Hill, M. O. (1979). *TWINSPAN: A FORTRAN Program for Arranging Multivariate Data in an Ordered Two-way Table by Classification of Individual and Attributes*: Cornell University Press.

[50] Ter Braak, C. T., & Smilauer, P. (1998). CANOCO reference manual and user's guide to Canoco for Windows: software for canonical community ordination (version 4).

[51] Chahouki, M. A. Z. (2013). Classification and ordination methods as a tool for analyzing of plant communities. Multivariate analysis in management, engineering and the sciences, 221.

[52] Dyakov, N. R. (2014). Gradient analysis of vegetation on the south slope of Vitosha Mountain, Southwest Bulgaria. Applied Ecology and

Environmental Research, *12*(4), 1003-1025.

[53] Khan, W., Khan, S. M., & Ahmad, H. (2015). Altitudinal variation in plant species richness and diversity at Thandiani sub forests division, Abbottabad, Pakistan. Journal of Biodiversity and Environmental Sciences, *7*(1), 46-53.

[54] Ahmad, S., Mohammad, N., & Khan, S. (2013). Evaluating the effecst of soil pH and moisture on the plant species of Changa Manga Forest using Van Dobben circles. Middle East Journal of Scientific Research, *17*(10), 1405-1411.

[55] Arifa, Z., Almas, J., Sheikh, S. A., & Zaheer-ud-Din, K. (2018). Floral and edaphic data recorded from Sahiwal district, Pakistan analyzed by multivariate techniques. Biologia (Pakistan), *64*(1), 173-177.

[56] Urooj, R., Ahmad, S. S., Ahmad, M. N., Ahmad, H., & Nawaz, M. (2016). Ordination study of vegetation analysis around wetland area: a case study of Mangla dam, Azad Kashmir, Pakistan. Pakistan Journal of Botany, *48*(1), 115-119.

[57] Urooj, R., Ahmad, S. S., Ahmad, M. N., & Khan, S. (2015). Ordinal classification of vegetation along Mangla dam, Mirpur, AJK. *Pak. J. Bot, 47*(4), 1423-1428.

[58] David, V. (2017). *Data treatment in environmental sciences*: Elsevier.

[59] Greig-Smith, P. (1983). *Quantitative plant ecology* (Vol. 9): Univ of California Press.

[60] Tüxen, R., & Whittaker, R. (1973). *Handbook of vegetation science*: Springer.

[61] Miyawaki, A., & Fujiwara, K. (1988). Vegetation mapping in Japan *Vegetation mapping* (pp. 427-441): Springer.

[62] Khan, S. M., Harper, D., Page, S., & Ahmad, H. (2011b). Species and community diversity of vascular flora along environmental gradient in Naran Valley: A multivariate approach through indicator species analysis. Pak. J. Bot, *43*(5), 2337-2346.

[63] Rieley, J. O., & Page, S. E. (1990). *Ecology of plant communities. A phytosociological account of the British vegetation*: Longman Scientific and Technical.

[64] Bhadra, A., & Dhal, N. (2010). Tree seedling diversity in the protected natural forest of Gandhamardan hill ranges. The Bioscan, *2*, 451-459.

[65] Pahwa, J. S., Brewer, P., Sutton, T., Yesson, C., Burgess, M., Xu, X., ... Fiddian, N. J. (2006). *Biodiversity World: a problem-solving environment for analysing biodiversity patterns.* Paper presented at the Sixth IEEE International Symposium on Cluster Computing and the Grid (CCGRID'06).

[66] Mucina, L. (1997). Classification of vegetation: Past, present and future. Journal of Vegetation Science, *8*(6), 751-760.

[67] Phillips, O. L., Vásquez Martínez, R., Monteagudo Mendoza, A., Baker, T. R., & Núñez Vargas, P. (2005). Large lianas as hyperdynamic elements of the tropical forest canopy. Ecology, *86*(5), 1250-1258.

[68] Baillie, J. E., Hilton-Taylor, C., & Stuart, S. N. (2004). 2004 IUCN red list of threatened species.

[69] Zou, D., He, Y., Lin, Q., & Cui, G. (2007). Evaluation of the level of threat and protective classification of the vegetation of Makehe Forest in Sanjiangyuan Nature Reserve, west China. Frontiers of Forestry in China, *2*(2), 179-184.

[70] Feola, S., Carranza, M. L., Schaminée, J., Janssen, J., & Acosta, A.

(2011). EU habitats of interest: an insight into Atlantic and Mediterranean beach and foredunes. Biodiversity and Conservation, *20*(7), 1457-1468.

[71] Vačkář, D., ten Brink, B., Loh, J., Baillie, J. E., & Reyers, B. (2012). Review of multispecies indices for monitoring human impacts on biodiversity. Ecological Indicators, *17*, 58-67.

[72] Khan, S. M., Page, S., Ahmad, H., & Harper, D. (2014). Ethno-ecological importance of plant biodiversity in mountain ecosystems with special emphasis on indicator species of a Himalayan Valley in the northern Pakistan. Ecological indicators, *37*, 175-185.

[73] Dasti, A. A., Saima, S., Mahmood, Z., Athar, M., & Gohar, S. (2010). Vegetation zonation along the geological and geomorphological gradient at eastern slope of Sulaiman Range, Pakistan. African Journal of Biotechnology, *9*(37), 6105-6115.

[74] Ren, H. B., Niu, S. K., Zhang, L. Y., & Ma, K. P. (2006). Distribution of vascular plant species richness along an elevational gradient in the Dongling Mountains, Beijing, China. Journal of Integrative Plant Biology, *48*(2), 153-160.

[75] Siddiqui, M. F., Ahmed, M., Wahab, M., Khan, N., Khan, M. U., Nazim, K., & Hussain, S. S. (2009). Phytosociology of Pinus roxburghii Sargent (chir pine) in lesser Himalayan and Hindu Kush range of Pakistan. Pak. J. Bot, *41*(5), 2357-2369.

[76] Haarmeyer, D. H., Schmiedel, U., Dengler, J., & Bösing, B. M. (2010). How does grazing intensity affect different vegetation types in arid Succulent Karoo, South Africa? Implications for conservation management. Biological Conservation, *143*(3), 588-596.

[77] Pinke, G., & Pal, R. (2008). Phytosociological and conservational study of the arable weed communities in western Hungary. Plant Biosystems-An International Journal Dealing with all Aspects of Plant Biology, *142*(3), 491-508.

[78] Lambers, H., Chapin III, F. S., & Pons, T. L. (2008). *Plant physiological ecology*: Springer Science & Business Media.

[79] Jongkind, C., & Hawthorne, W. (2005). A botanical synopsis of the lianes and other forest climbers. Forest climbing plants of West Africa: diversity, ecology and management, 19-39.

[80] Guisan, A., & Zimmermann, N. E. (2000). Predictive habitat distribution models in ecology. Ecological modelling, *135*(2-3), 147-186.

[81] Zhang, H., Qian, Y., Wu, Z., & Wang, Z. (2012). Vegetation-environment relationships between northern slope of Karlik Mountain and Naomaohu Basin, East Tianshan Mountains. Chinese Geographical Science, *22*(3), 288-301.

[82] Enright, N., Miller, B., & Akhter, R. (2005). Desert vegetation and vegetation-environment relationships in Kirthar National Park, Sindh, Pakistan. Journal of Arid Environments, *61*(3), 397-418.

[83] Jabeen, T., & Ahmad, S. (2009). Multivariate analysis of environmental and vegetation data of Ayub National Park Rawalpindi. Soil and Environment (Pakistan).

[84] Jones, M. M., Tuomisto, H., Clark, D. B., & Olivas, P. (2006). Effects of mesoscale environmental heterogeneity and dispersal limitation on floristic variation in rain forest ferns. Journal of ecology, *94*(1), 181-195.

[85] Tavili, A., ROSTAMPOUR, M., ZARE, C. M., & Farzadmehr, J. (2009). CCA application for vegetation-environment relationships evaluation in arid environments (southern Khorasan rangelands).

[86] Zhang, J.-T., & Zhang, F. (2011). Ecological relations between forest communities and environmental variables in the Lishan Mountain Nature Reserve, China. African Journal of Agricultural Research, *6*(2), 248-259.

[87] Hejcmanovā-Nežerková, P., & Hejcman, M. (2006). A canonical correspondence analysis (CCA) of the vegetation–environment relationships in Sudanese savannah, Senegal. South African Journal of Botany, *72*(2), 256-262.

[88] Eilu, G., Hafashimana, D. L., & Kasenene, J. M. (2004). Density and species diversity of trees in four tropical forests of the Albertine rift, western Uganda. Diversity and Distributions, *10*(4), 303-312.

[89] Heydari, M., & Pourbabei, H. (2010). Vegetation-environmental relationships and ecological species groups of the Ilam oak forest landscape, Iran. Caspian Journal of Environmental Sciences, *8*(2), 115-125.

[90] Nguyen, T. V., Mitlohner, R., Bich, N. V., & Do, T. V. (2015). Environmental factors affecting the abundance and presence of tree species in a tropical lowland limestone and non-limestone forest in Ben En National Park, Vietnam. Journal of forest and environmental science, *31*(3), 177-191.

[91] Boulangeat, I., Gravel, D., & Thuiller, W. (2012). Accounting for dispersal and biotic interactions to disentangle the drivers of species distributions and their abundances. Ecology letters, *15*(6), 584-593.

[92] Hotra, L., Kolasky, E., & Voss, J. (2003). Abiotic factors influencing the abundance and distribution of marsh marigolds along Carp Creek.

[93] Muzein, B. S. (2006). *Remote sensing & GIS for land cover, land use change detection and analysis in the semi-natural ecosystems and agriculture landscapes of the central Ethiopian Rift Valley*. Verlag nicht ermittelbar.

[94] Roy, P., & Ravan, S. A. (1994). Habitat Management for Biodiversity Maintenance using aerospace remote sensing. Tropical ecosystems: A synthesis of tropical ecology and conservation, 309-345.

[95] Roy, P. (1993). Remote sensing for forest ecosystem analysis and management. Environmental studies in India, 335-363.

[96] Schaepman, M. E. (2005). *Imaging spectroscopy in ecology and the carbon cycle: Heading towards new frontiers*. Paper presented at the Proceedings of the EARSeL 4th workshop on Imaging Spectroscopy, Warsaw, Poland.

[97] Salem, B. (2003). Application of GIS to biodiversity monitoring. Journal of Arid Environments, *54*(1), 91-114.

[98] Codjoe, S. N. A. (2007). Integrating remote sensing, GIS, census, and socioeconomic data in studying the population–land use/cover nexus in Ghana: A literature update. Africa Development, *32*(2).

[99] Milla, K. A., Lorenzo, A., & Brown, C. (2005). GIS, GPS, and remote sensing technologies in extension services: Where to start, what to know. Journal of extension, *43*(3), 2.

[100] Zinner, D. (1996). GIS and remote sensing techniques as tools for surveying primates. Ecotropica, *2*, 41-47.

[101] Cai, G., Ren, H., Yang, L., Zhang, N., Du, M., & Wu, C. (2019). Detailed urban land use land cover classification

at the metropolitan scale using a three-layer classification scheme. Sensors, *19*(14), 3120.

[102] Tapiador, F. J., & Casanova, J. L. (2003). Land use mapping methodology using remote sensing for the regional planning directives in Segovia, Spain. Landscape and urban planning, *62*(2), 103-115.

[103] Agrawal, A. A., Ackerly, D. D., Adler, F., Arnold, A. E., Cáceres, C., Doak, D. F., ... Mooney, K. A. (2007). Filling key gaps in population and community ecology. Frontiers in Ecology and the Environment, *5*(3), 145-152.

[104] Austin, M. (2007). Species distribution models and ecological theory: a critical assessment and some possible new approaches. Ecological modelling, *200*(1-2), 1-19.

[105] Osborne, P. E., Foody, G. M., & Suárez-Seoane, S. (2007). Non-stationarity and local approaches to modelling the distributions of wildlife. Diversity and Distributions, *13*(3), 313-323.

[106] Shahid, M. (2011). Monitoring deforestation and urbanization growth in rawal watershed area using remote sensing and gis techniques.

[107] Zafar, S. M., Baig, M. A., & Irfan, M. (2011). *Application of GIS/RS for management zoning of Margallah Hills National Park, Islamabad.* Paper presented at the International Conference on Environmental Science and Technology.

[108] Wardlow, B. D., Egbert, S. L., & Kastens, J. H. (2007). Analysis of time-series MODIS 250 m vegetation index data for crop classification in the US Central Great Plains. Remote Sensing of Environment, *108*(3), 290-310.

[109] Ashraf, M., Kahlown, M. A., & Ashfaq, A. (2007). Impact of small dams on agriculture and groundwater development: A case study from Pakistan. Agricultural Water Management, *92*(1-2), 90-98.

[110] Noss, R. F. (1990). Indicators for monitoring biodiversity: a hierarchical approach. Conservation biology, *4*(4), 355-364.

[111] Powell, G. V., Barborak, J., & Rodriguez, M. (2000). Assessing representativeness of protected natural areas in Costa Rica for conserving biodiversity: a preliminary gap analysis. Biological Conservation, *93*(1), 35-41.

[112] Ramesh, B., Menon, S., & Bawa, K. S. (1997). A vegetation based approach to biodiversity gap analysis in the Agastyamalai region, Western Ghats, India. *Ambio*, 529-536.

[113] Champion, S. H., Seth, S. K., & Khattak, G. (1965). Forest types of Pakistan. *Forest types of Pakistan.*

[114] CI-IAGHTAI, S., SHAH, S. H., & AKIITAR, M. (1978). PHYTOSOCIOLOGICAL STUDY OF 'THE GRAVEYARDS OF PESHAWAR DISTRICT, NWFP, PAKISTAN.

[115] Zeitler, P. K. (1985). Cooling history of the NW Himalaya, Pakistan. Tectonics, *4*(1), 127-151.

[116] Khan, A. N., Collins, A. E., & Qazi, F. (2011a). Causes and extent of environmental impacts of landslide hazard in the Himalayan region: a case study of Murree, Pakistan. Natural Hazards, *57*(2), 413-434.

[117] Farooq, S., & Malik, M. (1996). Landslide Hazard Management and Control in Pakistan-A Review.

[118] Khan, A. N. (1992). *Housing and landslides: a case study of Murree, Pakistan.* University of Nottingham.

[119] Saqib, Z., Mahmood, A., Malik, R. N., Mahmood, A., Syed, J. H., & Ahmad, T. (2014). Indigenous knowledge of medicinal plants in Kotli Sattian, Rawalpindi district, Pakistan. Journal of Ethnopharmacology, *151*(2), 820-828.

[120] Ahmad, T. (2011). Geology of the Himalayan Mountain Range, with special reference to the western Himalaya. Geologi, *63*, 142-147.

[121] Gansser, A. (1964). Geology of the Himalayas.

[122] Ishfaq, M. (1999). Application of GIS and RS to slope instability problems (Landsliding). *Northeastern Murree, NWFP, Pakistan*.

[123] Clift, P. D., Giosan, L., Blusztajn, J., Campbell, I. H., Allen, C., Pringle, M., ... Alizai, A. (2008). Holocene erosion of the Lesser Himalaya triggered by intensified summer monsoon. Geology, *36*(1), 79-82.

[124] Syed, F. S., Yoo, J. H., Körnich, H., & Kucharski, F. (2010). Are intraseasonal summer rainfall events micro monsoon onsets over the western edge of the South-Asian monsoon? Atmospheric research, *98*(2-4), 341-346.

[125] Archer, D. R., & Fowler, H. J. (2008). Using meteorological data to forecast seasonal runoff on the River Jhelum, Pakistan. Journal of Hydrology, *361*(1-2), 10-23.

[126] Association, W. M. (2013). World Weather Information Service. *Retrieved May, 30*.

[127] Shaheen, H., Khan, S. M., Harper, D. M., Ullah, Z., & Qureshi, R. A. (2011). Species diversity, community structure, and distribution patterns in western Himalayan alpine pastures of Kashmir, Pakistan. Mountain Research and Development, *31*(2), 153-159.

[128] Shaheen, H., Qureshi, R., Iqbal, S., & Qasem, M. F. (2014). Seasonal availability and palatability of native flora of Santh Saroola Kotli Sattian, Rawalpindi, Pakistan. African Journal of Plant Science, *8*(2), 92-102.

[129] Ullah, R. (2009). *Evaluation Of Soil Productivity Under Different Cropping Patterns On Sloppy Lands (Terraces) In Pothowar.* Arid Agriculture University, Rawalpindi.

[130] Hamayun, M., Khan, M. A., & Begum, S. (2003). Marketing of medicinal plants of Utror-Gabral valleys, Swat, Pakistan. Ethnobotanical Leaflets, *2005*(1), 44.

Chapter 12

Understanding Past and Present Vegetation Dynamics Using the Palynological Approach: An Introductory Discourse

Sylvester Onoriode Obigba

Abstract

Palynology is a multi-disciplinary field of science that deals with the study and application of extinct, [fossilised] and extant palynomorphs (pollen and spore) and other related microscopic biological entities in the environment. It is divided into palaeo- and actuo-palynology, and provides substantial proxies to understanding past and present vegetation dynamics respectively. With reference to the two geological principles of uniformitarianism and of the evolution of fauna/flora, the distribution of plant indicators across ecological zones, palynomorph morphology and pollen analysis, palynology can be used to identify the change in past and present local and regional vegetation and climate and humans impact on the environment. Other supportive areas of endeavour like radiocarbon dating, sedimentology, taphonomic processes and geomorphology can be used to triangulate inferences drawn from palynological data. Palynomorphs are made of outer cell walls embedded with an inert, complex and resistant biopolymeric signature (called sporopollenin) which helps to facilitate long term preservation in different environmental matrices under favourable conditions, hence its widespread applicability. Palynology have proven to very reliable in reconstructing past vegetation, decrypting essential honeybee plants and understanding the impact of climate on plant population using pollen analysis, for which is the basis for the application of palynology in environmental studies. The application of palynology in climate, vegetation and anthropogenic studies begins with the selection of matrix (sediments from lake, river, ocean, excavation, relatively intact soil profile, bee products), coring or collection of samples, subjection to a series of chemically aided digestion, separation, physical filtration, decanting, accumulating of palynomorphs, microscopic study and ends with the interpretation of recovered information. Literature review on the application of palynology for understanding vegetation and climate interactions is presented in this paper.

Keywords: Palynology, vegetation dynamics, pollen, spores, palynomorphs, palaeo-vegetation, pollen analysis, environmental reconstruction, climate, Quaternary

1. Introduction

In this review, vegetation dynamics is the succession pattern, spatial distribution, diversity and interaction of plants with humans traceable by pollen footprints in terrestrial ecosystems. Kim [1] opined that vegetation dynamics is greatly influenced by climatic factors and patterns of land use by humans. The changes in vegetation pattern occur rapidly or gradually, and palynology can be used to study these changes. Palynology provides fair playing ground for participatory and collaborative research bordering on understanding past and recent changes in the vegetation of an area. By virtue of the principles of palynology and of pollen analysis, pollen grains and spores are best applied to resolving environmental puzzles especially those related to vegetation change. It is important to emphasise that palynology is highly preferred and considered rich in providing indices on the change in vegetation in a place for several reasons; the chemically resistant compound embedded in the outer walls of pollen and spores facilitating preservation, the ubiquitous nature of pollen and spore, high pollen productivity, distinctiveness in the morphology of pollen types helping in identification of parent flora among others. The organic compound in the pollen is resistant to microbial attack, temperature regimes and pressure when buried in the soil and lastly because palynomorphs are produced in abundance, transported by wind, human, insects (and other animals) or water to different environments and ubiquitous in nature. The ubiquitous characteristic nature of pollen is a function of its productivity and there is currently very limited data on it. Unpublished data with Sowunmi [2] and Obigba [3] has revealed that the pollen productivity for *Tridax procumbens* L. is 116,270, *Ricinus communis* L. - 1.7 million, *Bombax buonopozense* P. Beauv. - 5.3 million, *Adansonia digitata* L. - 2.6 million, *Annona senegalensis* Pers. - 796,791, *Vitellaria paradoxa* C.F. Gaertn. - 793,529, *Elaeis guineensis* Jacq. - 111,640, *Vitex doniana* Sweet - 31,160, *Parkia biglobosa* (Jacq.) D. Don. - 6,306 and *Bridelia ferruginea* Benth. – 740 in a single flower irrespective of the size of the pollen grain and the flower. This is indicative of that fact that a single flower in a tree can produce millions of pollen grains and by this singular action; it is now possible to find pollen in different environmental matrices.

1.1 Palynology and its applications

The term palynology was first introduced in 1944 [4]. The word palynology was derived from two Greek words "*paluno*" meaning "to sprinkle" or 'dust' supposedly related to airborne or wind dispersed pollens and '*logos*' meaning 'study'. Palynology as simply pronounced as 'pal-uh-NOL-uh-jee' is the study of pollen, spores and microscopic sized entities of biological and uncertain origin (ranging from 5 to 500 μm). These entities have resistant cell wall capable of withstanding routine pollen analytical processes involving strong acids treatment. It is also referred to the study of fossilised and extant microscopic structures and their application in the environment [5]. These entities include pollen, spore, algae and their spores, dinoflagellates and their cysts, amoeba and acritarchs of unknown origin. Their ability to withstand the actions of strong acids (hydrochloric, hydrofluoric and sulphuric acids) is credited to the presence of a cellulosic chemical compound called sporopollenin [6], a compound word for the chemically similar CHO compound in spores – sporonin and pollen – pollenin [7].

Sowunmi [2] defined palynology as

> *'the study of extant or fossil microscopic-sized structures, palynomorphs, which cannot be dissolved by hydrofluoric and hydrochloric acids and which are generally resistant to degradation in acidic and non-oxidative sediments or deposits, their dispersal and the applications thereof'* (p. 2).

It is primarily divided into two (past – palaeopalynology and present - actuopalynology), and has become highly applicable in several other emerging fields but only those involving vegetation dynamics will be discussed here. Palaeovegetation (or palaeoecology), petrolipalynology, palynostratigraphy, archaeopalynology, forensic palynology, pharmaceutical palynology, melissopalynology, paleobotany, palynotaxonomy or systematic palynology, and aeropalynology are some of the areas of research in palynology. There are several other emerging fields of interdisciplinary research in palynology which has spanned into plant systematics, apiculture, public health, earth sciences, climatology, environmental reconstruction and archaeology, however, other supportive mechanisms like radiocarbon dating, sedimentology, taphonomic processes and geomorphology can be used to triangulate inferences drawn from recovered palynomorphs.

1.2 Palaeovegetation and environmental reconstruction

This area of research is centred on understanding past ecological dynamics in view to elucidating the past and present legacies of humans, the changes in the regional and local ecosystems, the impact of climate variability and how these information can be used in predicting future changes or current patterns. Palynology has been used for decades for understanding palaeovegetation dynamics and changes from analysing different substrates like guano deposits [8, 9], climate changes in forests [10], rock shelters [11], lakes sediments [12] or surface sediments [13]. The word 'palaeo' means 'past, ancient, old or prehistoric', so, it will be acceptable to say palaeovegetation is vegetation of the past or prehistoric vegetation. One of the way in understanding change in vegetation is by studying palynomorph abundance and variability in undisturbed stratified sediments. Two geological principles are used to support palaeovegetation studies. The first is the Principle of Uniformitarianism which proposes that the natural geologic laws or processes that exist in the present day are same or at one time were observed in the universe in the past, and these changes apply to every other area on earth. The inference is that, the earth has always had uniform changes and that the present changes can be used to uncover changes that occurred in the past and vice versa. Therefore, the changes that occurred in past in terms of vegetation are almost same as at today. The second law is the 'Evolution of fauna and flora' which says that in a vertically stratified sedimentary soil profile, the stratum on top is younger in age and formation than the one below. What this means is that, the farther the stratum down the earth, the older the soil and the closer the stratum to the surface, the younger it will be. This also implies that these strata are embedded with fossilised plant and animal remains preserved over time. In recent times, the law has been referred to as the 'principle of fauna/flora succession' [14, 15] where fossilised materials succeed themselves in the vertical strata. That is, the fossilised fauna and flora beneath evolved to the next stratum just next to it (on top) and so on till it gets to the earth surface or top soil. This order occurs in a reliable format except for disturbed and distorted soil profile. This principle is applied to paleo-vegetation up to what I called 'actuo-vegetation' using pollen analysis of each stratum referred to as sub sample. Since inception several studies have been conducted on this. Novello *et al.* [16] described how palynology was used to decipher last glacial (115,000 years before present) to Holocene (about 12,000 years before present) vegetation and environmental change in South America using cave deposits. Using certain pollen types, the palaeovegetation changes with respect to pollen abundance in sediments through routine pollen analysis are presented below. These reviews provide clues to the vegetation dynamics based on the presence or absence of pollen grains in the sediments and possible factors influencing their abundance in the sampled regions.

1.3 Melissopalynology and conservation of bee flora

Melissopalynology is the branch of palynology that deals with the study of palynomorphs in honey and other honeybee products like propolis, beewax and bee breed. The aim is to find out the botanical and geographical origin of the honey [17]. Honey is produced by the action of eusocial honeybees foraging for proteins and carbohydrates. They visit 'nectariferous' and 'polliniferous' flowers (**Figure 1**, No 6 & 8) for pollen and nectar and sweet fruits like pineapple, mango, water melon and others (**Figure 1**, No. 1–5 & 7) for natural sugars. Honeybees are regular visitors of very colourfully scented flowers for nectar or pollen because their larval and adult dietary requirements depend on it [18]. Pollen is the bee's major source of protein, fat, minerals and vitamins, while nectar is the major source of carbohydrates from which honeybees source for energy. In the course these foraging expedition, the honeybees collects pollen and other non-pollen materials (honeydew elements) co-incidentally for the production of honey in their hives. The pollen is the focus for this aspect of palynology and its usefulness for vegetation dynamics of the present day. Preliminarily, pollen grains are the male microgametophyte of either unicellular or multicellular form that is produced in the flowers with the primary responsibility of pollination and fertilisation. This invariably means they can be used to track flowering patterns for honeybee plants. Honey bees are major pollinators among flying insects and are so essential in conserving plant diversity.

Apiculture is the aspect of agriculture that covers this part of biological sciences. Melissopalynology thus deals the representation of pollen types (that is flowering plants visited by honeybees) in honeys collected and marketed for humans. Depending on the rate of foraging and seasonality of flowering in honeybees plants, honeybee farmers can collect or extract honey from honeycombs or artificially manufactured hives on a weekly or monthly basis. The taste, colour, texture and fragrance of the honey are dependent on the type of flora visited. If the honey is derived from a single flora, it is called unifloral and if it is from several floras, it is referred to as multifloral or polyfloral honey. In understanding flora dynamics, mulitfloral honeys

Figure 1.
Honey bee foraging for nectar and/or pollen: 1: Citrullus lanatus *(Thunb.) Matsum. & Nakal (water melon) fruit, 2–3:* Mangifera indica *(mango) fruit, 4–5:* Anacardium occidentale *L. (cashew) fruit. 7:* Antigonon leptopus *Hook. & Arn. (Mexican creeper) flower, 7:* Ananas comosus *(L.) Merr. (pineapple) fruit, 8:* Canthium danlapii *flower. (Source: Author original photos).*

are best for analysis. This can reflect the yearly pollen calendar for a locality where plants are cultivated or grown in the wild. Several studies have been carried out in this aspect by many researchers. Vegetation dynamics *vis-a-vis* floral diversity can be safely constructed using melissopalynology to show flowering pattern for important bee plants. Flora that needs to be conserved for enhancing health of honeybee colonies and the production of economically and medicinally important honey can be revealed through melissopalynology. A perfect example is presented in Lau *et al.* [19] paper where they studied the annual spatial and temporal dynamics in the vegetation of urban and suburban areas in Texas, Florida, Michigan and California by collecting pollen foraged by honeybees. Hence, pollen is an essential tool in the analysis of honey as it indicates the major and minor plant taxa utilised by honeybees.

1.4 Aeropalynology

The atmosphere is made of several airborne particulate matter of which pollen and spores are part of. Wind dispersed pollen and spores are released from lower green and flowering plants respectively at different times in the year and can be used to trace seasonality and presence of pollen in the atmosphere for public health reasons. Aeropalynology as the name implies is the study of airborne palynomorphs sampled through a pollen trap. This study is important if deleterious allergy triggers must be identified. One of the founding fathers of palynology, Erdtman, defined aeropalynology as the study of pollen and spores in the atmosphere [20]. In Ezike *et al.* [21], a monthly survey airborne palynomorphs in North Central Nigeria was carried for one year with the aim of finding the abundance, diversity and variation of wind pollinated flora in the region. The study attempted to the link pollen dispersal and meteorological environmental changes using a pollen trap. Fern spores, pollen types, algal cysts and diatoms were recovered with varying abundance across the year. Aeropalynology can be used to identify phytoecological groups in the atmosphere which is supposed to be a representation of the regional vegetation. In Anyigba, Kogi State, Essien and Nkang [22] recovered 47 airborne palynomorphs (pollen types) from 29 plant families through the pollen analysis after collecting samples for both dry and wet seasons. They found three major vegetation types (forest, savanna and human impacted). Thus, airborne palynomorphs can be used as indicators for regional flora or of the immediate environment. Monthly retrieval of airborne palynomorphs can be used to infer the flowering seasonality of wind dispersed or pollinated plants in that region.

1.5 Archaeopalynolgy

This area of research is in environmental archaeology where the interaction of humans with their environment (particularly the plants) in antiquity is deciphered by detail analysis of pit excavations. The archaeological materials alongside with the palynomorphs recovered are used to interpret past interactions of humans with their flora. There are anthropological studies available on this aspect involving the use of palynology e.g. farming history and prehistoric weapon production and furnace use [11, 23]. Johnston [24] mentioned that archaeologists in the course of their study find fossilised pollen and spores of different shapes in excavations; hence, archaeopalynology is the study of palynomorphs in archaeological sites in an attempt to reconstruct the ancient lifestyle (diets, farming practices, raw material sourcing), food sources, physical landscape, domestication attempts and the understanding the impact of humans on earth. The methods in the analysing archaeopalynological samples are outlined in the paper by Dontella and Federico [25]. It includes removal of organic and inorganic matter, microscopy, identification and counting. Based on the pollen types found, the vegetation types and interaction of humans with the flora can be elucidated. The law of geologic laws of uniformitarianism and flora/fauna succession is also applied here. It is important to succinctly note that pollen analysis remains the basis for the application of palynology in vegetation studies. Some important archaeo-palynological works has been carried out in Nigeria like those of Orijemie [11, 23].

2. Pollen analysis, identification and vegetation-climate interaction

2.1 Pollen (or palynomorph) analysis

The application of palynology in any field is basically dependent on the use of pollen analysis for deducing inference on vegetation-climate-human interactions. Pollen analysis is relatively laborious, time consuming and expertise demanding. The purpose for pollen analysis is to disintegrate the palynomorphs from their matrices and concentrate them for proper identification. It is only the series of analytical procedures commencing with the collection of palynomoprh embedded substrates (honeybee products – pollen pellet, honey, propolis, terrestrial and aquatic sediments, air borne particulate matter, excavations, anther from flowers, faecal matter, drug samples and rocks) from the field to laboratory processing of the substrates (**Figure 2**). Microscopy which helps to determine relative abundance of one palynomorph type in comparison to other follows immediately after the laboratory processing. Depending on the type of matrix and the aim of the analysis, different laboratory procedures are employed. For example, the qualitative study of palynomorphs requires acetolytic (chemical removal of protosplamic content using 9:1 of acetic anhydride and concentrated hydrochloric acid) processing for elucidation of exine ornamentations or patterns. This may not be necessary for quantitative study of pollen and spores in honey or sediments. They type of matrix determines the number of treatments to be used. For soil matrices, the numbers of chemical processes are more than other matrices like honey, drugs samples, or pollen pellet.

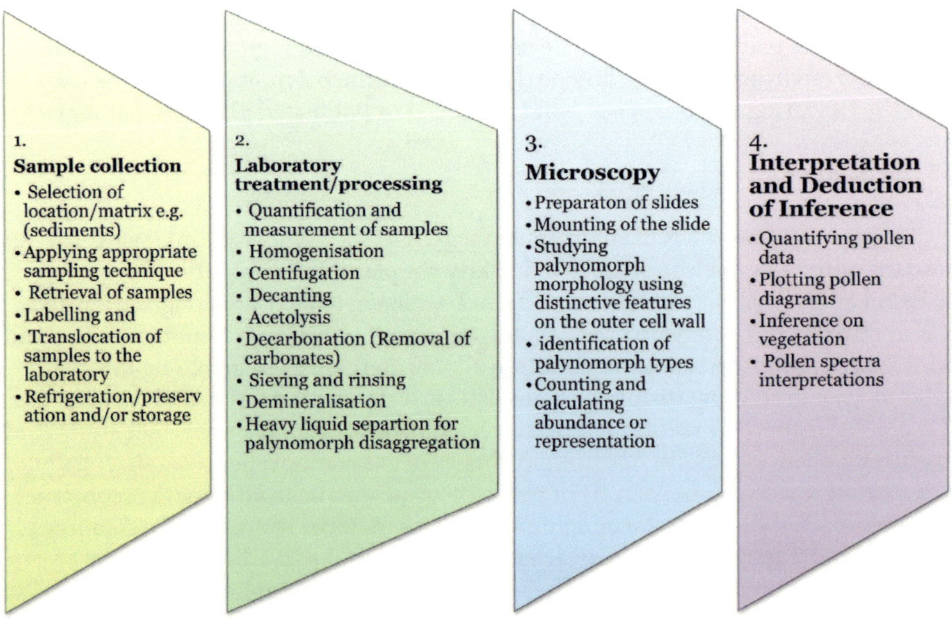

Figure 2.
Illustrating the stages involved in pollen analysis.

Categorisation	Quantification (in %)	Interpretation based on dominance of the plant species
Predominant	more than 45	Overwhelmingly abundant flora in the vegetation
Secondary	16–45	Major vegetation coverage in the area that is complimenting the 'predominant' plant species
Important minor	3–15	Very useful flora present in the vegetation
Minor	between 1–3	Scanty and insignificant in number
Present	less than 1	Rarely noticed in the vegetation

Table 1.
Palynomorph quantitative representation and interpretation in vegetation studies.

Erdtman [20] and Faegri and Iversen [26] gives full description of pollen analysis, however, the figure below shows a summarised procedure for pollen analysis of palynomorph embedded samples.

In the interpretation of results from pollen analysis, there are several limiting factors. These factors influence the representation of palynomorphs when recovered. Some of them are pollen dispersal mechanism, pollen productivity, and differential preservation capacity against environment induced deterioration. In interpreting of pollen analysis, microscopy, identification and counting are used as the quantification presented in **Table 1**. In counting and providing information about the abundance of a particular pollen type, Jones and Byrant [27] and Louveaux, *et al.* [28] formula is used as shown in **Table 1** above.

2.2 Pollen identification

The application of palynology on every other area of research is largely dependent on the accurate identification which is powered by the impeccable description of the morphological features of the pollen types recovered from the environmental matrix. Pollen grains are unicellular to multicellular units composed of a cell wall and protoplasm. The morphological features and their distinctiveness are found on the cell wall especially in the outer cell wall called the exine. The parameters used in describing pollen grains are polarity, symmetry, aperture types, pollen class, pollen size, and exine ornamentation (details in **Figure 3**). Expert experience is highly in identification since some pollen types may have close resemblance but represent different vegetation types e.g. *Lophira alata* Banks ex Gaertn. is found in tropical freshwater swamp forests and *Lophira lanceolata* is representative of wooded savanna. Pollen from the Melastomaceae and Combretaceae are almost indistinguishable, hence are classified into one group even though other gross morphological feature are different. Palynomorph identification is carried out using pollen albums, reference pollen collections and published atlases [29–31].

2.3 Late quaternary climate vegetation interaction in tropical West Africa

According to the geological time scale, the Late Quaternary is from 65,000 years BP (before present) to date. During the period, the plant communities were greatly influenced based on their response to climate change. Information on the fluctuation in mangrove and freshwater swamp forests, the relative dominance of other forms of forests and savanna is provided. As provided in details in Sowunmi [32] paper, the following were some of the vegetation-climate changes that had occurred basically in the expansion and contraction of forests and savanna and a wet-dry climate cycle. These changes were categorised into six time frames as presented in **Table 2**.

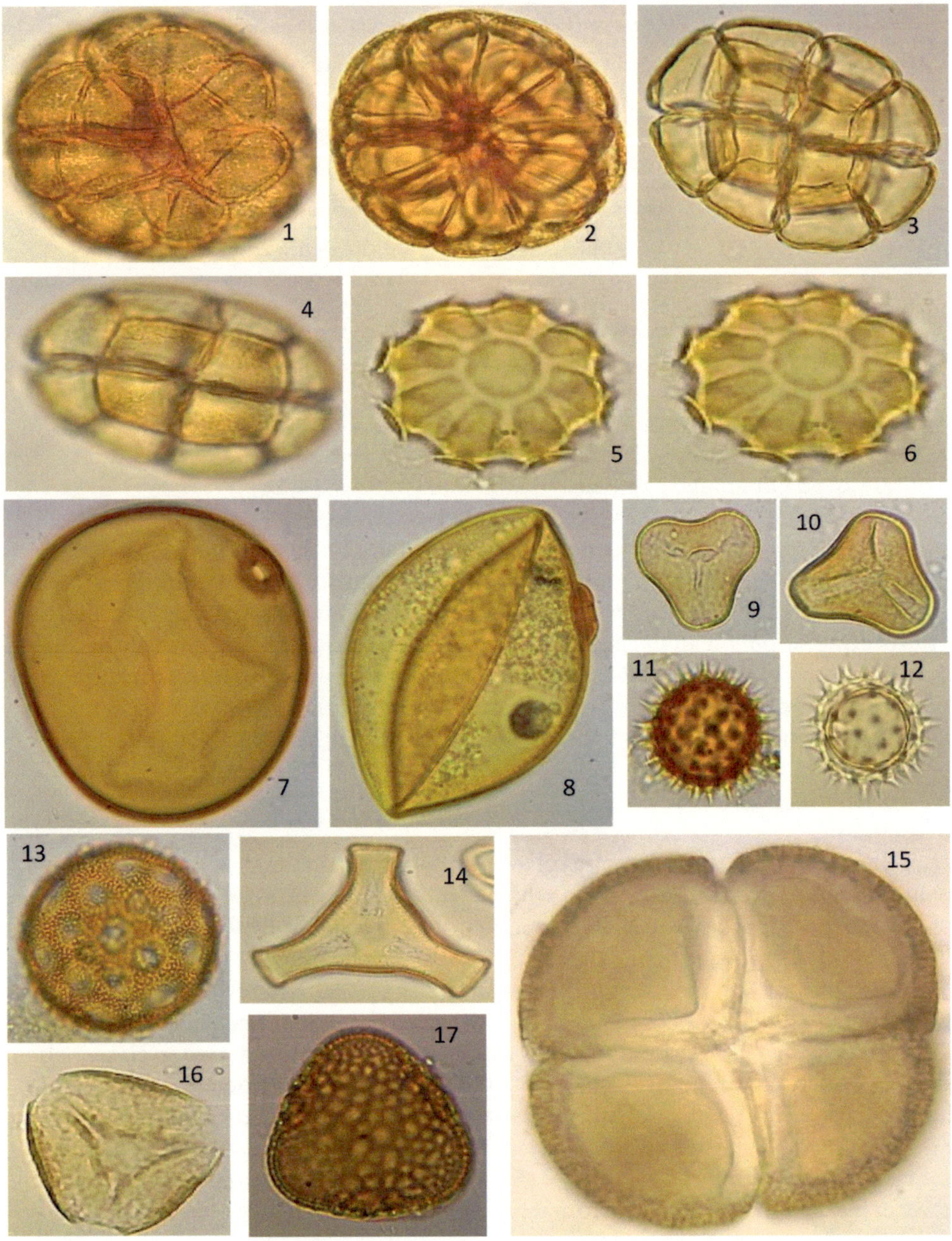

Figure 3.
Pollen grains (or types) of different flora and their morphological distinctiveness: 1–2: Parkia biglobosa, *3–4:* Acacia *sp. Martius, 5–6:* Securidaea *sp. 7–8:* Zea mays, *9–10:* Elaeis guineensis, *11–12: Asteraceae grain, 13:* Talinum triangulare, *14: Loranthaceae– cf.* Tapinanthus *sp. (Blume) Rchb. 15:* Annona sengalensis, *16:* Parinari *sp. 17:* Bombax costatum *Pellegr. & Vuill. (**Source:** Author original photomicrographs).*

3. Empirical studies

3.1 Palynomorphs, human-plant interactions and vegetation change

Anthropocentric (human cultural) and the anthropocene (climate/human induced) forces have altered ecosystems, plant growth response, habitat characteristics, and behaviour of plants in recent times. In West Africa, there are evident

Phases	Time frame	Climate	Savanna	Forest	Sea level
1	between 40,000 and 30,000 yrs. BP	Wet	Reduction of savanna	Extensive presence of freshwater swamp, distension of mangrove swamp farther south of the present day limits. Dry forests in the northern region became wetter (that is a gradual movement from drier to wetter type forest conditions)	Sea transgression which led to the re-establishment of the mangrove swamp (Inchirian transgression).
2	between 30,000 and 25,000 yrs. BP	Drier	Expansion of savanna southwards to cover forests. Deserted or Saharan landscape replaced by Sudanian vegetation farther north. Ergs formation in sahel and sudan	Destruction of forests or reduction of forests into pockets or refugia	Occurrence of the Ogolian regression, fall in sea levels
3	between 25,000–5,000 yrs. BP	Wetter		Forests re-established, became denser with increased species diversity, regrowth, expansion and extension of the Mangrove swamp	Nouakchottian transgression, rise in sea level
4	between 5,000 and 3,500 yrs. BP	Very wet and later dry	Expansion of savanna	Mangrove swamp forest disappeared (? 4000–3500 yrs. BP), forest reduction	Nouakchottian transgression continuation until 4000 yrs. BP from 5000 yrs. BP, later sea regression
5	between 3,500 and 3,000 yrs. BP	Wetter	Contraction of savannas	Expansion of forests	
6	From 3,000 yrs. to present day	Warm and dry	Sudan and Sahel savannas depreciated in vegetation.	Northern axis of the forests was replaced by woodland savanna	
	increase in human inference in the natural vegetation (the Anthropocene), burning of vegetation, animal grazing and extensive farming				

*Source: Sowunmi [32]. * BP = before present.*

Table 2.
Climate vegetation changes in West Africa in the late quaternary.

adaptive changes in certain plants including their survival and growth in diverse vegetation zones. Today, many savanna species are found growing favourably in residential areas in mangrove and fresh water swamp regions in West Africa. Although a few species are still considered useful in deciphering vegetation dynamics during pollen analysis as presented in **Table 3**. This is particularly noticed in

Vegetation types	Pollen/flowering plants						Spores/lower plants		
Savanna	*Berlinia grandiflora. Borassus aethiopum*	*Bombax costatum, Lophira lanceolata*	*Vitellaria paradoxa, Bridelia ferruginea*	*Syzygium guineensis, Hymenocardia acida, Pavetta crassipes*	*Adansonia digitata, Parkia biglobosa*	*Lannea* spp.			
Montane	*Justicia laxa*	*Justicia flava*	*Podocarpus milanjianus*	*Myrica arborea*	*Podocarpus latifolius*				
Open/disturbed forest/open vegetation	*Elaeis guineensis, Solanum torvum*	*Alchornea cordifolia*	Asteraceae, Poaceae (grasses)	*Ageratum conyzoides, Ludwigia erecta*	*Piper umbellatum*	Amaranthaceae/ Chenopodiaceae	*Sydowia polyspora*	*Alternaria* spp.	
Rainforest/ Freshwater swamp forest	*Milicia excelsa, Alstonia boonei,*	Cyperaceae, *Symphonia globulifera*	*Irvingia gabonensis*	*Piptadeniastrum africanum, Celtis brownii,*	*Uapaca* spp., *Typha dominigensis.*	*Mitragyna ciliata*	*Botryoccocus brauni, Pseudoschizaea* sp.,	*Ceratopteris* sp.	*Selaginella* sp. *Botryococcus* sp.
Mangrove swamp	*Rhizophora mangle*	*Rhizophora racemosa*	*Rhizophora harrisonii*	*Nypa fruticans, Avicennia germinans*	*Bruguiera* sp.	*Avicennia nitida*	*Acrostichum aureum*	*Pteris* sp.,	*Nephrolepis* sp., *Valsaria* sp.
Human interference/ Cultivation/ Farming/Fire	*Alchornea cordifolia,*	*Musa sapientum, Triumfetta rhomboidea*	*Alchornea laxiflora*	*Elaeis guineensis, Ipomoea* spp.	*Solanum* spp. *Momordica charantia*	*Vigna gracilis*	*Alternaria* spp.	*Pteris intricata*	*Ustilago* spp.
	Dioscorea spp., *Sorghum bicolor,*	*Arachis hypogea, Corchorus olitorius*	*Citrus* sp., *Manihot esculentus*	*Zea mays, Ocimum gratissimum, Sesamum indicum*	Poaceae, *Corchorus olitorius,*	*Aspilia africana, Chromolaena odorata*	*Cytobasidium* spp.	*Nephrolepis biserrata*	*Glomus* sp.,
Residential and Ornamentals	*Lagerstromia indica*	*Aspilia africana*	*Thunbergia grandiflora*	*Caesalpinia pulcherrima*	*Mangifera indica*	*Pinus* spp.	*Alternaria* spp.	*Cladosporium* spp.	*Fusarium* spp.

Source: Orijemie [11, 23], Adekanmbi et al. [33], Akinbola et al. [34], Redondo et al. [35], Numbere [36].

Table 3.
Palynomorph (plants) based on indication of vegetation types in tropical West Africa.

the way savanna species thrive and survive in forest regions in southern Nigeria, although the survival of forest species in savanna has not been convincingly proven. Anthropogenic factors has led to the opening of forests canopies and planting of savanna (including ornamental) plants species and the guinea savanna region and deciduous low land rain forest is fast becoming a forest-savanna mosaic sometimes referred to as forest savanna transition. The availability of water, human interference through cultivation, burning, soil spatial variations, or herbivory pressure characterise this transition zone. Depending on the microclimate and anthropogenic impact, savanna and forest plant species co-exist, hence the presence of other vegetation-specific plant indicators are used to make decisions. Spores are usually used as bio-indicator for microclimates hence; the percentage of pollen and spore are good indices for understanding vegetation change and climate variability. Correct pollen identification is crucial to this application. **Table 2** shows the plant or pollen indicators that can be used to different vegetation types if there abundance in sediments is measured on the basis of **Table 1** above.

3.2 Understanding vegetation dynamics using percentage representation of palynomorphs recovered from environmental matrices

As earlier established in this review, pollen and spore percentage representation can be used to understand vegetation changes. This section will focus on changes in the abundance of certain pollen types (inferably, the plant species) in different environments across the globe and the possible factors influencing these changes. Some of these plants in *Typha* spp., *Elaeis guineensis, Bridelia* spp. *Annona* spp., Cyperaceae, Chenopodiaceae, Asteraceae, and Amaranthaceae to mention but a few. Haung et al. [37] in their study on pollen distribution in a large freshwater lake (Boston Lake) in the arid regions of Xinjiang, China using 61 surface samples found *Typha* L. to have average percentage representation (8.6%) when Chenopodiaceae had *ca.* 50%. Remarkably, *Typha* and *Phragmites* plants were abundant on the west side of the lake, and they asserted that hydrodynamic conditions affect *Typha* pollen. This indicates that lithological factors could moderate the representation of palynomorphs in sediments for some plant species just like the factors influencing the dominance of some plants in a region over others.

In Cameroun, Assi-kaudjhis [38] studied vegetational evolution in the Crater Lake Bambili which lies in the volcanic zone through the pollen analysis of sediments cored from two sites around the lake, a region located in the Guinean-Congolian forest belt. An inventory of the plant biodiversity was taken, and *Annona senegalensis, Bridelia ferruginea,* and *Typha* were found from 1600 m – 800 m asl as savanna elements within the local vegetation. Two cores of 13.5 m and 14.01 m depths, respectively, were taken from a few meters from the lake in 2007 and 2010. The results from pollen analysis revealed that *Bridelia*-type pollen was recorded (7.82%), Amaranthaceae/Chenopodiaceae, Poaceae (75.68–34.44%) and undifferentiated Asteraceae pollen was recovered from the sediments (30.29%). There was generally low amount of tree pollen. The distinctive pollen of *Annona senegalensis* and *Typha* were not found in the core while *Bridelia* was under-represented. Njokuocha [12] studied a 116 cm core from Holocene deposits in Lake Obayi in Nguru, Nsukka, which yielded 78 pollen types from 47 families. Njokuocha [12] found that *Elaeis guineensis* was well represented n sediments at depth 88–116 cm (*ca.* 40%) with Poaceae which was continually abundant in the core with specific abundance ranging from 20 to 40% representation in depths between 25 and 45 cm.

Marlon *et al.* [39] carried out high resolution sedimentological, geochemical and pollen analysis on a 5.75 m sediment core from the coastal plains of the Doce River, southeastern Brazil, which was characterised by many valleys resulting from Quaternary deposition of silt. The region was composed of tropical rainforest like Annonaceae, pioneering freshwater species like *Cyperus* sp. L. (*ca.* 80%), Asteraceae (*ca.* 18%) and Amaranthaceae (*ca.* 2%). From depth 5.5–1.5 m, five ecological groups were observed where *Typha* was only *ca.* 2% with Cyperaceae (3–30%), Poaceae (30–80%). Within lake regions of 1.5–0.8 m depth, *Typha* was only *ca.* 2% represented while Poaceae was 14–40%. The herbaceous plain of depth 0.7–0 m on the surface also yielded under-representation of *Typha* and *Hydrocleis* pollen (< 2%). No detailed discussion was reported on why the *Typha* pollen was under-represented.

In India's Lonar Crater Lake, Riedel *et al.* [40] investigated modern pollen vegetation relationships using Holocene lacustrine sediments and surface samples. They found *Typha augustata* L. near the site of coring and *Annona squamosa* L. characterised the steep faces above the dry deciduous forest in around the site which made up *ca*. 30% of the local vegetation. Results revealed strong differences in pollen assemblages and studied trapping media samples although local arboreal vegetation was adequately represented in soil samples. The pollen of *Typha* and Cyperaceae pollen accounted for 33% of the total pollen present. Poaceae was over-presented while the pollen of *Annona* L. appeared scattered or with single grains even when it formed part of the local vegetation. Channel and surface run off water transport influence pollen assemblages.

Travedi *et al.* [41] reported the under-representation of *Annona cf. squamosa* whilst attempting to establish modern pollen rain vegetation relationship from ten surface samples from Chaudhari-Ka-Tal, Raebareli District, Utter Pradesh of India. They averred that low pollen productivity owing to its entomophilous mode of pollination may have been the factor responsible for the under-representation. There was sparse abundance of *Annona* plants in the local vegetation. Pollen spectra from three of the surface samples revealed that *Annona* pollen was merely 0.65% while *Typha* pollen ranged from 3.2 to 28% from the southern flank samples analysed from the lake. From the western flank, *Annona* pollen was merely 1.65% in the sediments and *Typha* ranging from 17.6 to 22%. The eastern flank of the lake recorded the highest occurrence of *Typha* pollen (ranging from 22.3 to 39%) probably due to the marshy nature of the lake. They however argued that the under-representation of tree taxa could be attributed to low pollen productivity when compared with grasses and herbaceous taxa. Chenopodiaceae/Amaranthaceae pollen was reported to be over-represented also. The increase in fungal spores may have indicated microbial attack, however, there was no certainty on the possible causes of under-representation mentioned and also no further studies were carried out on the exine chemistry to substantiate on the variations in percentage occurrence in the same lake.

3.3 Climate variability induced vegetation dynamism and interactions

The succession of one vegetation type by another is influenced by climatic, human or edaphic factors, or a combination of the triad. Across the globe, the phenomenal change and succession of vegetation in the past and present have been revealed through palynological studies. Few empirical studies are reported in this chapter for clearer understanding.

In Maya region of southern Mexico and Central America, Franco-Gaviria *et al.* [42] investigated the impact of climate variability and human activities on the vegetation communities from two sediment sequences collected from two lakes (Lakes San Lorenzo and Esmeralda) in the highlands of Chiapas, Mexico during the late Holocene. The records reveal a long-term trend towards drier conditions with superimposed centennial-scale droughts. A declining moisture trend from 3,400 to 1,500 cal yr. BP consistent with southward displacement of the Intertropical Convergence Zone was reported. According to them, the climatic conditions with dense human occupation converted the vegetation from forest to more open systems. From paleoecological records of the area, cultural abandonment of the area which occurred ca. 1500 cal yr. BP probably favoured the forest recovery process at that time. About 600 cal yr. BP, wetter conditions promoted the establishment of modern montane cloud forests, with a diverse mixture of temperate and tropical elements. Some of the palynomorphs found in abundance were *Pinus* sp. L. and Cyperaceae pollen (at 3400–3200 cal yr. BP), *Pinus* sp. (over 200%), *Myrica* sp. L. (5–12%), *Quercus* L. (22–68%) and *Alnus* sp. Mill (2–8%) pollen (3200–2400 cal yr. BP). *Alchornea* Sw., Poaceae and *Quercus* aboreal elements peaked at ca. 2500–1500 cal yr. BP. Herbaceous taxa like Amaranthaceae, Asteraceae, Poaceae decreased from 1200 to 600 yr. BP at Lake san Lorenzo. Charcoal concentrations were low generally, but had peaks at ca. 2,500, 1600 and 1100 yr. cal BP. They concluded that the importance of microhabitats is in the maintenance biodiversity through time, even under scenarios of high climate variability and anthropogenic pressure.

In south western region of Nigeria, Orijemie [43] investigated climate-vegetation dynamics using an 8 m-core drilled in Ikorigho with comparison with Ahanve to provided evidence of late Holocene mangrove dynamics and environmental changes. The vegetation was found to have changed from mangrove to low land rainforest. Mangrove swamp forest species were indicated by pollen and spore of *Rhizophora* spp. L., *Avicennia* spp. L. and *Acrostichum aureum* L.; freshwater swamp forest include *Uapaca* Baill., *Mitragyna ciliata* Aubrev. & Pellegr. and *Symphonia globulifera* L.f. and a few lowland rainforest taxa (*Celtis brownie* Rendle, *Pycnanthus angolensis* (Welw.) Warb). There were marked reduction in *Rhizophora* spp. at certain periods which almost always coincided with an upsurge in Poaceae and Cyperaceae pollen obviously indicative of prevailing drier climate, and lowered sea level. In constrast, Sowunmi [44] reported that the environment in Ahanve – Badagry, experienced a very reduced or complete disappearance of mangrove species ca. 3100 yrs. B.P. Non-pollen palynomorphs like few charcoal particles at the lowest sections, and significant increase in charcoal particles in the topmost sections were found of the core providing evidence of relatively recent history of human interactions like tree felling, bush burning and agriculture.

In central Gabon, Ngomanda *et al.* [45] investigated vegetation changes during the past 1300 years are reconstructed in western equatorial Africa using a high-resolution pollen record from Lake Kamalete. The Kamalete pollen data showed the persistence over the past 1300 years of a relatively stable forest-savanna mosaic, associated with significant changes of the forest component. Three successive stages of forest dynamics was found. First, at 1325 yrs. BP, moist semi-evergreen rainforest existed around the catchment of Lake Kamalete. There was consistent presence of above 70% Gramineae pollen that the site was always primarily in savanna. Secondly, from c. 1240 to 550 yr. BP, a noticeable increase in shade-intolerant plant species indicate openings in the rainforest canopy. Thirdly, at 550 cal BP, a mature forest was re-established, corresponding to progressive savanna colonisation by forest pioneer species such as *Aucoumea klaineana* Pierre, *Lophira alata* and *Fagara macrophylla* (Oliv.) Engl. This new phase of forest expansion coincided with a marked lithological change, indicating an increase in lake-level. It was concluded that the major vegetation changes observed were due to climatic variability, and anthropogenic action had limited influence.

In Benin Republic in West Africa, Tossou *et al.* [46] was able to present information on the coastal halophytic mangrove vegetation history based on palynological data collected. They found that the mangrove swamp went through several physiognomic changes from the middle to late Holocene. In the course of the middle Holocene that is from 7500 to 2500 yr. BP. They found that the mangrove was extensive, and with high density of monospecific mangrove species dominated by Rhizophora. During the late Holocene, the mangrove regressed around 3000 years BP which indicated a period of drop in sea level and disappeared about 2500 yr. BP. It has been replaced by swamp meadows dominated by *Paspalum vaginatum* Sw. and a fresh water environment colonised by taxa such as *Persicaria* Mill., *Typha*, *Ludwigia* L., and *Nymphaea lotus* indicating a return of wetter climate drop in sea level.

In Lake Chad region, Amaral *et al.* [47] reported palynological evidences on the climate and vegetation changes that occurred in the Sahara – Sahel boundary which shifted northwards during the termination period of the African Humid Period (AHP). Dates obtained for sediments where between ca. 6700 and ca. 5000 yr. BP which encompassed part of the termination of the AHP. Results showed that, between ca. 6700 and ca. 6050 yr. BP, the vegetation close to humid savanna woodland, including elements currently found further southward, thrived in the vicinity of the Mega-Lake Chad in place of the modern dry woodland, steppe and desert vegetation. At the same time, the montane forest populations extended further southward on the Adamawa Plateau. The high abundance of lowland humid pollen taxa, particularly of *Uapaca*, was interpreted as the result of a northward migration of the corresponding plants during the AHP. The data retrieved indicated that between ca. 6700 and ca. 5000 yr. BP vegetation and climate changes must have occurred progressively, but that century-scale climate variability was superimposed on this long-term mid-Holocene drying trend as observed around ca. 6300 yr. BP, where pollen data indicate more humid conditions.

In Ghana, West Africa, Miller and Gosling [48] presented a fossil pollen record from sediment cores extracted from Lake Bosumtwi. The record covered the last c. 520,000 yrs. BP making it a apart of the Late Quaternary. The fossil pollen assemblages revealed that there was a dynamic vegetation change which can be broadly characterised as indicative of shifts between savanna and forest which also reflected the glacial – interglacial period. Savanna elements which heavily dominated the vegetation included Poaceae pollen (>55%) and was associated typically associated with Cyperaceae, Chenopodiaceaee/Amaranthaceae and Caryophyllaceae. Forest formations are were more diverse than the savanna, with the key taxa occurring in multiple forest zones being Moraceae, *Celtis* sp., *Uapaca* sp., *Macaranga* and *Trema* spp. Lour. The fossil pollen data indicated that over the last c. 520,000 yr. BP, the vegetation of lowland tropical West Africa has mainly been savanna; however six periods of forest expansion were evident which most likely correspond to global interglacial periods. A comparison of the forest assemblage composition within each interglacial suggests that the Holocene (11,000 yrs) forest occurred under the wettest climate, while the forest which occurred at the time of Marine Isotope Stage 7 probably under the driest climate.

In Cameroun, Lebamba *et al.* [49] investigated the vegetation dynamics and human interference of the Adamawa Plateau which is in between the Guineo-Congolian rain forest and Sudanian savanna in Central Cameroun from African Humid Period to the present day through the analysis of pollen and spores. They presented a 4000-yr old pollen sequence derived from the Lake Tizong sediments that extended from the end of the. Pollen sequence were distinguished into two major short-duration forested phases that lasted between ca. 3900–3000 yr. BP, and ca. 1900–1450 yr. BP. Within 4000–3000 yr. BP, arboreal/montane forest plants (*Podocarpus* sp. L'Her ex Pers., *Olea* sp. L., and *Rubus pinnatus* Willd.) dominated the vegetation, with associated semi-deciduous forest elements like *Celtis* sp. indicative of dry climate. A decrease in tree taxa was noticed and increase in freshwater forest taxa (Cyperaceae) around 3000–1900 yr. BP indicating a wet climate at that time with a slight increase in arboreal plants. From 1450 yr. till present savanna elements (e.g. *Hymenocardia* sp.) become more dominant. It was also found that a critical ecological threshold occurred around 3000 yr. BP when Poaceae reached higher percentages than forest taxa. Savanna was established until the present day with a brief expansion of lowland semi-deciduous forest, dominated by *Myrianthus arboreus* P.Beauv.-type pollen, between ca. 1000–700 cal. yr. BP. Although, human impacts and climatic factors driving vegetation change were difficult to differentiate, the late Holocene on the Adamawa plateau was characterised by a variable climate that resulted in significant vegetation transitions.

3.4 Honey, pollen and vegetation representation

Pollen analysis of honey samples started a long time ago dating back to late 1970 with Sowunmi [50] as foremost. Several reports are available on the melissopalynological analysis of honeys from different parts of Nigeria as reviewed as follows. One of the earliest studies was that of Agwu and Akanbi [51] in view to understanding the botanical origin of the honeys from Bichi, Edem-Ani, Nanka, Nimo Nsukka and Ogbomosho including Ohafia and Port Harcourt. They found out that 56 pollen types were identified in all belonging to 14 families, genera and tribes. Ogbomosho had the highest (29) number of pollen types. The study revealed that most of the honeys were rich in pollen apart from the Port-Harcourt and Ohafia honeys that were adulterated. The species that were dominant or most preferred by honeybees include *Vitellaria paradoxa, Lannea* sp. A. Rich. in Guillem, *Elaeis guineensis, Parkia clappertoniana* Keay, *Prosopis africana* (Gull. & Perr.) Taub., *Crossopteryx febrifuga* (Afzel. ex G. Don) Benth. and *Nicotiana tabacum* L. The vendors claim were verified as the pollen representation showed the vegetation where the honeys were produced. After this time, several others have been published.

Agwu *et al.* [52] analysed honey samples from four honey samples from different localities in Kogi State (Olowa, Ajogoni, Itama and Ojowu), Nigeria to determine their floral sources, ecological origin and season of production. Grains counts of 532, 589, 1033 and 720 were recovered respectively. Thirty-two pollen types encountered, 23 were identified to family level and two were unidentified. The predominant pollen types include those of *Acanthus* spp. L., *Alchornea cordifolia* (Schum. & Thonn.) Mull.Arg., *Anacardium occidentale* L., *Cassia mimosoides* L., *Elaeis guineensis*, *Hymenocardia acida* Tul., *Phyllanthus niruri* L., *Mangifera indica* L., *Tridax procumbens*, and *Zea mays* L. Results suggested vegetation types reflecting the lowland rainforest and secondary grassland and of quality. Kayode and Oyeyemi [53] undertook a study to determine the pollen quality of eight honey samples collected from Odigbo and Okitipupa areas of Ondo State, Nigeria. The result showed diverse pollen types that were visited by worker bees for their nectar and pollen source. The most frequently represented families were Fabaceae and Euphorbiaceae with 14 taxa and eleven taxa of pollen grains respectively. Some important plant taxa identified that are frequently visited by the bees include, *Lannae* sp., *Nauclea* sp. L., *Pericopsis* sp. Thwaites, *Lophira lanceolata* Tiegh. ex Keay, *Phyllanthus discoides* (Baill.) Mull.Arg, *Elaeis guineensis*, *Nauclea diderrichii* (De Wild. & T.Durand) Merr., *Brachystegia eurycoma* Harms and members of Combretaceae or Melastomataceae. These plants were found to be of great importance to the bees for honey production.

Melissopalynological analysis of four honey samples from four localities of Kogi State (Idah, Ajaka, Igalamela–Odolu and Inachalo-Oforachi) was carried out by Aina *et al.* [54] in a bid to ascertain the species of plants which were incorporated into honey. Pollen grain counts of 2274 to 3,195 were recorded. A total of 21 pollen types were recorded, 19 of which were identified to family level. Some predominant pollen types included *Leuceanea glauca* (Lam.) de Wit, *Parkia biglobosa*, *Elaeis guineensis*, *Phyllanthus* sp. and *Bombax buonopozense*. The four honey samples were found to be unadulterated and certified to be of quality. From the pollen identifications, the botanical and geographical origin as well as the season of honey production of each sample were determined and also associated with definite vegetation types which reflected the vegetation of low land rainforest/Guinea savanna.

Oyeyemi [55] collected three honey samples from an apiary for three months to determine the change in pollen content for the months of October to December in Ado Ekiti. The study revealed that pollen count ranged from 106,962 to 171,487. The honey samples were found to be multifloral in source and had abundant *Nauclea* sp., *Entada* sp. Adans., *Vitellaria paradoxa, Lannea* sp. members of the Rutaceae and Combretaceae or Melastomaceae which indicated that these plants were available at the locality where the apiary is situated to serve as pollen and nectar sources. Other pollen types encountered include *Vitex doniana*, Poaceae, and *Irvingia* sp. Hook.f. There was variation in the total pollen count for the months of sampling; however, the honeys indicated a high richness in pollen diversity and quality.

Adekanmbi and Ogundipe [56] analysed three honey samples bought from open markets in Lagos, Nigeria. The proportion of pollen from each of the honey samples varied from 196 to 280. The most abundant taxa were *Tridax procumbens* and *Elaeis guineensis.* The pollen grains in the Palmae and Asteraceae plant families are of great importance to the bees for honey production; this can be seen in the abundance displayed. Other pollen taxa recovered belong to the families Mimosaceae, Euphorbiaceae, Sapotaceae and Anacardiaceae providing clue on the ecological origin of the pollen grains in the honey sample. Pollen analysis of honey proved to be useful in deciphering nectar sources of honeybee *Apis mellifera adansonii* Latreille.

4. Conclusion

The genetically conscripted chemical signature called sporopollenin in the outer cell walls enhances the ability of palynomorphs to retain their shapes even after subjection to heat and pressure in sediment treatment with concentrated acid during routine palynological procedures. This enables clear identification of pollen types and in extension the vegetation dynamics. The identification of pollen flora in honey is currently been considered as good means of understanding economically and traditionally important plants for conservation and reforestation. The daily dispersal of pollen (including the allergenic ones) in the atmosphere sheds light on the seasonally of vegetation, their flowering, productivity of pollen and contamination of atmospheric quality. Studying cores from lakes, rivers and soil profiles provides evidence for the change in vegetation in relation to climate, plant response to stress and human interference in the past. This is applied based to vegetation science by virtue of the percentage representation pattern of palynomorphs. Although caution must be exercised in interpreting palynological data due to the moderating factors like differential preservation of pollen grains based inertness of the chemical compound in its outer wall, limited knowledge on pollen productivity of flowering plants, dispersal mechanisms which affects the quantity of grains recovered from soil and honey or atmospheric samples, issues of pollen morphological similarities. Describing and attributing vegetation characteristics to a locality or region, the modern behaviour of plants is to be considered.

Acknowledgements

The author's special appreciation goes to Prof (Canon) M. Adebisi Sowunmi for laying the academic foundation for the understanding of palynology. Her academic,

DOI: http://dx.doi.org/10.5772/intechopen.97755

career-based and motherly advices were so helpful and she is so highly appreciated. Her contributions were highly transformative. The author is also grateful to Dr. E.A. Orijemie and Mr. P.C. Opara for their support, encouragement and support. To my parents Mr. and Mrs. Austin O. Kobi, I express my gratitude for the prayers and financial support.

Author details

Sylvester Onoriode Obigba
Palynology Unit, Department of Botany, University of Ibadan, Ibadan, Oyo State, Nigeria

*Address all correspondence to: obigbasylvester@gmail.com

References

[1] Kim, D. (2014). *Vegetation dynamics*. Retrieved February 12, 2021 online from https://www.accessscience.com/content/vegetation-dynamics/YB150518. DOI: 10.1036/1097-8542.YB150518.

[2] Sowunmi, M.A. (2015). *The philosophy and principles of palynology*. Paper presented at the maiden workshop of the palynology Association of Nigeria held on the may 4th, 2015 at the University of Ibadan, Ibadan.

[3] Obigba, S.O. (2018). *Under-Representation of Certain Pollen Types in the Holocene Palynomorph Record at Ahanve, Southwestern Nigeria: An Investigation*. Unpublished M.Sc. Thesis, Department of Botany, University of Ibadan, Ibadan, Nigeria. 140 p.

[4] Hyde, H.A. & Williams, D.A. (1944). The right word. Pollen Analysis Circular, 8: 6 – 7.

[5] Steele, A. & Warrny, S. (2013). Reconstructing earth's past climates: The hidden secrets of pollen. Science Activities, 50: 62 – 71. DOI: 10.1080/00368121.2013.792763

[6] Fraser, W.T., Scott, A.C., Forbes, A.E.S., Glasspool, I.J., Plotnick, R.E., Kenig, F. & Lomax, B.H. (2012). Evolutionary stasis of sporopollenin biochemistry revealed by unaltered Pennsylvannian spores. New Phytologist, 196: 397 – 401. DOI: 10.1111/j.1469-8137.2012.04301.x

[7] Shaw, G. & Yeadon, A. (1964). Chemical studies on the constitution of some pollen and spore membranes. Grana Palynologica, 5(2): 247 – 252. DOI: 10.1080/00173136409430017.

[8] Onac, B.P., Hutchinson, S.M., Geanta, A., Forray, F.L., Wynn, G.J., Giurgiu, A.M., & Coroiu, J., (2015). A 2500-yr late Holocene multi-proxy record of vegetation and hydrologic changes from a cave guano-clay sequence in SW Romania. *Quaternary Research*, YQRES - 03625, 12C DOI: 10.1016/j.yqres.2015.01.007

[9] Widga, C. & Colburn, M. (2015). Paleontology and paleoecology of guano deposits in mammoth cave, Kentucky, U.S.a. Quaternary Research, 83, 427 – 436.

[10] Oslisly, R. (2013). Climatic and cultural changes in the West Congo Basin forests over the past 5000 years. *Philos. Trans. R. Society of London. Series* B 368:20120304

[11] Orijemie, E.A. (2018). Late Holocene paleoenvironment of the Tse dura, a later stone age (LSA) rock shelter, north Central Nigeria. Studies Quaternaria, 35(1), 41 – 53. DOI: 10.2478/squa-2018-0003

[12] Njokuocha, R.C. (2012). A contribution to the Holocene vegetation changes of Nigeria: Palynomorphs from a 200 cm Core from Ohe pond Nsukka, Nigeria. *Paper presented at the West African Quaternary Research Association (WAQUA) Third International Workshop* held in Accra – Ghana on October 8 – 12, 2012.

[13] Adeleye, M.A., Adebayo, M.B., & Adeonipekun, P.A. (2018). Palynological study of surface sediments from coastal area of Lagos, Nigeria. Ife Journal of Science, 20(1), 199 – 206.

[14] Winchester, S. (2001). *The map that changed the world: Williams Smith and the birth of modern geology*. New York: Harper Collins (p. 59 – 91).

[15] Yu, M., Wu, P., Widelitz, R. & Chuong, C. (2002). The morphogenesis of feathers. *Nature,* 420(6913):308 – 312. DOI: 10.1038/nature01196.

[16] Novello, V.F., Cruz, F.W., McGlue, M.M., Wong, C.L., Ward, B.M., Vuille, M., Santos, R.A., Jaqueto, P., Luiz, C.R., Atorre, T., Ribeiro, M.A.L., Karmann, I., Barreto, E.S., Cheng, H., Edwards, R.L., Marcos, S.P., & Scholz, D. (2019). Vegetation and environmental changes in tropical South America from the last glacial to the Holocene documented by multiple cave sediments proxies. *Earth and Planetary Science Letters,* 524: Paper 115717. DOI: 10.1016/j.epsl.2019.115717.

[17] Layek, U., Mondal, R. & Karnmakar, P. (2020). Honey sample collection influence pollen composition in determining true nectar-foraging bee plants. Acta Botanica Brasilica, 34(3): 478 – 486. https://doi.org/10.1590/0102-33062020abb0086

[18] Willmer, P. (2011). *Pollination and Floral Ecology*. Princeton, New Jersey: Princeton University Press. 789p.

[19] Lau, P., Byrant, V., Ellis, J.D. & Huang, Z.Y. (2019). Seasonal variation of pollen collected by honey bees (*Apis mellifera*) in developed areas across four regions in the united state. PLoS ONE, 14(6): e0217294. DOI: 10.1371/journal.pone.0217294.

[20] Erdtman, G. (1968). *Handbook of Palynology: An Introduction to the Study of Pollen Grains and Spores*. New York: Hafner Publishing Inc.

[21] Ezike, D.N., Nnamani, C.V., Ogundipe, T.O. & Adekanmbi, O.H. (2016). Airborne pollen and fungal spores in Garki, Abuja (north-Central Nigeria). Aerobiologia, 32(4). DOI: 10.1007/s10453-016-9443-5.

[22] Essien, B.C. & Nkang, A.E. (2013). Evaluation of the phtyoecological indicator species in the airborne palynomorphs from Anyigba, Kogi State, Nigeria. *Standard Scientific Research and Essays,* 1(14): 398 – 402. Retrievable at https://doi.org/10.1007/978-3-319-19944-3_27

[23] Orijemie, E.A. (2017). Understanding ancient farming in Tivland, north-Central Nigeria: Palaeoecological perspective. Global Journal of Archaeology and Anthropology, 1(4), 1 – 8.

[24] Johnston, G. (2019). *Palynology*. Retrieved online on January 15, 2021 from http://www.archaeologyexpert.co.uk/palynology.html/

[25] Donatella, M. & Federico, D.R. (2015). Archaeopalynological preparation techniques. In, Eung. E., Stasolla, C., Summer, M. and Huang, B. (eds), *Plant Microtechniques and Protocols* (pp. 495 – 506). Cham: Springer International Publishing.

[26] Faegri, K. & Iversen, J. (1989). *Textbooks of Pollen Analysis*, 4th ed. Chichester: John Wiley and Sons.

[27] Jones, G.D. & Byrant, V.M. (2004). The use of ETOH for the dilution of honey. Grana, 43: 174 – 182.

[28] Louveaux, J., Maurizo, A., & Vorwohl, G. (1978). Methods of melissoplaynology. Bee World, **59**(4): 139 – 157.

[29] Gosling, W.D., Miller, C.S. & Livingstone, D.A. (2013). Atlas of the tropical west African pollen flora. Review and Palaeobotany and Palynology 199: 1 -135. DOI: 10.1016/j.revpalbo.2013.01.003

[30] Sowunmi, M.A. (1973). Pollen grains of Nigerian plants I: Woody species. Grana 13: 145 – 186.

[31] Sowunmi, M.A. (1995). Pollen of Nigerian plants II: Woody species. Grana 34(2): 120 – 141.

[32] Sowunmi, M. A. (1986). Change in vegetation with time. In: *Plant Ecology in West Africa: Systems and Processes*. G. Lawson (ed.) Wiley, New York: 273 - 307.

[33] Adekanmbi, H., Ajikah, L., & Ogundipe, O. (2018). Palynology and paleovegetational reconstruction of Lagos coastal environment, southwestern Nigeria. *Paper presented at the Botany Conference* on Thriving with diversity held July 21 – 25, 2018 at Mayo Civic Centre, Rochester, Minnesota.

[34] Akinbola, S., Awotoye, O., Adepoju, K & Salami, A. (2016). Floristic indicators of tropical land use systems: evidence from mining areas in southwestern Nigeria. *Global Ecology and Conservation*, 7: 141 – 147. Retrieved from DOI: 10.1016/j.gecco.2016.06.003

[35] Redondo, MA., Berlin, A., Boberg, J. & Oliva, J. (2020). Vegetation type determines spore deposition within a forest – Agriculture mosaic landscape. *FEMS Microbiology,* 96(6): fiaa082. Retrieved online from https://doi.org/10.1093/femsec/fiaa082.

[36] Numbere, A.O. (2018). Mangrove species distribution and composition, adaptive strategies and ecosystem services in the Niger River Delta, Nigeria. In, Sharma, S. (ed.), *Mangrove ecosystem ecology and function*. DOI: 10.5772/intechopen.79028.

[37] Haung, X., Zhou, G., Ma, Y., Xu, Q. & Chen, F. (2010). Pollen distribution in large freshwater lake of arid regions: A case study on the surface sediments from Boston Lake Xinjiang, China. Frontier in Earth Science China 4(2): 174 – 180.

[38] Assi-Kaudjhis, C. (2012). Vegetation evolution in the mountain of Cameroun during the last 20,000 years: Pollen analysis of Lake Bambili sediment. In, Blanco, J.A. (ed.), *Forest Ecosystems-more than Just Trees* (p. 138 – 164). Rijeka, Croatia: InTech Publishers.

[39] Marlon, C.F., Marcelo, C.L.C., Luiz, C.R.P., Dilu, F.R., Flavio, L.L., Antonio, A.B.J., Jose, T.F.G., Yuri, F. & Kita, M. (2013). Mangrove vegetation changes on Holocene terraces of the Doce River, southeastern Brazil. Catena 110: 59 – 69.

[40] Riedel, N., Stebich, M., Anoop, A., Basavaiah, N., Menzel, P., Prasad, S., Sachse, D., Sarkar, S. & Wiesner, M. (2015). Modern pollen vegetation relationships in a dry deciduous monsoon forest: A case study from Lonar crater Lake, Central India. Quaternary International 317: 268 – 279.

[41] Travedi, A., Saxena, A. & Chauhan, M.S. (2016). Studies on pollen rain *vis-à-vis* vegetation relationship in Chaudhari-Ka-Tal, Raebareli District, utter Pradesh. Journal of the Paleontological Society of India 61(1): 85 – 90.

[42] Franco-Gaviria, F., Corree-Metro, A., Cordero-Oviedo, C., Lopez-Perez, M., Cardenes-Sandi, G.M. & Romero, F.M. (2018). Effects of late Holocene climate variability and anthropogenic stressors on the vegetation of the Maya highlands. Quaternary Science Reviews, 189, 76 – 90.

[43] Orijemie, A.E. (2012). Holocene mangrove dynamics and environmental changes inthe coastal region of south western Nigeria. Quaternary International Abstracts, 279-280, 346 – 461.

[44] Sowunmi, M.A. (2004). Aspects of Nigeria coastal vegetation in the Holocene: some recent insights. *Past climate variability through Europe and Africa.* Eds. R.W. Battarbee, R.W., F. Gasse and C.E. Stickly. Dordrecht, The Netherlands: Springer. p. 199 – 218.

[45] Ngomanda, A., Chepstow-Lusty, A., Makaya, M., Schevin, P., Maley, J., Fontugne, M., Oslily, R., Rabenkogo, N. and Jolly, D. (2005). Vegetation changes during the past 1300 years in western equatorial Africa: A high resolution pollen record from Lake Kamalee, lope reserve, Central Gabon. The Holocene, 15(7): 1021 – 1031.

[46] Tossou, M.G., Akoègninou, A., Ballouche, A., Sowunmi, M. A., and Akpagana, K. (2008). The history of the mangrove vegetation in Bénin during the Holocene: A palynological study. Journal of African Earth Sciences 52: 167-174.

[47] Amaral, P.G.C., Vincens, A., Guiot, J., Buchet, G., Deschamps, P., Doumnang, J.-C., and Sylvestre, F. (2013). Palynological evidence for gradual vegetation and climate changes during the African humid period termination at 13^0 N from a mega-Lake Chad sedimentary sequence. Climate of the Past, 9: 223 – 241.

[48] Miller, C.S. and Gosling, W.D. (2014). Quaternary forest associations in lowland tropical West Africa. Quaternary Science Reviews, 84: 7 – 25.

[49] Lebamba, J., Vincens, A., Lezine, A.M., Marchant, R. and Buchet, G. (2016). Forest-savanna dynamics on the Adamawa plateau (central Cameroun) during the African humid period termination: A new high-resolution pollen record from Lake Tizong. Review of Palaeobotany and Palynology, 235: 129 – 139.

[50] Sowunmi, M.A. (1976). The potential value of honey in paleopalynology and archaeology. Review of Palaeobotany and Palynology, **21**: 171 – 185.

[51] Agwu, C.O.C. & Akanbi, T.O. (1985). A palynological study of honey from four vegetation zones of Nigeria. Pollen et Spore, 27 (3-4): 335 – 348.

[52] Agwu, C.O.C., Essien, B.C. and Badmus, S.A. (2013). Melissopalynological study of honey samples from four localities in Dekina Local Government Area of Kogi State, Nigeria. *Journal of Biological and Chemical Research: an International Journal of Life Science and Chemistry*, 30(2): 921-928.

[53] Kayode, J. & Oyeyemi, S.D. (2014). Physico-chemical study and analysis of pollen grains in some commercial honey samples found in Ondo state, Nigeria. International Journal of Basic and Applied Science, **3**(2), 63 – 73.

[54] Aina, D.O., Enefola, J.O. & Agwu, C.O.C. (2015). Palynological study of honey samples from four localities in Kogi east, Nigeria. Journal of Biology and Life Sciences, 6(1), 29 – 38.

[55] Oyeyemi, S.D. (2017). Pollen analysis of honey samples from an apiary in ado Ekiti, Ekiti state, Nigeria. Nigerian Journal of Botany, 30(2), 203 – 212.

[56] Adekanmbi, O. & Ogundipe, O. (2019). Nectar sources for the honey bee (*Apis mellifera adansonii*) revealed by pollen content. Notulae Botanicae Horti Agrobotanici Cluj-Napoca, **37**(2), 211 – 217.

Chapter 13

Forest Vegetation and Dynamics Studies in India

Madan Prasad Singh, Manohara Tattekere Nanjappa, Sukumar Raman, Suresh Hebbalalu Satyanatayana, Ayyappan Narayanan, Ganesan Renagaian and Sreejith Kalpuzha Ashtamoorthy

Abstract

Forests across the globe have been exploited for resouces, and over the years the demand has increased, and forests are rather exploited instead of sustainable use. Focussed research on vegetation and forerst dynamics is necessary to preserve biodiversity and functioning of forests for sustanence of human life on Earth.This article emphasis that the India has a long history of traditional knowledge on forest and plants, and explorations from 17th century on forests and provided subsequent scientific approach on classification of forests. This also explains the developments of quantitative approach on the understanding of vegetation and forest diversity. Four case studies viz., Mudumalai, Sholayar, Uppangala, Kakachi permanent plots in the forests of Western Ghats has been explained in detail about their sampling methods with a note on the results of forest monitoring. In the case of deciduous forests, the population of plant species showed considerable fluctuations but basal area has been steadily increasing over time, and this is reflecting carbon sequestration. In Sholayar, a total of 25390 individuals of 106 woody species was recorded for < 1 cm diameter at breast height in the first census of the 10 ha plot in the tropical evergreen forest. In Uppangala, 1) a 27- year long investigation revealed that residual impact of logging in the evergreen forests and such forests would take more time to resemble unlogged forests in terms of composition and structure; 2) across a similar temporal scale, the unlogged plots trees < 30 cm gbh showed a more or less similar trend in mortality (an average of 0.8% year-1) and recruitment (1%). The Kakachi plot study revealed that 1) endemic species showed least change in stem density and basal area whereas widely distributed species showed greater change in both; 2) The overall recruitment of trees was 0.86 % per year and mortality 0.56% per year resulting in an annual turnover of 0.71% ; 3) majority of the gap species had high levels of recruitment and mortality resulting in a high turnover.Such studies can be used as early warning system to understand how the response of individual plants, species and forests with the climatic variability. In conclusion, the necessity of implementation of national level projects, the way forward of two such studies: 1) impact of climate change on Indian forests through Indian Council of Forestry Research and Education (ICFRE) colloborations and 2) Indian long term ecological observatorion, including the sampling protocols of such studies. This will be the first of its kind in India to address climate change issues at national and international level and helps to trace footprints of climate change impacts through vegetation and also reveals to what extent our forests are resilient to changes in the climate.

IntechOpen

Keywords: climate change, episodic recruitment, monitoring, permanent plots, trees, Western Ghats

1. Introduction

Forests across the globe have been utilized and many times exploited by humans ever since life style changed from nomadism to settled agri-based system. Forests used to supply many resources including fuel wood, medicine, timber, food etc. But over the years the demand has increased and forests are rather exploited instead of sustainable use.

Historically Indian forests attracted traders for its diversity in spices. The discovery of sea route to India resulted in exodus of European traders. The Portuguese, English, Dutch, French and Germans arrived in India in their quest for the plant species that were important as spices. Records are available that Romans and Arabs were briskly trading with Kings especially along the Western Ghats for variety of spices such as pepper, ginger, cardamom and other condiments.

1.1 Vegetation studies in India

India is richly endowed with climatic, edaphic and orographic gradients. Geographical extent of India ranges from tropical latitude to temperate latitude with tropic of Cancer passing through India dividing into almost half with one part predominantly tropical nature while the other is subtropical and temperate in nature. Hence the natural vegetation of India is also ranging from rainforests in the Western Ghats and Eastern Himalayas to desert in Rajasthan.

Ancient description of vegetation largely confined to composition of species primarily of medicinal use and utilization in rituals. Ancient India had a rich tradition of life sciences. There are reviews relating to the knowledge of plant sciences in ancient India [1]. The study of plants in ancent India was mainly under two heads namely, 1. Plants utilized for medicinal purposes and 2. Plants relating to agriculture [1]. However, the study on description of plants and animals was not popular. Takshashila University encouraged the collection, identifying and description of plants found around Takshashila. The traditional medicine in India which is primarily plant based has description of several plant species occurring in different forest types. Though, the descriptions largely confined to medicinal properties and parts of the plants used.

1.2 Studies on plant systematics

Significant contribution to botany of India was made during the colonial period with a strong pursuit of harvesting botanical resources. There were several European botanists and explorers contributed to the knowledge of botanical resources of India. Van Rheede H. A. (17th Century) the then Dutch Governor of Cochin made an effort to scientifically document the wealth of plants and indigenous medical knowledge with the native Malabaris. He produced a well written book, '*Hortus Malabaricus*' published during 1678–1693 [2]. This book contained scientific description and life size illustrations of about 742 native plants in Malabar (Kerala) and of which about 650 are of medicinal value. This book was used by several European botanists such as Linnaeus, De Candole, De Jussieu, Adanson, Blume and Wight to describe many species from India and Asia [3]. Some of the important names include, Johann Gerhard Koenig (1768) who made extensive collections from India, Robert Kydd (1787) who was instrumental in starting the Royal Botaical Gardens at Calcutta and Sir Joseph Hooker who wrote monumental book "*Flora of British India*" and Gamble and Fischer (1915-36) who wrote a comprehensive flora for the region of the presidency of Madras [4]. There were several Europeans who made collections and described many plant species from different parts of India. The development of science under colonial period is well described by Kochhar [4, 5]. Important component in development of plant science in India under colonial rule was establishment of botanical gardens not only in Calcutta but in several places across India with major focus on breeding and maintenance of economically important species from different colonial parts of southeast Asia.

1.3 Indian forest types

First systematic classification of Indian forest types was by Sir. H.G. Champion in 1936 which was later revised in the year 1968 by Champion H.G. and S.K. Seth [6]. They identified 16 major forest types based on rainfall and temperature (**Table 1**). Contemporaneously, the French Institute of Pondicherry (IFP; http://www.ifpindia.org) produced vegetation maps at the one million scale for peninsular India (by publishing 12 sheets between 1959 and 1973) in collaboration with Indian Council of Agricultural Research. Subsequently published six vegetation maps (scale 1: 250000) covering south and central Western Ghats region in collaboration with state forest departments of Karnataka, Kerala and Tamil Nadu [7–12]. These maps were produced considering bioclimate of the region, floristic series (dominant species based on climax species, structural and floristic composition) of the forest types, limits of forest types, altitudinal zonation, degree of degradation of forests, relationships between different stages of succession and the potentiality of a disturbed forest to return to the climax. Since then there are studies to improve the classification of forest types considering the feasibility for the forest managers to manage their forests. Attempts have been made by the Forest Survey of India (FSI) to revisit different forest types and reassign the forest types based on ground survey [13] (**Table 2**). In 2014, Indian Council of Forestry Research and Education (ICFRE) has revisited Champion and Seth [6] forest type classification by rapid assessment mode.

1.4 Major developments in understanding vegetation of India

There was a discernible change in describing vegetation in India. The trend was not to describe species found in any vegetation type but quantitatively describe the vegetation of a locality. There were several studies on quantitative description of vegetation in India. It was initiated by Rai [16], who inventoried all trees ≥10 cm dbh in four plots of 2.7, 2.7, 2.63 and 1.09 hectares respectively at Devimane, Malemane, Kodkani and Katlekan areas of the Western Ghats. Most such studies in the Indian evergreen forests have been conducted during the last decade of the 20th century, and many of them are once census plots. Interest in tree mortality and forest dynamics has increased recently because forest dynamics is thought to be involved in determining tree species diversity [17], and also thought to be related to global climate change, in particular [18]. Phillips and Gentry (*l.c.*) concluded that tree turnover rates have increased in tropical forests during the latter part of the 20th century. This has proved to be a controversial finding (e.g. [19, 20]), but one which could have important implications for biodiversity and atmospheric change. Apart from the global plot networks such as CTFS and Rainfor, India also made

Category 1: Moist Tropical Forests	
Group 1: Tropical Wet Evergreen Forests	
Sub group 1A: Southern Tropical Wet Evergreen Forests	Includes giant evergreen forests, moist forests of Andamans, montane forests of southern India, west coast evergreen forests. Imortant species include *Palaquium ellipticum, Cullenia exarillata, Vateria indica, Hopea* sp., *Dipterocarpus indicus, Calophyllum* sp., *Diospyros* sp., *Mesua ferrea.*
Sub group 1B: Northern Tropical Wet Evergreen Forests	Forests in Assam Valley, Upper Assam Valley Evergreen forests, Cachar evergreen forests (Dima Hasao district, Assam). Important species include *Michelia montana, Mesua ferrea, Dysoxylon binecteriferum, Dipterocarpus* sp., *Shorea assamica, Litsea monopetala, Artocarpus chaplasha, Garcinia pedunculata.*
Sub group 1C: Edaphic and seral types of evergreen forests	Includes cane brakes, wet Bamboo and pioneer Euphorbiaceous scrub.
Group 2: Tropical Semi-evergreen Forests	
Sub group 2A: Southern tropical semi-evergreen forests	Includes Andaman's semi-evergreen forests, west coast semi-evergreen forests, west coast secondary evergreen Dipterocarp forests. Some of the dominant species include *Terminalia paniculata, Olea dioica, Knema attenuata, Litsea* sp., *Syzygium* sp., *Diospyros* sp., *Artocarpus hirsutus, Mesua ferrea.*
Sub group 2B: Northern tropical semi-evergreen forests	Assam valley and alluvial semi-evergreen forests, sub-Himalayan secondary et mixed forests, Cachar and Orissa semi-evergreen forests. Important species are *Shorea robusta, Haldina cordifolia, Mesua ferrea, Litsea monopetala, Casearia graveolens, Syzugium* sp.,*Terminalia bellerica, Tetrameles nudiflora.*
Sub group 2C: Edaphic and seral types of semi-evergreen forests	Moist bamboo brakes, lateritic semi-evergreen forests, secondary moist bamboo brakes. Important species are *Pterocarpus marsupium, Anogeissus latifolia, Syzygium cumini, Phoenix acaulis* etc.
Group 3: Tropical Moist Deciduous Forests	
Sub group 3A: Andaman's moist deciduous forests	Andaman's moist and secondary moist deciduous forests. Dominant species include *Pterocarpus dalbergioides, Diospyros oocarpa, Celtis wightii, Terminalia bialata, Lagerstroemia hypoleuca*
Sub group 3B: South Indian moist deciduous forests	Moist teak bearing forest, southern moist mixed and secondary mixed deciduous forest. Important species include *Terminalia paniculata, Xylia xylocarpa, Terminalia bellirica, Lagersroemia microcarpa, Grewia tiliifolia, Dillenia pentagyna, Holigarna* sp., *Diospyros* sp., *Tectona grandis.*
Sub group 3C: north Indian moist deciduous forests	Moist Sal forests founding east Himalayas, Khasi hills, Siwalik hills and Assam. Dominant species include *Shorea robusta, Buchanania lanzan, Garuga pinnata, Schima wallichii, Madhuca latifolia, Mallotus philippensis, Lagerstroemia speciosa* (Syn. *L. macrocarpa*), *Gmelina arborea.*
Sub group 3D: Edaphic and seral types of moist deciduous forests	*Terminalia tomentosa* forests. Some of the species are *Tectona grandis, Dalbergia latifolia, Shorea robusta, Anogeissus latifolia, Lagerstroemia* sp.

Category 1: Moist Tropical Forests	
Group 4: Littoral and swamp forests	
Sub-group 4A: Littoral forests	Some of the common species are *Manilkara littoralis, Thespesia, Spinifex, Littoreus, Casuarina equisetifolia, Pandanus, Borassus, Phoenix, Callophyllum littoralis, Barringtonia, Pongamia,* twiners, climbers etc.
Sub-group 4B: Tidal swamp forests	Mangrove scrub, Mangrove forests. Dominant species are *Rhizophora* sp., *Avicennia* sp., *Bruguieria gymnorrhiza, Excoecaria agallocha.*
Sub-group 4C: Tropical fresh water swamp forests	*Myristica* swamp forests. Important species are *Myristica fatua, Gymnocranthera canarica, Mastixia arborea, Semecarpus travancorica, Hopea whitiana, Lophopetalum whitiana, Holigarna graham, Syzygium laetum* etc.
Category 2: Dry Tropical Forests	
Group 5: Topicl dry deciduous forests	
Sub-group 5A: Southeren tropical dry deciduous forests	Dry teak bearing forest, dry teak forest, dry red-sanders bearing forest, southern dry mixed deciduous forest. Some of the important species include *Tectona grandis, Anogeissus latifolia, Terminalia* sp., *Shorea roxburghii, Schleichera oleosa, Diospyros* sp., *Pterocarpus marsupium, Pterocarpus santalinus, Lagerstroemia* sp.
Sub-group 5B: northern tropical dry deciduous forests	Dry Sal- bearing forest, dry Siwalik Sal forest, Dry plains Sal forest, Dry peninsular sal forest, Northern mixed dry deciduous forest. Some of the important species are *Shorea robusta, Terminalia* sp., *Anogeissus latifolia, Anogeissus pendula, Anthocephalus chinensis, Albizia* sp., *Careya arborea, Sterculia villosa, Lannea coromandelica, Boswellia serrata* etc.
Sub-group 5C: Degrdation of tropical deciduous forests, edaphic type of dry eciduous forests and general seral types.	Dry deciduous scrub, dry savanna forest, dry tropical riverain forest, secondary dry deciduous forest. Important species are *Anogeissus latifolia, Terminalia chebula, Albizia* sp., *Pterocarpus marsupium* etc.
Group 6: Tropical thorn forests	
Sub-group 6A: Southern tropical thorn forests	Southern thon forest, southern thorn scrub, Important species include *Acacia* sp., *Ziziphus* sp., *Albizia chinensis*, *Diospyros montana*, *Erythroxylon monogynum*, *Bauhinia racemosa*, *Premna tomentosa*, etc.
Sub-group 6B: Nothern tropical thorn forests	Desrt thorn forest, riverain thorn forest, tropical *Euphorbia* scrub. Dominant species include *Acacia* sp., *Albizia* sp., *Butea monosperma, Prosopis* sp.,
Sub-group 6C: Genral degraded edaphic seral types.	*Acacia senegal* forest, *Salvdora* scrub, desert dune forest. Important species include *Salvadora oleoides, Acacia* sp., *Prosopis* sp., *Capparis* sp., etc.
Category 3: Montane subtropical forests	
Sub-group 8A: Southern subtropical broadleaved hill forests	Nilgiri subtropical hill forest, western subtropical hill forest, central Indian subtropical hill forests. Important species include *Persea macrantha, Neolitsea zeylanica, Litsea* sp., *Turpinia malabarica, Symplocos* sp., *Ilex denticulata* etc.

Category 3: Montane subtropical forests	
Sub-group 8B: Northern subtropical broadleaved hill forests	East Himalayan subtropical wet hill forest, Khasi subtropical wet hill forest. *Rhododendron arboreum, Myrica esculenta, Symplocos paniculata, Quercus leucotrichophora, Lyonia ovalifolia, Quercus lanata, Engelhardia spicata, Prunus nepalensis* etc.
Group 9: Subtropical Pine forests	Himalayan subtropical pine forest, Assam subtropical pine forests. Dominant species are *Cedrus deodara, Quercus* sp., *Alnus nepalensis, Acer oblongum, Pinus kesiya, Larix griffithii, Schima wallichii* etc.
Group 10: Subtropical dry evergreen forests	Subtropical dry evergreen forest. Dominant species are *Acacia* sp., *Olea cuspidata, Canthium dicoccum, Diospyros ebenum, Holoptelea integrifolia.*
Category 4: Montane temperate forests	
Sub-group 11A: Southern montane wet temperate forests	Southern montane wet temperate forest.Dominant species are *Lindera neesiana, Litsea* sp., *Cinnamomum* sp., *Isonandra candolleana, Symplocos* sp., *Beilschmiedia wightii, Daphniphylum neilgherrense, Syzygium* sp., etc
Sub-group 11B: Northern montane wet temperate forests	East Himalayan wet temperate forests, Naga hills wet temperate forests. Dominant species are *Pinus kesiya, Acer oblongum, Schima wallichii, Quercus serrata, Castonopsis* sp., *Magnolia campbellii.*
Group 12: Himalayan moist temperate forests	Lower western Himalyan tempeate forest, upper Himlayan temperate forest, East Himalayan temperate forest, East Himalayan mixed coniferous forest. Some of the important species include *Acer campbellii, Magnolia campbellii, Quercus pachyphylla, Michelia doltsopa, Betula alnoides, Actinodaphne microptera, Gordonia obtusa* etc.
Group 13: Himalayan dry temperate forests	1. Western type: Dry broad-leaved and confireous forest, dry temperate coniferous forest, west Himalayan dry temperate deciduous forest, west Himalayan high level dry blue pine forest, west Himalayan dry Juniper forest, east Himalayan dry temperate conifer forest, east Himalayan Juniper/birch forest. Dominant species are *Cedrus deodara, Pinus gerardiana, Pinus wallichiana, Picea smithiana, Rhododendron campanulatum, Juniperus macropoda, Larix griffithii, Tsuga dumosa* etc.
Group 14: Sub-alpine forests	West Himalayan sub-alpine birch/fir forest, east Himalayan sub-alpine birch/fir forest. The subalpine forests occur throughout the Himalaya above 3000 m elevation up to the tree limit., rainfall 83-600 mm. The forests are mainly evergreen, *Rhododendron* is common constituent. Tall trees are conifers; *Betula utilis* is present as the largest deciduous tree and associated with genera like *Quercus semecarpifolia*, *Sorbus* and *Rhododendron* sp. Western Himalaya sub-alpine forests are found in Jammu and Kashmir, Himachal Pradesh, and Uttrakhand. In the western Himalaya there are two types of forests (i) *Abies spectabilis* and *Betula utilis*, (ii) west Himalayan sub-alpine birch/fir forest.

Category 6: Alpine scrub	
Group 15: Moist alpine scrub	Birch/Rhododendron scrub forest, deciduous alpine scrub, alpine pastures Moist Alpine Scrub occurs throughout Himalaya, above timber line to 5500 m altitude, composed entirely of species of *Rhododendron* with some birch (*Betula*) and other deciduous trees. The tree trunks are short and highly branched, moss and ferns cover the ground. A thick layer of humus is present and soil is generally wet. In Kumaun, Uttrakhand, *Betula utilis* and *Rhododendron campanulatum* scrub forest occur. *Rhododendron-Lonicera* association occurs in Uttrakhand, in inner Himalaya. In eastern Himalaya, dense Rhododendron thickets occur at 3350-4600 m altitude. These forests are reported from Arunachal Pradesh, Sikkim and West Bengal.
Group 16: Dry alpine scrub	They are found in Jammu and Kashmir, Himachal Pradesh, Uttrakhand, and Arunachal Pradesh. Vegetation predominantly xerophytic dwarf shrubs; rainfall < 370 mm per year. Important species *Juniperus wallichiana*, *Lonicera* sp. *Potentilla* sp. Vegetation along the streams is composed of *Salix, Myricaria*, and *Hippophae rhamnoides*. In eastern Himalaya *Juniperus recurva* and *Juniperus wallichiana* occur at an altitude ranging from 3000 to 4600 m.

Table 1.
Major forest types in India with their sub types according to Champion and Seth [6] and characterstic species composition of different forests. Species composition follows Reddy et al. [14, 15].

efforts to understand the vegetation and its dynamics. There are some examples describing the efforts to understand the vegetation.

2. Forest dynamics study by Indian Institute of Science (IISc), Bengaluru, Karnataka, India

India's first biosphere reserve the Nilgiri Biosphere Reserve (NBR) was established in 1986 and the responsibility of conducting research was given to Center for Ecological Sciences (CES), Indian Institute of Science (IISc). As a principle institute responsible in setting NBR, IISc has a commitment towards ecological research in the biosphere. The climatic and altitudinal gradient in NBR harbours different vegetation types ranging from dry thorn forests to rainforests. The altitudinal range has dry forests in the lower elevation to high altitude montane forests. Hence there is a tremendous variation in species composition across both climatic and altitudinal gradient.

When IISc began its research in NBR there were several issues regarding the choice of study area. Based on both logistical and academic reasons, IISc decided to join the international network of 50-ha forest dynamics plots promoted by Prof. Hubbell [21]. CES selected species poor deciduous forests of Mudumalai for variety of reasons. Firstly, Mudumalai would complement plot at Barro Colorado Island (BCI), Panama (tropical semi-evergreen forest, neotropics) and Malaysian plot, FRIM, Malaysia (equatorial rainforest). Secondly the factors influencing the

Sl No	Forest types	Area in Sq. Km	Percent area
1	Group 1: Tropical wet evergreen forests	20,054	2.61
2	Group 2: Tropical semi-evergreen forests	71,171	9.27
3	Group 3: Tropical mist deciduous forests	135,492	1.65
4	Group 4: Littoral and Swamp forests	5596	0.73
5	Group 5: Tropical dry deciduous forests	313,617	40.86
6	Group 6: Tropical thorn forests	20,877	2.72
7	Group 7: Tropical dry evergreen forests	937	0.12
8	Group 8: Subtropical broadleaved hill forests	32,706	4.26
9	Group 9: Subtropical pine forests	18,102	2.36
10	Group 10: Subtropical dry evergreen forests	180	0.02
11	Group 11: Montane wet temperate forests	20,435	2.66
12	Group 12: Himalayan moist temperate forests	25,743	3.35
13	Group 13: Himalayan dry temperate forests	5627	0.73
14	Group 14: Sub alpine forests	14,995	1.96
15	Group 15: Moist alpine scrub	959	0.13
16	Group 16: Dry alpine scrub	2922	0.38
17	Plantation/TOF	64,839	8.45
	Total forest cover+ scrub	754,252	98.26
18	Grassland in different forest type groups (without forest cover)	13,329	1.74
	Grand total	767,581	100.00

Table 2.
Standard forest types of India according Forest Survey of India (FSI) classification.

dynamics in dry forests are totally different from factors influencing dynamics of forests at both Panama and Malaysia [22].

2.1 Choice of the site

IISc has selected 50 hectare area in 17th compartment of the Kargudi range in Mudumalai Tiger Reserve as because, 1. The area is relatively free of anthropogenic disturbances as settlements are far off, 2. This area was selectively felled during late 1960s and we could identify the trees that were removed from the plot. They could identify the species of the stumps left behind and map them spatially in the plot, 3. This area lies in the transition zone between dry and moist deciduous vegetation and has both elements represented in the plot.

2.2 Methods

Establishment of plot involved two steps a. gridding and b. enumeration of the plot. Gridding involves dividing the plot into blocks of 20 X 20 metres after correcting for slope. Correction for slope is important to give equal opportunity for all individuals to compete for resources.

Enumeration involves measuring of all woody individuals and mapping them. Block of 20 X 20 meters is further divided into blocks of 10 X 10 meters temporarily by laying ropes. All woody individuals >1 cm dbh (diameter at breast height) are identified, measured for size, marked with unique number and mapped by taking X and Y coordinates. Point of measurement (POM) was marked where size measurement was made.

2.3 Results

There were 25,929 stems duing the first census belonging to 71 species. Most abundant species was understorey tree *Kydia calycina* (Malvaceae) with 5175 individuals accounting for 20% of total abundance. Dominant canopy species such as *Anogeissus latifolia* (2280), *Lagerstroemia microcarpa* (3980). *Tectona grandis* (2143) and *Terminalia crenulata* (2776) accounted for 55.9% of total abundance. The genus *Ficus* (Moraceae) had five species and followed by *Terminalia* (Combretaceae) with three species. Relative abundance cumulative abundance of top ten species is listed in the **Table 3**. Top ten species accounted for 87.52% of the total abundance. There were 21 species with less than 10 individuals in the plot and 7 species had one individual in the entire 50 hectare plot (**Table 3**).

Most species showed clumped dispersion. *Kydia calycina, Lagerstroemia microarpa, Terminalia elliptica, Helicteres isora, Anogeissus latifolia, Catunaregam spinosa* and *Shorea roxburghii* are species that showed highly clumped distribution at 1 hectare scale while species such as *Tectona grandis, Emlica officinalis, Grewia tiliifolia, Syzygium cumini, Diospyros montana, Schleichera oleosa, Terminalia chebula* and *Gmelina arborea* showed lesser degree of clumping at one hectare scale.

2.4 Forest dynamics (1988–2016)

Population of woody species in the plot has shown considerable fluctuations over different census periods. Population has grown from 25,935 individuals >1 cm dbh in 1988 to 48,360 individuals in 2016 (**Figure 1**). The population across different census years has shown fluctuations. There was a negative trend during

Species	Family	Abundance (50 ha)	Relative abundance (%)	Cumulative abundance (%)
Kydia calycina Roxb.	Malvaceae	5175	19.96	19.96
Lagerstroemia microcarpa Wt.	Lythraceae	3980	15.35	35.31
Terminalia elliptica Willd. (syn. *Terminalia crenulata* Roth)	Combretaceae	2776	10.71	46.01
Helicteres isora L.	Malvaceae	2571	9.91	55.93
Anogeissus latifolia (Roxb. ex DC.) Wall. Ex Guill & Perr.	Combretaceae	2280	8.79	64.72
Tectona grandis L.f.	Lamiaceae	2143	8.26	72.99
Cassia fistula L.	Fabaceae	1881	7.25	80.24
Catunaregam spinosa (Thunb.) Tirveng. [syn. *Randia dumetorum* (Retz.) Lam., *Xeromphis spinosa* (Thunb.) Keay, *Randia spinosa* (Thunb.) Poir.]	Rubiaceae	770	2.97	83.21
Phyllanthus emblica L.	Phyllanthaceae	577	2.22	85.44
Grewia tiliifolia Vahl	Tiliaceae	539	2.08	87.52

Table 3.
Top ten dominant species in the plot during the first census (1988–1989).

1988–1992 (−31.9%), 1992–1996 (−13.2%) and 2000–2004 (−14.2%). There was huge surge of 138.9% in populations between 2004 and 2008 census. There was a positive trend during 1996–2000 (18.3%), 2008–2012 (1.1%), 2012–2016 (28.7%).

There is an overall increase of 86.4% with the total population during 1988–2016. However, individual species showed an interesting trend in the populations (**Table 4**). Among canopy species *Dalbergia latifolia* has shown a large change of 469.3% (population changed from 75 individuals in 1988 to 427 individuals in 2016). Among the four dominant canopy species except for *Lagerstroemia microcarpa* (13.4%) others such as *Anogeissus latifolia* (−25.3%), *Tectona grandis* (−17.4%) and *Terminalia elliptica* (22.2%) have shown negative change. Other species that have significant change include *Gmelina arborea* (−72.8%), *Ougeinia oojensis* (89.3%) and *Shorea roxburghii* (82.0%).

Among the understorey species, *Helicteres isora* (776.1%) and *Randia dumetorum* (679.0%) registered exceptionally higher population growth. *Helicteres isora* increased from 2573 individuals in 1988 to 22,543 individuals in 2016 while *Randia dumetorum* from 770 individuals in 1988 to 5999 individuals in 2016. There was a significant reduction in population of *Eriolaena quinquelocularis* from 249 individuals in 1988 to 7 individuals in 2016.

2.4.1 Patterns in mortality and recruitment

Entire plot (50 hectares) was annually censused for mortality and recruitment till 2008. Since 2009, annual census was done in sample plots of 40 meters X 40 meters inside the plot. There were 100 such randomly placed plots accounting for little more than 1/3rd of the total area. The reports on annual mortality for the plot from sample plots from 1989 to 2016 were published [23].

The community wide mean annual mortality rate was 7.67±5.75% (range 1.57–21.5%, N = 28) while mean annual recruitment rate was 11.1±14.0% range = 0.65–58.5%, N = 28). Though recruitment rate was higher than the mortality

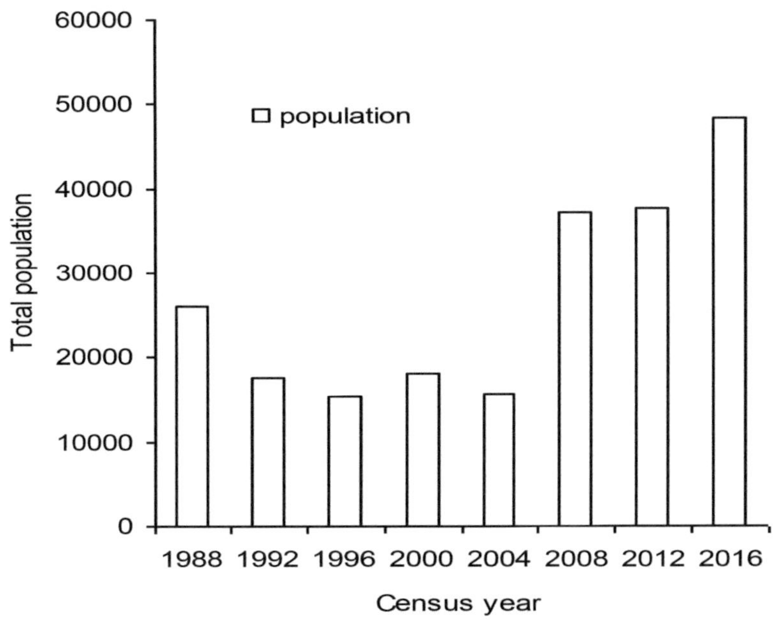

Figure 1.
Total population in different census years (all woody stems >1 cm dbh).

Species	Family	Census 1988	Census 2016	% Change
Canopy species				
Anogeissus latifolia (Roxb. ex DC.) Wall. Ex Guill& Perr.	Combretaceae	2281	1702	−25.38
Dalbergia latifolia Roxb.	Fabaceae	75	427	469.33
Diospyros montana Roxb.	Ebenaceae	130	135	3.85
Gmelina arborea Roxb.	Lamiaceae	59	16	−72.88
Grewia tiliifolia Vahl	Tiliaceae	540	592	9.63
Lagerstroemia microcarpa Wt.	Lythraceae	3982	4516	13.41
Ougeinia oojeinensis (Roxb.) Hochr.	Fabaceae	113	12	−89.38
Schleichera oleosa (Lour.) Oken	Sapindaceae	75	96	28.00
Shorea roxburghii G. Don (Syn. *Shorea talura* Roxb.)	Dipterocarpaceae	78	14	−82.05
Stereospermum personatum (Hassk.) Chatterjee	Bignoniaceae	112	129	15.18
Syzygium cumini (L.) Skeels.	Myrtaceae	414	442	6.76
Tectona grandis L.f.	Lamiaceae	2141	1768	−17.42
Terminalia elliptica Willd.	Combretaceae	2771	2154	−22.27
Understory species				
Cassia fistula L.	Fabaceae	1887	3025	60.31
Cordia oblique Willd.	Bignoniaceae	192	281	46.35
Eriolaena quinquelocularis (Wight & Arn.) Wight	Malvaceae	249	7	−96.39
Helicteres isora L.	Malvaceae	2573	22,543	776.14
Kydia calycina Roxb.	Malvaceae	5167	1018	−80.30
Lagerstoemia parviflora Roxb.	Lythraceae	94	51	−45.74
Phyllanthus emblica L.	Phyllanthaceae	576	296	−48.61
Catunaregam spinosa (Thunb.) Tirveng. [syn. *Randia dumetorum* (Retz.) Poir.]	Rubiaceae	770	5999	679.09
Radrmachera xylocarpa (Roxb.) K. Schum.	Bignoniaceae	358	356	−0.56
Terminalia chebula Retz.	Combretaceae	62	22	−64.52

Table 4.
Population changes observed among several canopy and understorey species in the 50 ha plot, Mudumalai.

rate, recruitment rate had high variability compared to mortality rate. There was considerable fluctuation in the mortality and recruitment rate across census periods (**Table 5**). Mean annual mortality due to fire across different census years was 2.87±5.75% (range = 0–20.6%, N = 28). Elephants resulted in mortality rate of 2.33±2.04% (range = 0.28–7.56%, N = 28) while mortality due to other causes was 2.47±2.49 (range = 0.46–12.58%). Mortality rates due to fire spiked in the years of dry season fire (1991, 1992, 1996, 2002 and 2010), resulting in mortality rates over 10%. Highest mortality of 12.58% was seen in the year 2016 where there was no dry season fire. Elephant related mortality was high in year 2008 after the massive "episodic recruitment" observed in one of the favoured speies *Helicteres isora*.

The community wide mean mortality rate was high during the first census (1989–1992) which had two years of consecutive dry season fires across different

Parameter	Census 1	Census 2	Census 3	Census 4	Census 5	Census 6	Census 7
Mortality (%)	11.57 ± 3.48	7.91 ± 5.45	3.38 ± 1.45	8.98 ± 8.18	6.91 ± 3.74	9.85 ± 8.79	5.14 ± 5.80
Recruitment (%)	2.30 ±1.17	3.98 ± 1.62	9.19 ± 4.25	5.75 ± 5.54	38.76 ± 15.74	11.87 ± 10.98	10.70 ± 12.29

Table 5.
Mean overall mortality and recruitment rates in the 50 hectare forest dynamics plot, Mudumalai.

Parameter	Census 1	Census 2	Census 3	Census 4	Census 5	Census 6	Census 7
Fire (%)	4.48 ± 2.86	4.92 ± 6.08	0.03 ± 0.05	5.13 ± 9.64	0.00 ± 0.00	5.54 ± 10.10	0.00 ± 0.00
Elephant (%)	4.62 ± 1.88	1.73 ± 0.69	1.38 ± 0.69	0.76 ± 0.43	4.22 ±2.55	2.52 ± 2.77	1.09 ± 0.19
Others (%)	2.48 ± 0.75	1.26 ± 0.47	1.97 ± 0.81	3.08 ± 2.76	2.70 ± 1.74	1.78 ± 2.06	4.05 ± 5.75

Table 6.
Mean mortality rate due to different causes in the 50 hectare forest dynamics plot, Mudumalai.

census period. Lowe mortality rate with low variability was seen during the third census (1996–2000). Recruitment rate also showed considerable fluctuation across different census periods ranging from as low as 2.3% during first census (1989–1992) to 38.76% during fifth census (2004–2008). Variability was high during the seventh census (2012–2016) (**Table 5**).

Mean mortality rates due to different causes across census periods is tabulated in the **Table 6**. Mean rate of mortality due to fire during the census period between (2004–2008) and (2012–2016) was zero suggesting fire did not result in the mortality of any individual. There was a considerable variability in mortality rates across other census periods (**Table 6**). Elephant related mortality rate was high during the first census period (1989–1992) and 5th census period (2004–2008) owing to abundance of elephant favoured species such as *Kydia calycina* and *Helicteres isora*. The mortality rate inflicted by other causes also showed considerable fluctuation with high variability during the 7th census period (2012–2016).

2.4.2 Basal area changes and biomass across census periods

The basal area in the plot has been steadily increasing over time (**Figure 2**). Above Ground Biomass (AGB) and hence carbon stock also shows a similar trend (**Figure 3**), with both the native woody vegetation and invasive ground vegetation showing increment. Basal area changes do not necessarily translate to AGB changes: for instance, the slight decline in basal area during 1992–1996 is not reflected in AGB, which may be partly due to differences in wood densities (e.g. hardwoods growing more than softwoods). Native woody vegetation biomass in 2004 shows a slight reduction owing to a severe drought in the preceding years and a large fire in 2002. However, the invasive *Lantana camara* L. increased substantially following the drought, and therefore the total biomass remained at the 2000 census level. Large increment in biomass were seen in all subsequent censuses: 2008, which followed a period of higher than average precipitation and no fires, 2012, despite a fire in 2010 and only 807 mm precipitation in 2012 (compared to the long-term average of 1260 mm), and 2016, which follows a fire-free census interval. The

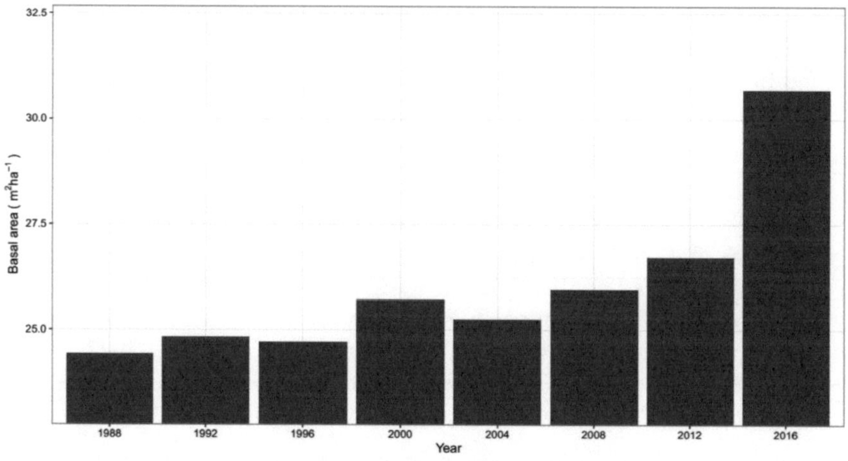

Figure 2.
Native woody-plant basal area (per hectare) in the 50-ha plot.

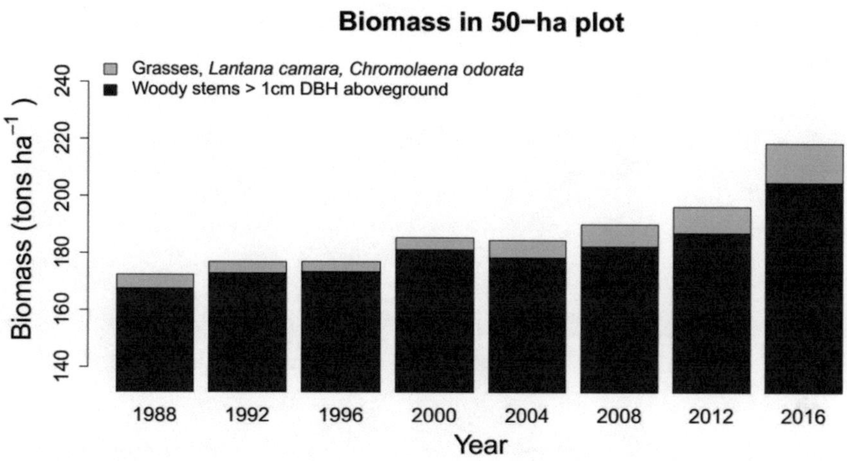

Figure 3.
Aboveground biomass (per hectare) in the 50-ha plot, showing contributions from native woody-plants as well as other ground vegetation.

estimates for basal area and biomass for the census period 2016 to 2020 is based is based on first 10 Ha.

3. Long-term monitoring programme of Kerala Forest Research Institute (KFRI), Peechi, Thrissur, Kerala

Being a part of Western Ghats range of mountains, one of the global biodiversity hotspots, Kerala has bestowed with diverse forest ecosystem with high degree of endemism. Kerala Forest Research Institute (KFRI) currently having more than 40 permanent plots across the state representing all major forest ecosystems (**Figure 4**) and more plots are coming up as a part of various ongoing research projects. KFRI Long-term monitoring programme represents all major ecosystems like mangrove, moist deciduous, dry deciduous, wet evergreen, montane shola forests and grasslands. As of now, the programme covers 50,309 woody individuals of more than 350 species. Majority of our plots are smaller in size (≤1 ha) in which survey would be conducted at five year intervals. These plots were established in different time periods under various research projects undertaken by KFRI. The oldest set of plots was established was during 2000–2002. Recently a large 10 Ha permanent plot was established in wet evergreen forests, Sholayar range. Vazhachal Forest Division, Kerala (**Figure 4**).

3.1 Sholayar 10 ha plot

A permanent plot of 10 ha (500 × 200 m^2) size was established in a tropical evergreen forest at Karadichola, Sholayar Range, Vazhachal Forest Division, Kerala in Southern Western Ghats. Plot establishment and baseline data collection were done based on the Forest-GEO [24] (CTFS) protocol during 2016–2017. Comparison of Sholayar plot with other sites which are following Forest-GEO protocol is summarized as **Table 7**. Complete inventory of woody individuals ≥1 cm dbh were

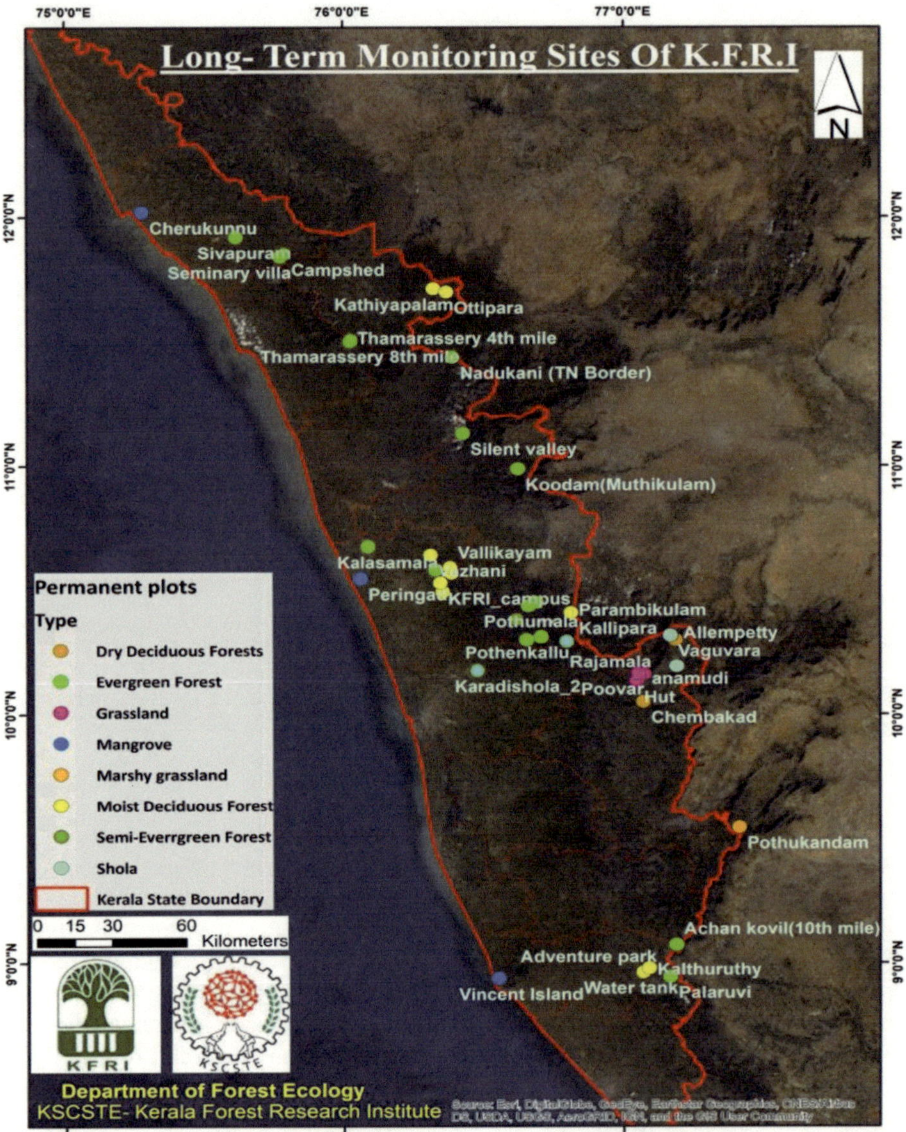

Figure 4.
Distribution of Long-term monitoring plots of KFRI in Southern Western Ghats.

CTFS plot location	Latitude	Mean elevation (m)	Land type*	Plot size (ha)	No. of species	No. of families	Trees/ ha	Basal area (m^2/ha)
Lambir, Malaysia	4.19	170	I	52	1182	83	6915	43.5
Yasun, Ecuador	−0.69	230	M	50	1114	81	3026	33.0
Pasoh, Malaysia	2.98	80	M	50	814	82	6708	31.0
Khao Chong, Thailand	7.54	140	M	24	593	Na	5063	Na
Korup, Cameroon	5.07	200	M	50	494	62	6580	32.0
Ituri, Dem. Rep. of Congo	1.44	780	M	404	445	Na	7200	Na
Palanan, Philippines	17.04	110	I	16	335	60	4125	39.8
Bukit Timan, Singapore	1.25	150	I	2	329	62	5950	34.5
BCI, Panama	9.15	140	M	50	299	58	4168	32.1
Mo Singto, Thailand	14.43	770	M	30.5	262	Na	Na	Na
HuaiKhaKhaeng, Thailand	15.63	590	M	50	251	58	1450	31.2
La Planada, Colombia	1.16	1840	M	25	240	54	4216	29.8
Dinghushan, China	23.16	350	M	20	210	Na	3581	Na
Sinharaja, Sri Lanka	6.4	500	I	25	204	46	7736	45.6
DoiInthanon, Thailand	18.52	1700	M	15	192	Na	4913	Na
Luquillo, Puerto Rico	18.33	380	I	16	138	47	4194	38.3
Nanjenshan, Taiwan	22.06	320	I	3	125	41	12,133	36.3
Ilha do Cardoso, Brazil	−25.1	5	M	10	106	Na	Na	Na
Mudumalai, India	11.6	1050	M	50	72	29	510	25.5
Laupahoehoe, USA	19.93	1150	I	4	21	15	3078	67.36
Palamanui, USA	19.74	240	I	4	15	15	3487	8.6
Sholayar, India (KFRI PLOT)	10.29	950	M	10	106	44	2539	47.7

**I - Island, M - Mainland, Na - not available.*

Table 7.
Comparison of vegetation parameters of 10 ha plot at Sholayar, Kerala with other Forest-GEO (CTFS) sites around Globe.

done and each individual was permanently tagged with sequentially numbered aluminium tags. In the 10-ha plot, a total of 25,390 individuals of 106 woody species were recorded [25]. These individuals were belonging to 44 families and 81 genera.

Small-diameter class (1 cm ≤dbh<10 cm)were 3.6 times more abundant than large-diameter (dbh ≥10 cm) ones. There were 19,975 small-diameter plants (78.67% of all stems), averaging 1997 individuals/ha, while large-diameter trees had an abundance of 5415 plants and density of only 546 individuals/ha. The family Rubiaceae, is the most abundant, with densities >900 individuals/ha, followed by Euphorbiaceae, Urticaceae, Sapotaceae, Meliaceae, Malvaceae, and Putranjivaceae. Among tree genera *Palaquium, Cullenia*, of family Sapotaceae and Malvaceae respectively, were the most abundant, with >100 individuals/ha, followed by *Drypetes, Mesua, Aglaia, Vateria, Syzygium* and *Agrostistachys*. At tree species level, *Palaquium ellipticum, Cullenia exarillata, Mesua ferrea* were the most abundant, followed by *Aglaia tomentosa* and *Vateria indica.* Out of the 106 species, 38 species are endemic to Western Ghats and 20 are listed in IUCN categories. Among trees, *Palaquium ellipticum*, *Cullenia exarillata* and *Vateria indica* were dominant in terms of Importance Value Index (IVI) while among shrubs *Psychotria nudiflora* was the dominant, followed by *Dendrocnide sinuata* and *Psychotria anamalayana*. Total basal area (in 10 ha) was 477.24 m^2 (47.72± 7.58 m^2/ha) and family wise it was higher in Malvaceae (9.69 m^2/ha) and Sapotaceae (9.6 m^2/ha) followed by Dipterocarpaceae, Calophyllaceae, Putranjivaceae and Sapindaceae. The dominant genera by basal area were *Cullenia, Palaquium, Vateria, Mesua, Drypetes* and *Dimocarpus*. At species level, *Cullenia exarillata* and *Palaquium elipticum* were the most important followed by *Vateria indica, Mesua ferrea, Dimocarpus longan, Drypetes wightii* and *Holigarna nigra*. Small-diameter trees contributed 4.89% of the total basal area. Species level contribution to density and importance value index is summarized in **Table 2**. Girth class distribution pattern indicates it as a relatively undisturbed forest patch with healthy regeneration. Ecological and Ecophysiological studies are ongoing on the structure, function and dynamics of this system in the context of climate change (**Table 8**).

Sl. No.	Species	No. of Individuals (10 Ha)	IVI
1.	*Actinodaphne bourneae* Gamble	11	0.1939
2.	*Actinodaphne malabarica* N.P.Balakr.	42	0.9148
3.	*Aglaia lawii* (Wight) C.J. Saldanha	48	0.7948
4.	*Aglaiaper viridis* Hiern	30	0.515
5.	*Aglaia simplicifolia* (Bedd.) Harms	39	0.6217
6.	*Aglaia* sps.	22	0.4034
7.	*Aglaia tomentosa* Teijsm. & Binn.	706	7.2858
8.	*Agrostistachys borneensis* Becc.	594	6.7618
9.	*Alstonia scholaris* (L.) R.Br.	1	0.0206
10.	*Ancistrocladus heyneanus* Wall. ex J.Graham	62	0.6279
11.	*Antidesma montanum* Blume	249	4.3758
12.	*Aphanamixis polystachya* (Wall.) R.Parker	3	0.0590
13.	*Aporosa acuminate* Thwaites	163	1.8734
14.	*Aporosa cardiosperma* (Gaertn.) Merr.	12	0.1705
15.	*Aporosa* sp.	229	2.9408
16.	*Ardisia pauciflora* B.Heyne ex Roxb.	246	3.1307
17.	*Artocarpus heterophyllus* Lam.	1	0.0221
18.	*Atlantia racemosa* Wight ex Hook.	2	0.0391

Sl. No.	Species	No. of Individuals (10 Ha)	IVI
19.	*Baccaurea courtallensis* (Wight) Müll.Arg.	2	0.0434
20.	*Bhesa indica* (Bedd.) Ding Hou	109	2.4132
21.	*Bischofia javanica* Blume	2	0.1796
22.	*Boehmeria glomerulifera* Miq.	55	0.5937
23.	*Calophyllum polyanthum* Wall. ex Planch. &Triana	15	0.6838
24.	*Canarium strictum* Roxb.	22	0.5249
25.	*Canthium angustifolium* Roxb.	7	0.1210
26.	*Caryota urens* L.	1	0.0358
27.	*Chassalia curviflora* var. *ophioxyloides* (Wall.) Deb & B. Krishna	15	0.2601
28.	*Cinnamomum malabatrum* (Burm.f.) J. Presl	12	0.2231
29.	*Cipadessa baccifera* (Roth) Miq.	3	0.0603
30.	*Clausena anisata* (Willd.) Hook.f. ex Benth.	4	0.0854
31.	*Clausena austroindica* B.C.Stone & K.K.N. Nair	2	0.0392
32.	*Coscinium fenestratum* (Goetgh.) Colebr.	2	0.0416
33.	*Croton zeylanicus* Müll.Arg.	514	4.0406
34.	*Cullenia exarillata* A.Robyns	1027	27.9277
35.	*Dendrocnides inuata* (Blume) Chew	1376	11.0712
36.	*Dichapetalum gelonioides* (Roxb.) Engl.	1	0.0194
37.	*Dimocarpus longan* Lour.	373	8.2856
38.	*Diospyros paniculata* Dalzell	118	2.6948
39.	*Diospyros* sp.	27	0.6139
40.	*Drypetes malabarica* (Bedd.) Airy Shaw	413	6.8282
41.	*Drypetes wightii* (Hook.f.) Pax & K. Hoffm.	600	9.2968
42.	*Dysoxylum malabaricum* Bedd. ex C. DC.	390	5.2566
43.	*Elaeocarpus tuberculatus* Roxb.	32	1.3709
44.	*Elaeocarpus variabilis* Zmarzty	63	1.6316
45.	*Eugenia mooniana* Wight	130	1.4946
46.	*Fagraea ceilanica* Thunb	3	0.0853
47.	*Garcinia morella* (Gaertn.) Desr.	190	3.0605
48.	*Glycosmis macrocarpa* Wight	47	0.7725
49.	*Goniothalamus rhynchantherus* Dunn	106	1.5670
50.	*Holigarna arnottiana* Hook.f.	1	0.0403
51.	*Holigarna nigra* Bourd.	115	4.6342
52.	*Isonandra lanceolata* Wight	33	0.4739
53.	*Kunstleria keralensis* C.N. Mohanan & N.C. Nair	134	1.6989
54.	*Lasianthus rostratus* Wight	396	4.0462
55.	*Leea indica* (Burm. f.) Merr.	139	1.7541
56.	*Lepisanthes erecta* (Thwaites) Leenh.	3	0.0596
57.	*Litsea bourdillonii* Gamble	328	4.8171
58.	*Litsea* sp.	14	0.3211

Sl. No.	Species	No. of Individuals (10 Ha)	IVI
59.	*Litsea wightiana* (Nees) Hook.f.	208	2.8858
60.	*Luvunga eleutherandra* Dalzell	63	0.9326
61.	*Macaranga peltata* (Roxb.) Müll.Arg.	28	0.8126
62.	*Mallotus atrovirens* Wall. ex Müll.Arg.	167	2.5887
63.	*Mallotus aureopunctatus* (Dalzell) Müll.Arg.	180	2.2228
64.	*Mallotus philippensis* (Lam.) Müll.Arg.	35	0.7248
65.	*Mallotus* sps.	13	0.2298
66.	*Mangifera indica* L.	15	0.3446
67.	*Mastixia arborea* (Wight) C.B.Clarke	148	2.1348
68.	*Meiogyne pannosa* (Dalzell) J. Sinclair	360	3.8657
69.	*Meiogyne ramarowii* (Dunn) Gandhi	29	0.4852
70.	*Meliosma simplicifolia* (Roxb.) Walp.	145	2.6122
71.	*Memecylon* sp.	27	0.4366
72.	*Mesua ferrea* L.	878	14.255
73.	*Micrococca beddomei* (Hook.f.) Prain	8	0.1290
74.	*Microtropis stocksii* Gamble	1	0.0195
75.	*Myristica malabarica* Lam.	28	0.6579
76.	*Myristica attenuata* Wall.	362	5.6289
77.	*Nothopegia* sp.	1	0.0221
78.	*Ocotea lancifolia* (Schott) Mez	354	4.6706
79.	*Oreocnide integrifolia* (Gaudich.) Miq.	108	1.2404
80.	*Otonephelium stipulaceum* (Bedd.) Radlk.	61	0.9228
81.	*Palaquium ellipticum* (Dalzell) Baill.	1257	28.6825
82.	*Pavetta* sp.	1	0.0194
83.	*Persea* sp.	1	0.0201
84.	*Piper* sp.	6	0.1162
85.	*Polyalthia coffeoides* (Thwaites) Hook.f. & Thomson	23	0.5240
86.	*Psychotria anamalayana* Bedd.	604	5.0707
87.	*Psychotria nudiflora* Wight &Arn.	8340	37.6802
88.	*Sabia limoniacea* Wall. ex Hook.f. & Thomson	28	0.3739
89.	*Salacia fruticosa* Wall.	1	0.0196
90.	*Saprosma glomeratum* (Gardner) Bedd.	1	0.0193
91.	*Semecarpus travancorica* Bedd.	13	0.4371
92.	*Smythea bombaiensis* (Dalzell) S.P.Banerjee & P.K. Mukh.	43	0.7766
93.	*Spondias pinnata* (L.f.) Kurz	1	0.1700
94.	*Strobilanthes barbatus* Nees	37	0.2613
95.	*Strobilanthes* sp.	660	3.3360
96.	*Strychnos dalzellii* C.B.Clarke	188	2.1646
97.	*Symplocos macrophylla* Wall. ex A. DC ssp. *rosea* (Bedd.) Nooteb.	131	1.7905

Sl. No.	Species	No. of Individuals (10 Ha)	IVI
98.	*Syzygium lateum* (Buch.-Ham.) Gandhi	517	5.2191
99.	*Syzygium munronii* (Wight) N.P.Balakr.	73	1.2097
100.	*Syzygium* sp.	5	0.2121
101.	*Tetrastigma leucostaphylum* (Dennst.) Alston	22	0.3552
102.	*Thottea siliquosa* (Lam.) Ding Hou	121	1.4741
103.	*Toddalia asiatica* (L.) Lam.	11	0.1927
104.	*Turpinia malabarica* Gamble	145	4.1535
105.	*Vateria indica* L.	605	16.4577
106.	*Xanthophyllum arnottianum* Wight	1	0.0195

Table 8.
Species- level contribution to the community in the 10 ha. plot, Sholayar, Kerala.

4. Forest dyanmics study by the French Institute of Pondicherry

4.1 A case study: Uppangala Permanent sampling plots

Since the early 1980s, the French Institute of Pondicherry has been in collaboration with the Forest Department of Kerala and Karnataka to explore structure and diversity of wet evergreen forests of the Western Ghats. In 1979–80, a total of 147 trees ≥ 30 cm girth at breast height was monitored untill 1982 for growth (with a precision of 0.02 mm) at monthly intervals in a 0.2 ha plot at Attapadi. Monitoring the plot in the region was stopped for logistic reasons. Subsequently, IFP has established two sets of sample plots in low elevation wet evergreen forest in Kadamakkal RF, Sampaje Range, Kodagu (ca. 12°32′15″N, 12° 33' N, 75°39'4″E; **Figure 1a**). Currently this area comes under the Pushpagiri Wildlife Sanctuary in Kodagu district. The study area, the Uppangala was subjected to selective logging, between 1974 and 1983 [26]. During the logging operation, the forest was divided into compartments of 28 ha each, 237 to 359 large trees (stems ≥ 180 cm) of medium wood (> 0.5 but ≤ 0.72 g cm^{-3}) *Dipterocarp* species viz., *Dipterocarpus indicus* and *Vateria indica* were logged per compartment. An average of 8 to 13 dipterocarp trees ha^{-1} were logged manually and hauled using elephants locally, a method that causes much less damage than mechanized skidding. A few patches of forest remain unlogged. The elevation ranges between 400 and 600 m a.s.l. It belongs to the *Dipterocarpus indicus-Kingiodendron pinnatum-Humboldtia brunonis* type of wet evergreen forests and is a part of the West Coast Tropical Forests of Champion and Seth's classification. Uppangala receives slightly more than 5100 mm per year and the dry season lasts 4.5 months.

The first set of sample plots was installed in 1984 to study the post-logging effects on the forest dynamics of a once logged 30-ha compartment (**Figure 5b**). It consists of 14 plots of 600 m^2 (20 x 30 m). All trees ≥ 10 cm girth at breast height (gbh) were recorded during the first census. All the plots were recensused (except 4 plots, which were recorded as burnt) in 1988 and 1993 for recruitment and mortality. In 1989, a second set of sample plots was established in another 30-ha compartment (**Figure 5c**), which had escaped logging operation due to the ban on selective felling from 1987 in the forest of Western Ghats.

The unlogged compartment probably represents the last example of old-growth low-elevation *Dipterocarp* forest in the entire Western Ghats. Five north–south oriented transects (viz., A, B, C, D and E; **Figure 5c**) of 20 m wide, 180 to 370 m

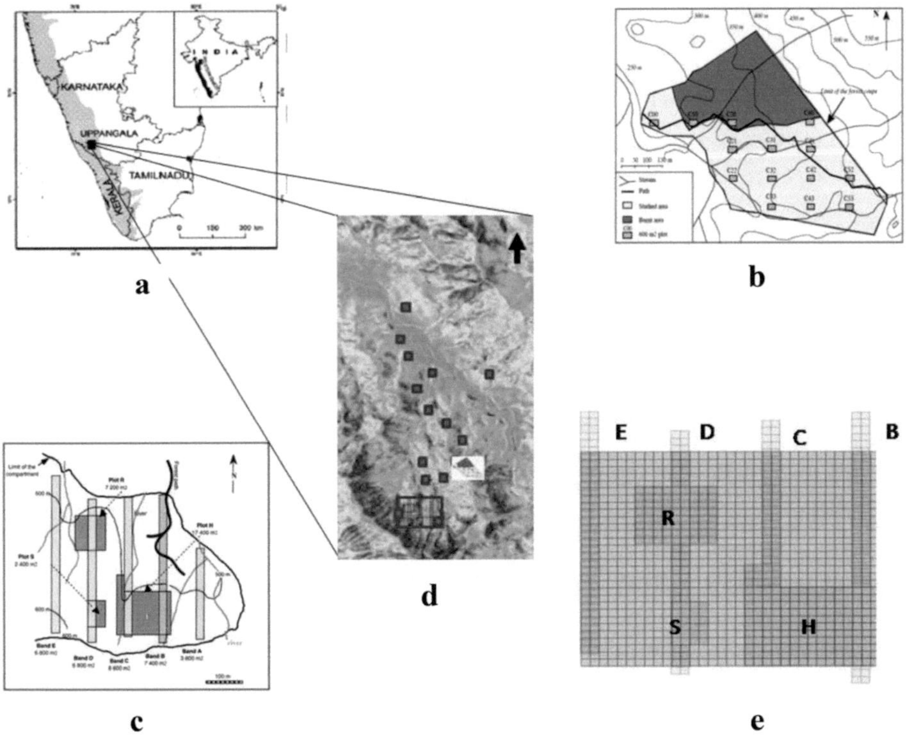

Figure 5.
(a-e) Uppangala study site, (a) Location of Uppangala study site in the central Western Ghats, Karnataka, India. Sampling designs for the inventory of trees: (b) systematic sampling in logged and (c) sampling plots and transects in unlogged (d) fifteen 1 ha plots in both compartments and (e) contiguous 9.9 ha plot in the unlogged compartment of the low elevation wet evergreen forests.

long and 100 m apart from each other were established to inventory trees ≥30 cm gbh. Collectively they represent a 3.12 ha^{-1} systematic sample. Subsequently, additional rectangular sampling plots viz., H, R and S, which overlap with sampling area of the transects and represent an additional area of 1.95 ha, were established between 1990 and 1993 to study the forest dynamics according to topography (slope and more or less flat terrain). Totally 3870 trees were identified, mapped and installed with dendrometric belt (precision of 0.2 mm) for growth monitoring, which has no equivalent in any other tropical forest in the world. The sampling area has been monitored annually for recruitment and mortality. In 2010–2011, fifteen 1 ha plots was established to appraise allometric relationship of tree diameter and tree crown (for trees ≥ 30 cm gbh) and to estimate above ground biomass. These 1 ha plots were sampled randomly in both the logged and unlogged forests. Of these, four plots were selected to understand the residual impact of logging on species composition, population structure and biomass. In 2013–2014, the sampling area of the unlogged compartment has been increased to 9.9 ha and all trees ≥ 30 cm gbh were inventoried within a 330 x 300 m^2 area (**Figure 5d**). All the trees were identified, girth measured and mapped.

4.2 Results of three decade long research in Uppangala

4.2.1 Tree density and diversity

The systematic sampling plots of logged compartment was recorded with a total of 2748 trees ≥ 10 cm girth at breast height (gbh) during the first census in 1985 [27].

Similarly, a total of 1981 trees ≥ 30 cm gbh were recorded with 91 species in the first census of 3.14 ha area of the unlogged compartment in 1990 [28]. Pronounced species hierarchy is another characteristic feature of the forest. Just 10 most abundant species contributed 71% for the forest stand (**Table 9**). Subsequent additional sampling area allow us to monitor more number of trees in the unlogged compartment. Totally 3870 trees were enumerated in the 5.07 ha during 1994 census and all those trees fitted with stainless dendrometric belts for measuring growth with a precision of 0.2 mm. At present we are monitoring 6672 trees (of which 3127 trees were installed with dendrometer bands) representing 111 species in the unlogged forest plot. The forest is characterized by high tree density and basal area (661 stems ha^{-1}; 43 m^2 ha^{-1}). Pronounced species hierarchy is another characteristic feature of the forest. Just four species namely, *Dipterocarpus indicus* (emergent layer), *Vateria indica* (upper canopy and emergent), *Myristica dactyloides* (intermedidate) and *Humboldtia brunonis* (understorey) dominate the forest stand, and they collectively account for greater than 50% of density and basal area of the forest.

4.2.2 Impact of logging on tree diversity

A decade long monitoring of logged and unlogged forest for trees ≥ 30 cm gbh revealed the logged compartment had 347 trees and 54 species in 0.6 ha (1986) whereas the unlogged compartment had 1891 trees and 88 species in 3.12 ha in 1990 [29]. Initial stand density and basal area of the trees were slightly lower in the logged forest (578 stems ha^{-1}; 34.8 m^2 ha^{-1}) than in the unlogged forest (606 stems ha^{-1}; 39.3 $m^2 ha^{-1}$). Mean density and basal area for the 20 × 30 m^2 samples of the two compartments displayed no significant difference (t-tests, P>0.25). The mortality rate was more or less similar for the compartments (0.89% for logged and 0.87% for unlogged), which is lower than the rates observed in other tropical forests. Annual recruitment rate of logged (1.68%) and unlogged forests (1.34%) were not significantly different. Mean diameter increment was 2.1 mm and 2.9 mm $yr.^{-1}$ for unlogged and logged compartments. In the logged forest, *Antiaris toxicaria*, *Aphanamixis polystachya*, *Beilschmiedia wightii* disappeared while *Cinnamomum malabatrum*, *Holigarna arnottiana*, *Microtropis stocksii*, *Sterculia*

Species (Family)	Family	Abundance	Relative abundance (%)	Cumulative abundance
Vateria indica L.	Dipterocarpaceae	329	17	23
Humboldtia brunonis Wall.	Caesalpiniaceae	294	15	31
Myristica dactyloides Gaertn.	Myristicaceae	263	13	45
Knema attenuata (J.Hk. & Th.) Warb.	Myristicaceae	118	6	51
Palaquium ellipticum (Dalz.) Baill.	Sapotaceae	98	5	56
Drypetes elata (Bedd.) Pax. & Hoffm.	Euphorbiaceae	79	4	60
Dipterocarpus indicus Bedd.	Dipterocarpaceae	67	3	63
Reinwardtiodendron anamalaiense (Bedd.) Mabb.	Meliaceae	62	3	66
Mesua ferrea L.	Clusiaceae	56	3	69
Garcinia morella (Gaertn.) Desr.	Clusiaceae	44	2	71

Table 9.
Top-ten dominant species in the unlogged compartment plot during the first census 1990.

guttata and *Vitex altissima*. In the same time, unlogged plots showed no disappearance of species and appearance of three new species namely, *Agrostistachys borneansis*, *Clerodendrum viscosum* and *Syzygium hemisphericum*. This decade long investigation suggested that the logged compartment gradually recovered and resemble unlogged forest within 20 years. However the recent inventories of the logged compartment at 1 ha scale shows the residual impact of logging even after 27 years (**Table 10**; [30]). Logged plots had low floristic similarity between them (0.45 to 0.56%) and also with the unlogged plots (0.41 to 0.43%). Mantel and partial Mantel tests proved that logging was the main driver for the species composition rather than the elevation and spatial distance. Higher abundance of species belonging to canopy, intermediate and light wood categories and lower density of emergent, understory and medium wood types were recorded in the logged plots. As compared to unlogged plots, logged plots had 20–59% less above ground biomass (AGB) due to paucity of large trees, especially in the emergent and medium wood types. However, the logged plots had higher AGB in canopy and hardwood categories. These findings indicated that the compositional shifts has occurred in the logged patches and the recovery process may depend on the resurgence of emergent and medium wood categories (**Figure 6**).

4.2.3 Forest dynamics in unlogged forest

In the unlogged forest, over the study period of 1990–2016, mortality rates ranged from 0.7 to 1.2% $yr.^{-1}$ with an average of 0.8% $yr.^{-1}$ while the recruitment ranged from 0.4 to 1.2% $yr.^{-1}$ with an average of 1% $yr.^{-1}$ (**Figure 7**). The basal area of the stand showed a loss of 13.8% due to tree death and an addition of 21.6% basal area by growth of trees. Overall, it shows an increment by 7.8% of the stand basal area. During the period of 26 years, four species *Memecylon wightii* Thwaites, *Goniothalamus cardiopetalus* (Dalz.) J. Hk. & Thoms., *Clerodendrum viscosum* Vent. and *Walsura trifolia* (A. Juss.) Harms. disappeared by tree deaths and one species *Diospyros assimilis* Bedd. appeared by new recruits. A total of 73 species have registered either recruitment and/or mortality, while population density of the remaining 18 species was unchanged during 26 year period. Of these, 44 species showed decline in population density. Notable among them includes *Myristica dactyloides, Humboldtia brunonis, Palaquium ellipticum* and *Knema attenuata*, lost more than 10 individuals during the study period. Nineteen species showed increase in population density. They include *Kingiodendron pinnatum, Holigarna nigra*, *Diospyros bourdillonii* and *Leptonychia caudata* each was recorded with increase in population density of 5 individuals.

Variables	Logged plots				Unlogged plots		Total (mean)
	LP1	LP2	LP3	LP4	UP1	UP2	
Number of species	59	72	66	56	53	57	126 (60.5)
Number of families	32	28	25	23	24	25	41 (26.2)
Stem density	572	513	636	672	665	680	3738 (623.0)
Basal area (m^2 ha^{-1})	28.15	39.3	41.37	48.23	51.19	51.01	259.24 (43.21)
AGB (Mg ha^{-1})	268.05	396.71	454.84	491.44	611.59	649.82	2872.45 (478.74)

Table 10.
Impact of selective logging on tree species richness, composition and structure, after 27 years in comparison with unlogged forest at 1 ha scale (based on census 2010–2011; [30]).

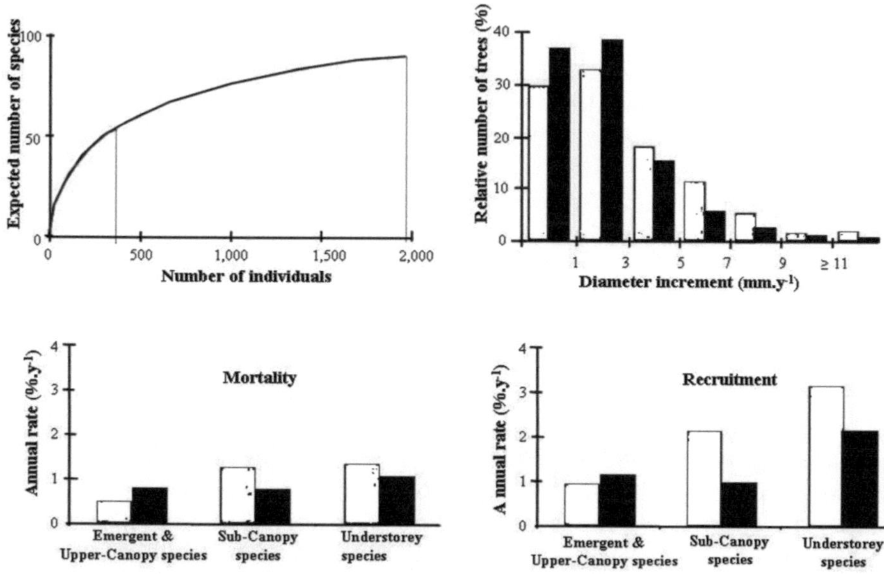

Figure 6.
Summarized results of a decade long monitoring study of logged and unlogged plots [31].

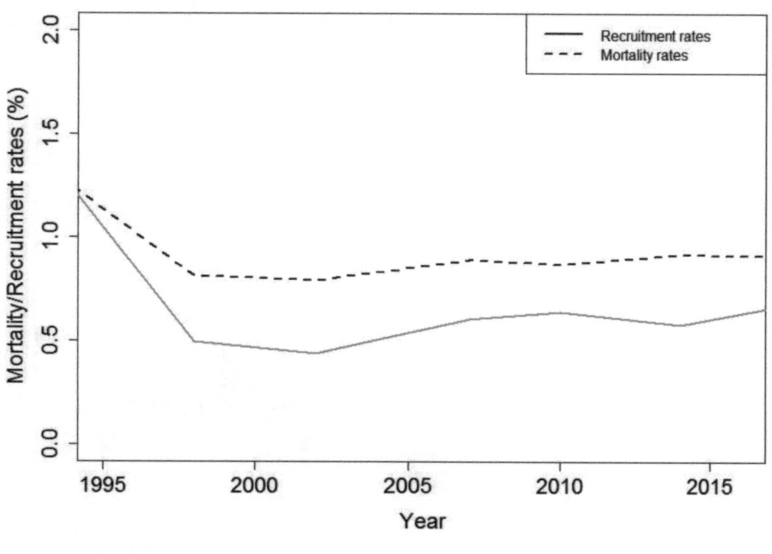

Figure 7.
Recruitment and mortality rates for trees ≥ 30 cm gbh in the unlogged forest.

4.2.4 Biomass estimation

Live Above Ground Biomass (AGB) of individual trees were determined using the regression equation of tropical moist forest stands: AGB = exp.(−2.977 + ln (ρD2H)).

Where D is the diameter at breast height in cm, H is total height in m and ρ is wood density in g cm^{-3} [32]. The estimated value ranged from 268 to 491 Mg ha^{-1} for those plots in the logged compartments and 611 to 649 Mg ha^{-1} for the unlogged compartment **Table 10**. The AGB ha^{-1} of unlogged plots of the present study is high compared to the available data on Indian forests and the other tropical forests across the continents: a mean of 287.8 ± 105.0 Mg ha^{-1} for South America, 393 ± 109.3 Mg ha^{-1} in Asia and 418 ± 91.8 Mg ha^{-1} in Africa for trees ± 10 cm dbh [33].

In summary, the continued monitoring of the plot will enhance our capacities to understand the forest dynamics in space and time, and response of the forest to the influence of climate change.

5. Forest dynamics plots in Kakachi forest, Kalakad Mundanthurai Tiger Reserve, Western Ghats

Small scale forest dynamic plots were established in the wet evergreen forest at Kakachi in Kalakad Mundanthurai Tiger Reserve (hereafter KMTR) (8° 33' N. Lat. 77°23'E. Long, **Figure 8**). It covers an area of 887 km^2 along the eastern slopes of Agasthyamalai range. The altitude ranges from 100 to 1890 m with generally steep slopes and deep valleys. KMTR supports large stretches of evergreen forests, which are contiguous to the rest of the WG, and endowed with large number of endemic and rare plant species, and provide habitats for rare animals such as Lion Tailed Macaque, Nilgiri langur, Tiger, Elephants etc. KMTR receives both South-West and North-East monsoons and being a major watershed, seven major rivers originates from the forest. These rivers meet the water requirements of the arid regions of south Tamil Nadu. Kakachi is located at 1300 m amsl and receives an annual rainfall of over 3500 mm. The rainfall is spread over 8 to 10 months in a year. The spread out of the rainfall in the study site is due to Southwest monsoon and Northeast monsoon rains. Mean maximum temperature is 24° C and minimum about 16° C [34].

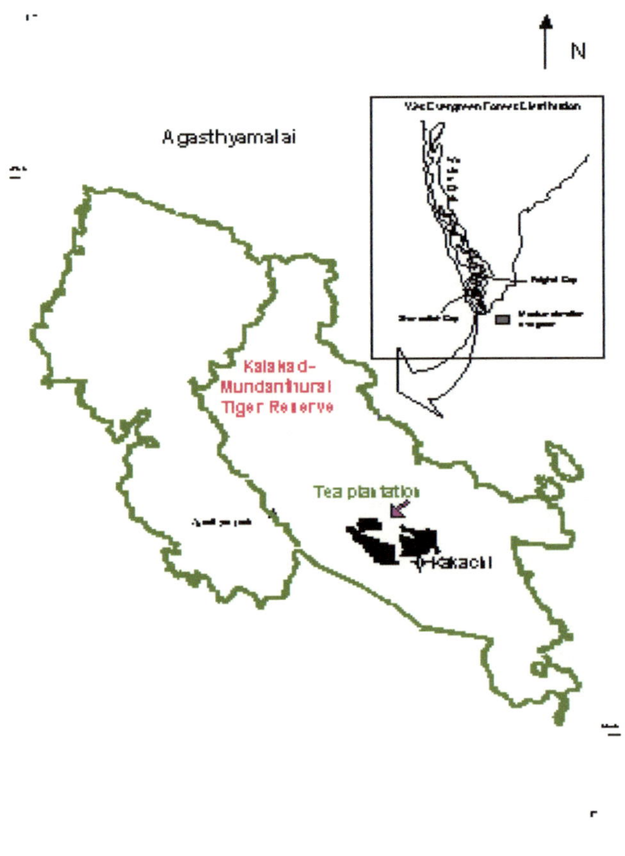

Figure 8.
Forest Dynamics Plots in Kakachi forest, Kalakad Mundanthurai Tiger Reserve, Western Ghats.

The terrain is highly undulating and is traversed by numerous mountain streams. The vegetation is characterized by three dominant tree species, *Cullenia exarillata, Palaquium ellipticum* and *Aglaia bourdillonii* [34]. Between 1993 and 1994 three 1 ha forest dynamic plots were established in an undisturbed wet evergreen forests at Kakachi, Kalakad Mundanthurai Tiger Reserve (KMTR) of Agasthyamalai range.

The principal objective of this study was to measure the changes in diversity, structure, recruitment and mortality of tree species compared to other forests within WG as well as globally. Following are the specific objectives:

(1) Determine the diversity and population structure of trees at Kakachi (2) Compare diversity and density of endemic with the widely distributed species in the site (3) Determine the overall recruitment, mortality and turnover rates of tree species.

Three 1-ha plots of 250 m x 40 m dimension were established within the wet evergreen forest during 1993 and 1994. A minimum distance of 1 km, spatially separated these plots. The plots were permanently marked using PVC pipes and all trees above 10 cm dbh at 1.3 m above ground were enumerated and tagged.

5.1 Floristics and species diversity

A total of 68 tree species >10 cm dbh were recorded from the 3 ha. The sixty-eight tree species belonged to 52 genus and 31 families. The most species rich family was Lauraceae with 12 species followed by Euphorbiaceae (7 sp.) and Myrtaceae (6 sp.). Seventeen families had only one species. *Syzygium* was the most common genus with 6 species followed by *Litsea* with 4 species. Genus with single species represented over 90% of the total genus. Shannon diversity index was 2.79 (=4.02 $\log^2$) and the evenness index was 0.66. The number of species recorded per ha was 46.

5.2 Stem density

A total of 2116 live stems were encountered in the 3 ha at an average of 705 stems ha^{-1}. Three species *Agrostistachys borneensis* (19%), *Cullenia exarillata* (16%) and *Palaquium ellipticum* (13%) represented 45% of the stems, while the other 65 species accounted for the remaining 55%. The species abundance relationship shows that majority of the species had a density between 4 and 8 individuals per 3 ha. Seventy percent of the species had less than 10 individuals in the 3 ha plot and only 10% (7 species) had over 100 individuals. Over 3.5% of the stems were dead during the first enumeration.

5.3 Basal area

Total basal area of all the trees was 60.51 m^2ha^{-1}. Two dominant species *Cullenia exarillata* and *Palaquium ellipticum* accounted for over 58% of the total basal area, while all other species individually accounted for less than 5%. Mean basal area of individual trees was 0.0769 m^2 ha^{-1} and ranged from 0.026 m^2 ha^{-1} to 0.3233 m^2 ha^{-1}.

5.4 Life forms

Of the 68 tree species 42 were canopy trees and 23 (35%) were understorey trees. Maximum height of canopy trees was 40 m. The tallest species is *Cullenia exarillata* with a mean height of 22 m. Canopy trees were at a higher density than understorey trees. Many of the common canopy trees were dominant component of the stand. Canopy species such as *Cullenia exarillata, Aglaia bourdillonii* and *Palaquium ellipticum* accounted for 66% (808) of the canopy trees. Among the understorey tree species *Agrostistachyis borneensis, Gomphandra coriacea,* and *Epiprinus mallotiformis* accounted for over 80% (663) of the total stems. Canopy trees were also larger in girth besides being more abundant, therefore contributed to higher total basal area 49.04 m^2 ha^{-1} compared to understorey trees 8.4 m^2 ha^{-1}.

5.5 Habitat

In terms of habit preferences there were 36 closed forest species and 33 gap species. Nineteen of the closed forest trees were canopy species and remaining 17 were gap invaders. Similarly in the understorey 14 were closed canopy species and 15 were gap species. The closed forest species were 11 times (1840) more abundant than gap species (144) and majority of the gap species contributed to less than 7 individuals per ha. Basal area of closed forest species was 22 times greater than gap species.

5.6 Endemic species richness gradient

5.6.1 Species

Thirty-three of the total 68 identified tree species (49%) in the plots were endemic to the Western Ghats. The endemic species richness increases from localised endemics to more widely distributed endemic species. Greater proportion (76%) of the species were endemic to the entire Western Ghats (EWGE, Entire Western Ghats Endemic), 18% to southern Western Ghats, (SWGE Southern Western Ghats Endemic comprising of Nilgiris and south of the Palghat gap) and 6% to Agasthyamalai (AGME Agasthyamalai Endemic) region alone (localized endemic). Some of the common endemic species are *Palaquium ellipticum*- endemic to whole of Western Ghats, *Litsea keralana* - restricted to southern Western Ghats and *Aglaia bourdillonii* - endemic to Agasthyamalai.

5.6.2 Density

Endemic species of the Western Ghats accounted for 51% (1079) of the total stems encountered in the 3 ha. EWGE were the most numerous, and accounted for 83% (893) of the stems followed by 16% (172) for AGME species and only 1.3% (14) to SWGE. The density of trees under the 3 endemic gradients is significantly different ($KW = 9.84$, $p < 0.01$). The WGE species were significantly more abundant than SWG species (Dunn's test $p < 0.01$). The median density value was 8 for EWGE species and 2 for SWGE species. Localized AGME species such as *Aglaia bourdillonii* was at high density. Contrary to species richness, density is high for local endemic species and highest for EWGE species but very low for SWGE species.

5.6.3 Basal area

Basal area of endemic species accounted for 94% of the total basal area, of which 95% were EWGE, 0.6% SWGE endemic and 5% AGME. Though there were only two species endemic to Agasthyamalai, one of them *Aglaia bourdilonii* was a high-density species but accounted for only 3.3% of the basal area. Trend in basal area was also similar to density; EWGE highest followed by AGME and finally SWGE.

	Total	Plot 1	Plot 2	Plot 3
Recruitment	0.86	0.05	1.84	0.96
Mortality	0.56	0.59	0.61	0.49
Turnover	0.71	0.32	1.22	0.73

Table 11.
Demographic parameters across the 3 one ha plots.

5.7 Changes in recruitment, mortality and turnover of tree species over time

Endemic species showed least change in stem density and basal area whereas widely distributed species showed greater change in both. The overall recruitment of trees was 0.86% per year and mortality 0.56% per year resulting in an annual turnover of 0.71% (**Table 11**). Thirty-three species did not show any recruitment and mortality. Forty species showed no recruitment and 37 species no mortality. The dominant species such as *Cullenia exarillata, Palaquium ellipticum, Agrostistachys borneensis* and *Aglaia bourdillonii* had low recruitment and mortality rate.

Majority of the gap species had high levels of recruitment and mortality resulting in a high turnover. Some closed forest and canopy species such as *Nageia wallichiana* (Podocarpus), *Elaeocarpus tuberculatus* and understorey species such as *Antidesma menasu, Syzygium mundagam* and *Miliusa wightiana* showed high levels of recruitment. Gap species had higher mortality and recruitment than closed forest species. Recruitment and mortality was not significantly different between canopy and understorey species. In general gap species was the major contributor to the turnover in the forest.

6. Way forward

Long-term data is essential for undestanding vegetation dynamics. Vegetation dynamics is directly related to climate variability that an ecosystem experience. Extreme events such as floods, drought and snowfall forms part of long-term variability in climate. Vegetation response to such extreme events depends upon the type and intensity of an event. Government of India has initiated two major national projects to undestand and combat the impacts of varibility in climate through understanding natural vegetation dynamics. They are, (a) Indian Long-Term Ecological Obsevatories (ILTEO) and (b) Studies of impact of climate change on Indian Forest System through long-term monitoring, an All India Coordinated Research Project (AICRP) managed by Indian Council for Forestry Research and Education (ICFRE), Dehradun. ICFRE with its nine Institutes, Forest Research Institute (FRI), Dehradun, Institute of Forest Genetics and Tree Breeding (IFGTB), Coimbatore, Institute of Wood Science and Technology (IWST), Bengaluru; Institute of Forest Biodiversity (IFB), Hyderabad; Rain Forest Research Institute (RFRI), Jorhat; Tropical Forest Research Institute (TFRI), Jabalpur; Himalayan Forest Research Institute (HFRI), Shimla; Institute of Forest Productivity (IFP), Ranchi; and other Institutions like Ashoka Trust for Research in Ecology and Environment (ATREE), Bengaluru; Indian Institute of Science (IISc) Bengaluru; French Institute, Pondicherry; and Kerala Forest Research Institute (KFRI), Peechi, Thrissur, Kerala have initiated long-term ecological monitoring studies on the effects of climatic variability on the forest ecosystem. This will be the first of its kind in India to address climate change issues at national and international level and helps to trace footprints of climate change impacts through vegetation and also reveals to what extent our forests are resilient to change in the climate. Further it will also address the issues flagged by UNFCCC, IPCC, NAPCC, SAPCC etc.

Major objectives of the programme includes establishment of Permanent Preservation Plots (PPP) to observe and understand the changes in species diversity, composition and growth pattern due to climate change over a period of time. The methodology for selection and laying of sample plots, assessment, identification and tagging of plants is based on Centre for Tropical Forest Science (CTFS) protocol. Aimed at precision, uniformity, and large scale of international acceptance, it was decided to laydown country wide permanent plots (preferably 10 Hectare size) in major forest types of the country wherein woody individuals >1 cm diamter at breast height (DBH) would be monitored for vital parameters such as recruitment, mortality and growth in relation to climate. The study also includes dendrochronology, edaphic factors, survivality, regeneration, invasive species, dynamics of soil microflora, phenological studies, insect-pest incidence, disease infection, pollinators etc in the pemanent plot and surrounding forest area.

Complementary to this initiative Indian Government launched a new pan India research program- Indian Long Term Ecological Observatories (ILTEO) with a larger goal of assessing the influence of climate change on the biodiversity at national level. To address issues related to climate change the Government of India has set up Indian Network for Climate Change Assesment (INCCA) to provide frame work to monitor impacts of climate change, assess the drivers of climate change and to develop decision support system. It is been recognzed that climate change of one of major drivers, Long-term ecological monitoring is required to identify pattern and drivers of change. Moreover long term monitoring is required to frame the national policies and signing international conventions such as United Nations Framework Convention on Climate Change (UNFCC). There are several isolated programs monitoring the changes. However, there is a need for unified multidisplinery national level program to address the issues of climate change. All India Coordinated Research project under ICFRE is one such national level effort to address encouragement and research to climate change.

Acknowledgements

Authors are thankful to Director General, ICFRE, Dehradun, for granting All India Coordinated Research Project on Climate Change (AICRP-31). Authors also thankful to Shri Nirmalya Bala, FRI, Dehradun, National Project Coordinator for the encouragement and support. We also thank State Forest Departments of Karnataka, Kerala and Tamil Nadu for encouragement and research permissions.

Author details

Madan Prasad Singh[1]*, Manohara Tattekere Nanjappa[1], Sukumar Raman[2], Suresh Hebbalalu Satyanatayana[2], Ayyappan Narayanan[3], Ganesan Renagaian[4] and Sreejith Kalpuzha Ashtamoorthy[5]

1 Institute of Wood Science and Technology (IWST), Bengaluru Karnataka, India

2 Centre for Ecological Sciences and Divecha Center for Climate Change, Indian Institute of Science, Bangalore, Karnataka, India

3 Research Scientist, French Institute, Pondicherry, India

4 Ashoka Trust for Research in Ecology and the Environment, Bangalore, Karnataka, India

5 Kerala Forest Research Institute, Peechi, Thrissur, Kerala, India

*Address all correspondence to: mpsinghifs1989@gmail.com

References

[1] Uma Shaanker, R. and Ganeshaiah, K.N. 1993. Pollination biology in India – surveying the past. Curr. Sci., 65 (3): 195-198.

[2] Manilal, K.S. 2003. On the English edition of Van Rheed's Hortus Malabaricus. University of Kerala.

[3] Manilal, K.S. 2008. Medicinal plants described in Hortus Malabaricus, The first Indian regional flora published in `1678 and its relevance to the people of India today. In Proceedings of International seminar on "Multidisciplinary Approaches in Angiosperm Systematics".

[4] Kochhar, R.K. 1993. Ardaseer Cursetjee (1808-1877), the first Indian Fellow of the Royal Society of London. Notes and Records: The Royal Society Journal of the History of Science. 47 (1): 33–47. http://doi.org/10.1098/rsnr.1993.0004.

[5] Kochhar, R.K. 1992. Science in British India. I. Colonial tool. Current Science, 63 (11): 689-694.

[6] Champion H.G. and S.K. Seth. 1968. A revised survey of the forest types of India. Delhi. Manager of Publications. 404 pp.

[7] Pascal, J.P., Ramesh, B.R. and Kichenassamy, K. 1992. Vegetation Map of South India - Map 4: Bangalore – Salem. Institut Français de Pondichéry.

[8] Pascal, J.P., Shyam Sunder, S. and Meher-Homji, V.M. 1984. Forest Map of South India-sheet: Belgaum-Dharwar-Panaji. Institut Français de Pondichéry.

[9] Pascal, J.P., Shyamsundar, S. and Meher-Homji, V.M. 1982. Vegetation Map of South India–Map 3: Mercara - Mysore. Institut Français de Pondichéry.

[10] Ramesh, B.R., De Franceschi, D. and Pascal, J.P. 1997a. Forest Map of South India – sheet: Thiruvananthapuram-Tirunelveli. Institut Français de Pondichéry.

[11] Ramesh, B.R., De Franceschi, D. and Pascal, J.P. 1997b. Vegetation Map of South India - Map 6: Thiruvananthapuram – Tirunelveli. Institut Français de Pondichéry.

[12] Ramesh, et al. 2002. Forest Map of South India: Mercara-Mysore Institut Français de Pondichéry.

[13] Senthilkumar, N., Prakash, S., Rajesh Kannan, C., Arun Prasath, A. and Krishnakumar, N. 2014. Revisiting forest types of India (Champion and Seth, 1968): A case study on Myristica swamp forest in Kerala. International Journal of Advanced Research, 2(2): 492-501.

[14] Pascal, J.P. and Pelissier, R. 1996. Structure and floristic composition of a tropical evergreen forest in south-west India. Journal of Tropical Ecology, 12 (2): 191-214.

[15] Reddy, C.S., Jha, C.S., Dhadwal, V. K., Hari Krishna, P., Pasha, S.V., Satish, K.V., Dutta, K., Saranya, K.R.L., Rakesh, F., Rajashekar, G. and Diwakar, P.G. 2015. Quantification and monitoring of deforestation in India over eight decades (1930-2013). Biodiversity and Conservation. Doi 10.1007/s 10531-015-1033-2

[16] Rai, S.N. 1981. Rate of growth of some evergreen species. Indian Forester, 107(8): 513-518.

[17] Phillips, O.L., Hall, P., Gentry, A.H., Sawyer, S.A. and Vasquez, R. 1994. Dynamics and species richness of tropical rainforests. Proc. National Acad. of Sciences of the USA., 91: 2805-2809.

[18] Phillips, O.L. and Gentry, A.H. 1994. Increasing turnover through time in tropical forests. Science. 263: 954-958.

[19] Phillips, O.L. 1995. Evaluating turnover in tropical forests. Science, 268: 894-895.

[20] Sheil, D. 1995. Evaluating turnover in tropical forests. Science, 268: 894-895.

[21] Hubbell, S.P. and Foster, R.B. 1983. Diversity of canopy trees in a neotropical forest and implications for conservation. In: Sutton S.L., Whitmore, T.C. and Chadwick, A.C. (eds.) Tropical rain forest: ecology and management, pp. 25-41. Blackwell Scientific, Oxford, UK.

[22] Sukumar, R., Dattaraja, H.S., Suresh, H.S., Radhakrishnan, J., Vasudeva, R., Nirmala, S. and Joshi, N. V. 1992. Long term monitoring of vegetation in a tropical deciduous forest in Mudumalai, southern India. Current Science, 62: 608-616.

[23] Pulla, S., Suresh, H.S., Dattaraja, H. S. and Sukumar, R. 2017. Multi dimensional tree niches in a tropical dry forest. Ecology. 98: 1334-1348.

[24] https://forestgeo.si.edu

[25] Gamble, J.S. and Fischer, C.E.C. 1915–1935. Flora of the Presidency of Madras Vols. I–III. Adlord and Sons Ltd., London.

[26] Loffeier, M.E. 1988. Reconstitution après exploitation sélective en forêt sempervirente du Coorg Inde I: Méthodes et résultats préliminaires d'une étude floristique et structurale Acta Ecologica, 9: 69-87.

[27] Kershaw, K.A. 1973. Quantitative and Dynamic Plant-Ecology. 3rd Edn., ELBS and Edward Arnold Ltd, London.

[28] Philips, O.L. and Gentry, A. 1994. Increasing turnover through time in Tropical forests. Science, 263: No. 5149. pp. 954-958.

[29] Sasidharan, N. 2004. Biodiversity Documentation for Kerala Part 6: Flowering Plants. Kerala Forest Research Institute, Peechi.

[30] Jeyakumar, S., Ayyappan, N., Muthuramkumar, S. and Rajarathinam, K. 2017. Impacts of selective logging on diversity, species composition and biomass of residual lowland dipterocarp forest in central Western Ghats, India. Tropical Ecology, 58(2): 315-330.

[31] Pélissier, R., Pascal, J.P., Houllier, F. and Laborde, H. 1998. Impact of selective logging on the dynamics of a low elevation dense moist evergreen forest in the Western Ghats South India. Forest Ecology and Management, 105: 107-119.

[32] Chave, J., Andalo, C., Brown, S., Cairns, M.A., Chambers, J.Q., Eamus, D., Fölster, H., Fromard, F., Higuchi, N., Kira, T., Lescure, J.P., Nelson, B.W., Ogawa, H., Puig, H., Riéra B. and Yamakura, T. 2005. Tree allometry and improved estimation of carbon stocks and balance in tropical forests. Oecologia, 145: 87-99.

[33] Slik, J.W.F., Paoli, G., McGuire, K., Amaral, I., Barroso, J., Bastian, M., Blanc, L., Bongers, F., Boundja, P., Clark, C., Collins, M., Dauby, G., Ding, Y., Doucet J.L., Eler, E., Ferreira, L., Forshed, O., Fredriksson, G., Gillet, J.F., Harris, D., Leal, M., Laumonier, Y., Malhi, Y., Mansor, A., Martin, E., Miyamoto, K., Araujo-Murakami, A., Nagamasu, H., Nilus, R., Nurtjahya, E., Oliveira, A., Onrizal, O., Parada-Gutierrez, A., Permana, A., Poorter, L., Poulsen, J., Ramirez-Angulo, H., Reitsma, J., Rovero, F., Rozak, A., Sheil, D., Silva-Espejo, J., Silveira, M., Spironelo, W., ter Steege, H., Stevart, T., Navarro-Aguilar, G.E., Sunderland, T., Suzuki, E., Tang, J., Theilade, I., van der Heijden, G., van Valkenburg, J., Van Do, T., Vilanova, E., Vos, V., Wich, S., Wöll, H., Yoneda, T., Zang, R., Zhang, M.G. and Zweifel, N. 2013. Large trees drive forest aboveground biomass variation in moist lowland forests across

the tropics. Global Ecology and Biogeography, 22: 1261-1271.

[34] Ganesh, T., Ganesan, R., Devy, M. S., Davidar, P. and Bawa, K.S. 1996. Assessment of plant biodiversity at a mid-elevation evergreen forest of Kalakad–Mundanthurai Tiger Reserve, Western Ghats. India. Current Science, 71(5): 379-392.

Section 2

Ecophysiology

Chapter 14

Photosynthetic Antenna Size Regulation as an Essential Mechanism of Higher Plants Acclimation to Biotic and Abiotic Factors: The Role of the Chloroplast Plastoquinone Pool and Hydrogen Peroxide

Maria M. Borisova-Mubarakshina, Ilya A. Naydov, Daria V. Vetoshkina, Marina A. Kozuleva, Daria V. Vilyanen, Natalia N. Rudenko and Boris N. Ivanov

Abstract

The present chapter describes the mechanisms of reactive oxygen species formation in photosynthetic reactions and the functional significance of reactive oxygen species as signal messengers in photosynthetic cells of plants. Attention is given to the acclimation mechanisms of higher plants to abiotic and biotic factors such as increased light, drought, soil salinity and colonization of plants by rhizosphere microorganisms. Special attention is paid to the reactions of reactive oxygen species with the components of the chloroplasts plastoquinone pool leading to production of hydrogen peroxide as a signal molecule, which is involved in acclimation of plants to these stress conditions. The chapter also presents the data demonstrating that regulation of the size of the light-harvesting antenna of photosystem II is one of the universal mechanisms of the structural and functional reorganization of the photosynthetic apparatus of higher plants exposed to the abiotic and biotic factors. These data were obtained for both model Arabidopsis (*Arabidopsis thaliana*) plants as well as for agricultural barley (*Hordeum vulgare*) plants. It is hypothesized that hydrogen peroxide, produced with involvement of the plastoquinone pool components, plays the role of a signaling molecule for regulation of the photosystem II antenna size in higher plants when environmental conditions change.

Keywords: photosynthesis, photosynthetic antenna, higher plants, acclimation, hydrogen peroxide

1. Introduction

Studies of such a global biological process as photosynthesis are always relevant. This is confirmed by a huge number of laboratories around the world studying different aspects of this process. Over the history of photosynthesis studies, various changes in photosynthetic parameters were constantly observed under the influence of external conditions, environmental factors, such as light intensity, temperature, the content of carbon dioxide in the air, drought, and soil salinity. For practical human needs, investigation of the mechanisms of these changes may give some clues that can help to maintain high productivity of food crops and other economically important photosynthesizing organisms.

Currently, one of the hot spots in this field is the problem of signal transmission from the photosynthetic apparatus to other systems of the photosynthetic cell, primarily to the systems responsible for the composition and structure of photosynthetic apparatus. It turns out that all signaling mechanisms are interconnected and in many cases represent a network of signals transferred from one signal messenger to another. An important milestone in the early 1990s was the discovery that nuclear systems react on the redox state of the chloroplast plastoquinone (PQ) pool and that the PQ pool plays an important role in protection of the photosynthetic apparatus of plant cells under various environmental conditions [1]. For instance, it was shown that PQ pool is involved in regulation of the size of photosynthetic light-harvesting antenna of photosystem II (PS II) in plants. Photochemically active PQ pool is situated within thylakoid membranes of chloroplasts. This fact raises a question: how exactly can it affect the expression of nuclear-encoded antenna genes in leaf cells when the environmental conditions change [2]?

Apart from that, for over 40 years, scientists have been studying the signaling functions of reactive oxygen species (ROS). One of the ROS, hydrogen peroxide (H_2O_2), was shown to play a major role in many signaling pathways in plants [3]. In the present chapter, we summarize the evidence that the PQ pool components are involved in H_2O_2 formation in chloroplast and that the hydrogen peroxide plays the essential role in regulation of photosynthetic antenna size of PS II under biotic and abiotic factors.

2. General structure of photosynthetic electron-transport chain

In the photosynthetic electron-transport chain (PETC) of thylakoid membranes, the absorption of energy of photons and the subsequent photochemical transformation of energy is performed by pigment-protein membrane complexes: photosystem II (PS II) and photosystem I (PS I). When plants are illuminated, the electrons originating from water decomposition are transported from PS II via the plastoquinone pool (PQ pool) to the cytochrome b_6/f complex and subsequently, via plastocyanin, to PS I for reduction of ferredoxin (Fd), which serves as an electron donor for $NADP^+$ reduction catalyzed by ferredoxin-$NADP^+$ reductase (FNR). Electron transport is accompanied by protons pumping into the thylakoid lumen, a slit-shaped intrathylakoid space, resulting in the creation of a transmembrane electrochemical proton gradient required for ATP synthesis.

Each photosystem consists of a reaction center (RC) and a light harvesting complex (LHC), or "antenna". The energy of photons is captured by the antenna complexes of both photosystems, with LHC I capturing excitation energy for PS I, and LHC II capturing excitation energy for PS II, however LHC II can function as antenna for both photosystems (see further) [4]. The PS II core complex of higher plants is surrounded by complexes that include polypeptides encoded by the *lhcb* (light-harvesting complex b) gene family and contain chlorophylls *a* and *b* as well as carotenoids in different proportions [5]. The inner antenna is comprised of three small monomeric proteins CP29, CP26, and CP24 (encoded by the *lhcb4*, *lhcb5*, and *lhcb6* genes, respectively). The outer peripheral LHC II antenna is mainly formed by three types of heterotrimers of proteins encoded by genes *lhcb1*, *lhcb2*, and *lhcb3:* strongly bound heterotrimers of two Lhcb1 and one Lhcb2 proteins (S-type), moderately bound heterotrimers of two Lhcb1 and one Lhcb3 proteins (M-type), and loosely bound heterotrimers of two Lhcb1 and one Lhcb2 proteins (L-type) [6–8]. CP29, CP26, and CP24 proteins bind directly to the core complex of PS II; the M-trimers bind to PS II via the CP29 and CP24 proteins, and the S-trimers bind via CP26 [7, 9]. The resulting PS II-LHC II complex can additionally associate with L trimers.

LHC I consists of four separate polypeptides (Lhca1–4, which are encoded by the *lhca* (light-harvesting complex a) gene family, combined into two heterodimers: Lhca1-Lhca4 and Lhca2-Lhca3. PS I has a docking site for LHC II consisted of PsaH, PsaL, and PsaI subunits [10, 11].

3. Superoxide anion radical and hydrogen peroxide formation in chloroplasts

Photosynthetic apparatus does not only evolve molecular oxygen (O_2) via water oxidation in PS II in the light, but also reacts with O_2 molecules that leads to O_2 consumption and formation of reactive oxygen species (ROS). In the ground state, O_2 molecule has two unpaired electrons localized on different anti-bonding orbitals (triplet state). Most organic molecules exist in a singlet state that limits their spontaneous reactions with O_2, and this is a favorable factor for biological life. When O_2 is reduced by the PETC components or other cell components, such ROS as superoxide anion radical ($O_2^{\bullet-}$) and hydrogen peroxide (H_2O_2) are initially formed. $O_2^{\bullet-}$ has been found to be the primary product of O_2 reduction in PETC [12]. *In vivo*, O_2 reduction and CO_2 assimilation occur simultaneously, with oxygen accounting for 5 to 50% of the electrons from the PETC in various plants under various conditions [13]. Another ROS, singlet oxygen ($^1O_2$), emerges because of spin inversion of one electron on the outer orbital of O_2 molecules. In the PETC, $^1O_2$ is formed in PS II, and $O_2^{\bullet-}$ is formed predominantly by the components of PS I. In chloroplasts, other ROS are also formed: perhydroxyl radical ($HO_2^{\bullet}$), hydroxyl radical ($HO^{\bullet}$), as well as hydroperoxides (ROOH) and radicals of organic molecules: peroxide radical ($ROO^{\bullet}$) and alkoxyl radical ($RO^{\bullet}$).

A stromal protein Fd, which is an electron carrier from the terminal cofactors of PS I to FNR, has long been considered a major O_2 reducing agent in chloroplasts. A number of studies have shown that addition of Fd to isolated thylakoid membranes devoid of stroma components significantly increased oxygen reduction [14, 15]. However, Kozi Asada, a pioneer researcher in the field of ROS formation in chloroplasts, observing a very low stimulation of $O_2^{\bullet-}$ production following the addition of Fd to the thylakoid suspension, concluded that Fd is not involved in O_2 reduction *in vivo* [16]. Indeed, the rate constant of Fd oxidation by O_2 is as low as 10^3 M^{-1} s^{-1} [17] meaning that a significant O_2 reduction with Fd can only be achieved with a significant increase in the amount of Fd itself [18]. In the presence of $NADP^+$, *i.e.*, under optimal conditions preventing the accumulation of the reduced Fd in large amounts, the Fd-dependent reduction of oxygen was limited to 5–7% of the total oxygen reduction rate [19]. The fraction of Fd-dependent oxygen reduction can increase when $NADP^+$ is deficient, for example when CO_2 is limited or Calvin-Benson cycle enzymes are inhibited. Low rates of O_2 reduction by Fd has a great physiological meaning since *in vivo* reduced Fd is used as an electron donor not only for $NADP^+$ reduction but for numerous reactions in the chloroplasts stroma.

Based on direct [20, 21] and indirect [21] evidence it has been shown that PS I is the main site of O_2 reduction to $O_2^{\bullet-}$. It has long been accepted that O_2 is reduced by the terminal cofactors of PS I, [4Fe-4S] clusters F_A/F_B, located in the protein subunit PsaC exposed to stroma. However, the rapid electron flow from F_A/F_B to O_2 *in vivo* may prevent efficient reduction of Fd. The contribution of another [4Fe-4S] cluster of PS I, F_X, to O_2 reduction was suggested in the study showed that light-induced H_2O_2-dependent iodination of thylakoid proteins occurred primarily in PS I subunits harboring F_X cluster [20]. However, no direct experimental confirmation of F_X involvement has been presented. It has recently been shown that removal of the F_A/F_B cofactors by chemical treatments, making F_X the terminal cofactor in the treated PS I complexes, results in a decreased rate of oxygen reduction [22], which argues against the assumption of a key role of F_X in oxygen reduction.

The involvement of another PS I cofactor, phylloquinone (PhQ) in the quinone-binding sites (A_1-sites), in O_2 reduction was suggested based on the stimulation of flash-induced O_2 uptake when PhQ was added to thylakoid membranes devoid of quinones [23]. O_2 reduction in PS I under steady-state illumination was investigated in isolated PS I complexes from the cyanobacterium *Synechocystis* sp. PCC 6803 of wild-type and a mutant strain with blocked PhQ biosynthesis (the mutant *menB*), in which PQ is incorporated into the A_1-sites [24]. It was shown that O_2 reduction rate in high light was lower in the PQ-containing PS I than in PhQ-containing, while the steady-state electron transport from quinone to F_A/F_B and then to O_2 was barely changed under studied conditions. This effect was attributed to the greater ability of phyllosemiquinone, $PhQ^{\bullet-}$, to reduce O_2 due to the lower redox potentials of PhQs at the A_1-sites compared to PQ. Moreover, unlike $PhQ^{\bullet-}$, plastosemiquinone, $PQ^{\bullet-}$, molecules at the A_1- sites are easily protonated [25] that also reduces the probability of their reaction with oxygen, while increases the probability of $PQ^{\bullet-}$ double reduction to PQH_2. In work with isolated PS I complexes from green alga *Chlamydomonas reinhardtii*, the involvement of PhQ in oxygen reduction has been also shown [22]. It was found that PhQ is the main site of oxygen reduction under increasing illumination, even under conditions of concurrent $NADP^+$ reduction, i.e., when Fd, FNR, and $NADP^+$ are present. Moreover, the principal role of $PhQ^{\bullet-}$ of one of the asymmetric branches of PS I (A-branch) was suggested based on the results with PS I complexes, in which the lifetime of $PhQ^{\bullet-}$ in the A-branch is ~2 orders of magnitude longer.

When considering the possible components involved in the reduction of O_2 in the membrane, the shift in the redox potential of the $O_2/O_2^{\bullet-}$ pair from −160 mV at 1 M O_2 in water to approximately −550 mV in a hydrophobic environment must be considered. The E_m values of $PhQ/PhQ^{\bullet-}$ pairs in the A- and B-branches of cyanobacterial PS I are −671 and − 844 mV, respectively [26] that makes their reaction with oxygen thermodynamically favorable even in such a hydrophobic environment as that of PhQ in its binding sites in PS I. One consequence of this reaction can be the appearance of $O_2^{\bullet-}$ within the thylakoid membrane. Indeed, this appearance has been shown in a number of studies [18, 27, 28]. Addition of Fd and $NADP^+$ did not suppress formation of $O_2^{\bullet-}$ within thylakoid membranes [18] emphasizing the physiological significance of the observed process.

$O_2^{\bullet-}$, formed in the chloroplast stroma, are disproportionated there with the formation of H_2O_2 (reaction 1).

$$O_2^{\bullet-} + O_2^{\bullet-} + 2H^+ \rightarrow H_2O_2 + O_2 \quad (1)$$

In chloroplasts, disproportionation reaction is catalyzed by superoxide dismutase (SOD) with the rate constant ~10^7 M^{-1} s^{-1}. The SOD content in the stroma is high, and it is concentrated mostly at the stromal surface of the thylakoids [29]. In the stroma, ascorbate and glutathione also can reduce $O_2^{\bullet-}$ to H_2O_2.

However, using isolated thylakoids it was shown that H_2O_2 is produced not only outside the thylakoid membrane (implying the stroma phase *in vivo*) but also within the membrane; this H_2O_2 was called the "membrane" H_2O_2 [30]. Using various approaches [30–32], it was shown that the rate of the "membrane" H_2O_2 production correlated with increasing illumination and reached 60% of the total H_2O_2 produced in the light. We suggest that the "membrane" H_2O_2 is formed not due to the disproportionation reaction (since this reaction is hampered in aprotonic membrane phase) but due to the reaction between reduced plastoquinone PQH_2 and the $PhQ^{\bullet-}$ -generated $O_2^{\bullet-}$ (reaction 2):

$$PQH_2 + O_2^{\bullet-} \rightarrow PQ^{\bullet-} + H_2O_2 \quad (2)$$

This reaction is thermodynamically favorable in aqueous solutions because of the high difference between the Em_7 values of the redox pairs $PQ^{\bullet-}/PQH_2$ (370 mV) and $O_2^{\bullet-}/H_2O_2$ (940 mV). This reaction can occur predominantly at the membrane/stroma interfaces since PQH_2 molecules and water molecules tends to form hydrogen bonds. The first indirect evidence in favor for the reaction of PQH_2 with $O_2^{\bullet-}$ was obtained by studying the oxidation of the PQ pool after swithing off the light cessation [33]. The PQ pool in thylakoids, which was reduced during illumination and consisted of PQH_2 molecules, was not oxidized in the dark in the absence of oxygen [33, 34]. Under aerobic conditions, the PQ pool oxidation exhibited two-phase kinetics [33, 35], with the fast component being attributed to oxidation of PQH_2 molecules by $O_2^{\bullet-}$ that accumulated in the membrane in the light. Further, the occurrence of reaction 2 was confirmed by evaluating the redox state of the PQ pool when $O_2^{\bullet-}$ was artificially supplied to the thylakoid membrane from a xanthine-xanthine oxidase system [36]. The higher apparent electron capacity of the PQ pool in the presence of external $O_2^{\bullet-}$ showed the additional electron leakage from the PQ pool during $O_2^{\bullet-}$ generation and confirmed the occurrence of reaction 2.

4. Mechanisms of acclimation of higher plants to light conditions and the regulatory role of the redox state of the plastoquinone pool

Effective adaptation to changing environmental conditions is a prerequisite for plant survival and competitiveness. Plant acclimation to different light conditions has been a subject of interest of many scientists for a long time [37, 38]. Plants are divided into shade-tolerant and light-demanding plants, although there is also an intermediate category. Studies show that shade-tolerant plants are generally less adaptable to changing light conditions than light-demanding plants [39].

Under varying light conditions, plants require different amounts of photosynthetic products such as ATP and NADPH for normal metabolism. The synthesis of ATP and NADPH is regulated by changes in the functioning of the photosynthetic electron-transport chain. Light intensity is the most rapidly and frequently changing abiotic factor, but it is also one of the most important, given that it is the energy of photons that is required to activate photosynthesis. "Proper" adaptation of plants to light conditions is essential to ensure efficient use of light under low light conditions and to prevent photo-oxidative damage under high light conditions.

The following typical characteristics of plants adapted to high light intensity compared to plants growing at low light intensity are recognized:

1. leaves are thicker with higher amount of cell layers, larger cells [40–42];
2. higher ratio of Chl *a* to Chl *b* [42, 43];
3. increase in the number of chloroplasts in the cell [38];
4. decrease in granular structure [43, 44];
5. high content of β-carotene and xanthophyll cycle pigments [38];
6. higher PS II/PS I ratio [45];
7. higher electron transfer rates, higher CO_2 assimilation rates, and higher starch content; transition of the photosynthetic electron transfer chain to a reduced state [40, 42, 46];
8. higher energy dissipation capacity [47–49];
9. changes in the activity of carbonic anhydrases, enzymes that catalyze the reversible reaction of carbonic acid formation from carbon dioxide and water [50];
10. changes in the ratio of alternative electron transport pathways, accumulation of ROS and induction of corresponding signaling pathways influencing the gene expression [46].

It was found that during changes in plant illumination, the optimization of photosynthetic activity at the level of light energy absorption occurs due to activation of mechanisms leading to changes in the size of photosynthetic antenna complexes. When the spectral composition of light changes, the photosynthetic antenna complexes of thylakoids can be reorganized by reversible migration of the outer part of LHC II between PS II and PS I that leads to changes in the antenna sizes of both photosystems. This process is called state transitions and represents the reorganization of light-harvesting pigment-protein complexes of thylakoid membranes by the action of light through phosphorylation/dephosphorylation of LHC II proteins [51, 52]. Only weakly bound LHC II trimers, which consist of products of the *lhcb1* and *lhcb2* genes, have been shown to migrate between photosystems, whereas strongly and moderately bound trimers remain bound to PS II [53]. Phosphorylation of LHC II proteins by the enzyme STN7 kinase [54] leads to LHC II migration from PS II to PS I under red light illumination that excites predominantly the reaction centers of PS II. Dephosphorylation by the phosphatase enzyme TAP38/PPH1 [55] enables LHC II to return from PS I to PS II when illuminated with far-red light that excites predominantly PS I reaction centers, or in the dark. The STN7 kinase is a transmembrane protein [56, 57]. The involvement of the PQ pool in state transitions is not disputed; however, this does not appear to be directly related to the redox state of the pool, but rather to the appearance of PQH_2 molecules in the light. Interaction of STN7 kinase with the stromal loop subunit of the complex results in the kinase activation, while binding of PQH_2 molecules to the oxidation site of the cyt b_6/f complex leads to dissociation of the kinase from the cyt b_6/f complex [58, 59].

It is generally accepted that state transitions occur at low light intensities (100–200 μmol photons m^{-2} s^{-1}) [60, 61]. There is evidence that state transitions do not occur at high light intensity [62, 63] because of the inhibition of STN7 kinase by reduced thioredoxin [64]. We found that state transitions proceed in barley at higher light intensities than in Arabidopsis [65].

Not only LHC II proteins but also PS II core proteins can exhibit reversible phosphorylation [66, 67]. Phosphorylation of core proteins occurs by STN8 kinase activity, whereas dephosphorylation occurs by PBCP phosphatase. In high light, as previously described, STN7 kinase activity is inhibited, whereas STN8 kinase becomes active, resulting in phosphorylation of PS II core proteins only [60, 68]. Apparently, the described reversible phosphorylation of PS II and LHC II proteins plays a major role in the distribution of energy between the photosystems when the light intensity changes [60]. Phosphorylation of the PS II core proteins in high light facilitates the "unpacking" of PS II-LHC II complexes that is necessary to repair damaged PS II centers [69, 70]. Reversible phosphorylation of PS II RC proteins affects the ultrastructure of the thylakoid membrane and regulates the PS II repair cycle [71–73].

A change in the expression of chloroplast genes such as *psbA*, *psaA*, and *psaB*, which encode PS II and PS I reaction center proteins, is also a mechanism of plant acclimation to light conditions [74]. This process has been shown to be slower than state transitions, but faster than the long-term response (see further) [75]. This response is necessary for rapid changes in the biosynthesis of photosynthetic electron transport chain proteins that are encoded by chloroplast genome, in particular, as described above, the RC subunits that allows the regulation of the stoichiometry of the photosystems to be adjusted [74]. This stoichiometry can be different depending on environmental conditions, ensuring optimal equilibrium in the photosynthetic electron transport chain. The chloroplast sensor kinase, CSK, has been shown to play an important role in adjusting the PS II to PS I stoichiometry and the PQ pool was considered to be the main site of the photosynthetic redox control of chloroplast gene expression [76]. However, the signal that affects the chloroplast sensor kinase is still unknown.

Another mechanism of plant acclimation to light is the regulation of the PS II antenna size. Higher plants can increase the size of LHC II in shade and, conversely, decrease the size of LHC II in high light [38], thus optimizing photosynthetic activity and protecting the photosynthetic electron transport chain from photoinhibition. Since PS II is unstable in high light, the change in the PS II antenna size is one of the important mechanisms of plants to adapt to high light intensities. When irradiance is increased for a prolonged period (days), the PS II antenna size is reduced by suppressing the biosynthesis of the peripheral LHC II proteins, disassembling of the PS II pigment-protein complexes, which include these proteins, and subsequently by proteolysis the proteins. Such changes are necessary to reduce the amount of absorbed light energy and, as a consequence, to adapt to long-term high light. The "adaptive" antenna size reduction is seen as the decrease in the levels of Lhcb1, Lhcb2, Lhcb3, and Lhcb6 proteins [37, 77]. The minimal antenna unit in high-light-adapted higher plants contains Lhcb4, Lhcb5, and S-type LHC II trimers in addition to the PS II core complex [77, 78]. It has been shown that proteolysis of Lhcb proteins is triggered after the first 24 hours in high light [79, 80].

PS II antenna size has been found to be regulated by the redox state of the PQ pool [81–83]. A detailed analysis of the mechanism of this regulation was performed using *viridis zb63*, a barley mutant devoid of PS I, but with an actively functioning PS II [84]. It was found that even at low light intensities, the PQ pool was reduced as much as possible, and the PS II antenna size was reduced in the mutant but not in wild type.

It can be concluded that the chloroplast PQ pool is involved not only in the regulation of state transitions, but also in the regulation of gene expression of both chloroplast- and nuclear-encoded genes [85].

5. Establishing the signaling role of H_2O_2 in the regulation of the size of the photosystem II light harvesting antenna in higher plants

As presented above, the redox state of the PQ pool plays an important role in triggering signaling pathways, which are necessary for regulating plastid and nuclear gene expression [86]. It is generally accepted that a high level of PQ pool reduction is the chloroplast signal to reduce the antenna size of PS II under high light conditions. Since the 1960s and 1970s, scientists have been involved in understanding the molecular signal about the redox state of the plastoquinone pool. However, the question about the nature of the signal indicating the redox state of the PQ pool remained open for a long time.

In Section 2.2 the data showing the involvement of the PQ pool in production of the membrane H_2O_2 in thylakoids were presented, therefore it was proposed that this H_2O_2 can be a candidate by which the redox state of the PQ pool imposes its regulatory effect in acclimation of higher plants to light conditions. In [86], using barley plants (*Hordeum vulgare*), several approaches were developed to change the H_2O_2 content in leaves at both low and high light intensities without affecting the redox state of the PQ pool allowing the correlation between the level of hydrogen peroxide, the PQ pool redox state, and the PS II antenna size to be assessed. The effect of hydrogen peroxide on composition of the thylakoid pigment-protein complexes, in particular on the PS II antenna composition, was revealed. The downsizing of the PS II antenna was suppressed in high light in leaves possessing high reduction level of the PQ pool, but low hydrogen peroxide content; at the same time, a decrease in the antenna size was observed in low light in the presence of elevated amount of hydrogen peroxide in leaves in spite of low reduction level of the PQ pool. The data obtained in that work indicate that it is the H_2O_2 content that determines the size of the PS II antenna, *i.e.,* the amount of pigment-protein complexes of the PS II antenna in leaves. This work was the first direct evidence, confirming the involvement of H_2O_2 in the signaling pathway leading to adjustment of the PS II antenna size in higher plants.

6. Acclimation mechanisms of plants to drought and soil salinity conditions and the discovery of the PS II antenna size regulation under these conditions

Abiotic and biotic stress factors affect many physiological processes, especially photosynthetic activity. Drought conditions are one of the major factors limiting plant growth and productivity, leading to significant changes in plant cell metabolism [87, 88]. Under drought conditions, photosynthesis is slowed down due to both a reduction in leaf area and a decrease in rate of photosynthesis per unit of leaf surface. The decrease in metabolism results from closure of stomata, reduced CO_2 availability and carbon metabolism that was shown for model *Arabidopsis thaliana* plants as well for agricultural plants such as *Vitis vinifera*, *Oryza sativa* and others [89].

In arid regions plants have developed xeromorphic traits to reduce transpiration. To do this, plants may shed leaves, reduce the number and the size of new leaves [90]. Another adaptive response is sclerophyllia (stiff-leafiness): stiff leaves are less affected by drought and easily regain functionality under normal conditions [91].

To reduce the negative effects of drought on photosynthesis, the photosynthetic apparatus increases thermal dissipation of absorbed energy and changes the activity of the xanthophyll cycle and ROS production/detoxification. Biochemical efficiency of photosynthesis under drought conditions depends on ribulose-1,5-bisphosphate regeneration and ribulose bisphosphate carboxylase/oxygenase activity [88, 92]. C4-photosynthesis is considered as a major adaptation, necessary to limit water loss, to reduce photorespiration, and to increase photosynthetic efficiency under drought conditions [93]. However, many crop plants, including rice, buckwheat, soybeans, and potatoes, use C3-photosynthesis. Drought is known to affect electron transport along the PETC [94, 95], to lead to decreased activity of the PS II oxygen-evolving complex, as well as PS II and PS I RCs [96–98], and to lead to lipid peroxidation of thylakoid membranes and hence to membrane damage [99–101].

C3-plant adaptation to drought involves multiple interactions of hormones, ROS, sugars, and many metabolic pathways. Computational models integrating data on gene expression, physiological and metabolic processes, as well as modern transgenic crossbreeding techniques, allow to improve photosynthesis as well as crop yield. Key phytohormones such as abscisic acid (ABA), cytokinins, gibberellic acid (GA), auxin, and ethylene control drought adaptation processes [102]. If plants are exposed to drought, ABA is synthesized in the roots and transported to the leaves to increase plant tolerance to this stress through stomatal closure and slower growth [103].

In many plants, drought primarily affects the root system [104]. The growth of tap roots is usually unaffected by drought, but lateral roots grow much slower due to suppression of lateral root meristem activity [105]. Small lateral roots provide an absorptive surface for water that also represents an adaptive strategy. Special tissues, such as rhizoderma, with thickened walls or corked exoderm, or reduced cortical layers are also considered to play role in acclimation to drought [90]. Hydrotropism also helps plants to adapt to drought stress [106, 107]. The interaction of auxin, cytokinins, GA, and ABA could be a potential way for changes in root architecture [108].

Under drought conditions, osmotic regulation is responsible for stomatal conductance, photosynthesis, leaf water capacity, turgor, and plant growth [109, 110]; at the same time both salt content and mechanical resistance change [111]. Sugars (sucrose, glucose, fructose, trehalose) are osmolytics affecting osmotic regulation [112, 113]. Substances such as proline, glycine, and betaine help protecting the plants from the damaging effects of drought [114].

Soil salinity is another major issue that negatively affects crop productivity of agricultural plants. The physiological response of plants to soil salinity includes many aspects that have not yet been fully characterized [115]. Under salinity conditions, plant growth and development are impaired due to water deficit, cytotoxicity is increased due to excessive ion uptake that consequences in an imbalance in plant metabolism. In addition, salinity is accompanied by oxidative stress due to increased ROS formation in plant cells [116, 117]. It was shown that in *Romaine lettuce*, a low salt-tolerant plant, long-term salt treatment led to enhancement of total carotenoid content [118].

Plant responses to salinity have been divided into two main stages [119, 120]. The first stage is ion-independent and occurs within minutes and first days, causing closure of the stomata and inhibition of cell expansion, primarily in the shoots, and results in limitation of plant growth [121–123]. The second stage occurs over several days or even weeks and is associated with an increase in the levels of cytotoxic ions, which slows down metabolic processes, causes premature senescence and eventually cell death [120, 124]. Salt stress causes outflow of water through aquaporins, which increases intracellular ion concentration, inactivating the photosynthetic apparatus [125]. Tolerance to both types of salt stress is regulated by multiple physiological and molecular mechanisms such as osmotic tolerance, ion tolerance, tissue tolerance, etc. [120, 123].

Salinity has been shown to change the ultrastructure of chloroplasts in higher plants: thylakoids swell [126], chloroplast structure changes [127], the number and size of plastoglobules increase [128]. The interaction of organelles, especially chloroplasts, mitochondria, and peroxisomes, is important for plant adaptation to stress conditions, particularly salinity [129]. Mitochondria, peroxisomes, and other organelles localize near chloroplasts for more efficient metabolite exchange [127].

For a large number of species [130, 131], it was shown that effects of drought and salinity lead to a more efficient conversion of starch into sucrose, which functions as an osmoprotector to reduce negative effects of environmental factors.

In Section 5 the data demonstrating that the amount of hydrogen peroxide determines the size of the PS II light-harvesting antenna were presented [86, 132]. Since the increase in hydrogen peroxide production also occurs in response to other stress conditions, *e.g.*, to drought and salinity, we hypothesized that the reduction of PS II antenna size may be one of the universal mechanisms of changing the structural and functional organization of photosynthetic apparatus during plant adaptation to various stress conditions. Acclimatory responses of *Arabidopsis thaliana* and barley (*Hordeum vulgare*) plants to drought and salinity conditions and the variations in the PS II antenna size were studied in detail recently [133, 134]. The main objective of these studies was to investigate the course of acclimatory changes before the manifestation of the negative effects of the selected stressors. The changes indicating an increase in the reduction level of the PQ pool were detected several days after introduction of these stress factors. After 7–14 days (depending on plant species), a decrease in the size of PS II light harvesting antenna was observed in plants under conditions of drought and salinity that was confirmed by a decrease in content of PS II antenna proteins and by downregulation of gene expression of these proteins under the stress conditions. Drought and salinity resulted in almost two-fold increase in the content of hydrogen peroxide in leaves compared to control leaves. Therefore, theses data demonstrated that the reduction of the size of PS II antenna represents one of the universal mechanisms acclimation of higher plants to mild stress conditions. The PQ pool reduction state along with the hydrogen peroxide content were proposed to be the important factors needed for the observed structural rearrangement [133, 134].

7. Changes in the functioning of higher plants and in the PS II antenna size in response to colonization by rhizosphere microorganisms

Environmental fluctuations and low soil fertility determine low yields and, consequently, low profitability of the agricultural sector of the economy. It is possible to avoid dependence of agricultural production on external conditions by increasing the resistance of plants to adverse environmental factors. The problem of low photosynthetic efficiency of agricultural crops (*e.g.*, of corn) when growing in the field is associated with the fact that most of the plant biomass is in shade [135]. Therefore, approaches to improve the acclimation of agricultural plants to suboptimal light conditions are being actively developed to improve their yield [136–138]. To increase crop yields a complex of approaches is required, including methods of biological protection of plants under stress conditions, under both biotic and abiotic environmental factors. At the moment, biotechnological development of methods to increase plant resistance is being actively pursued. One of such approaches is the use of rhizosphere microorganisms that are non-pathogenic for plants and should be non-pathogenic for animals and humans.

The plant rhizosphere is a narrow region of soil that is closest to the root system of plant, where the roots secrete large amounts of metabolites from the root hairs.

These metabolites act as chemical signals for motile bacteria, some of which can stimulate plant growth, increase plant productivity, and resistance to phytopathogens and abiotic factors. The plant microbiome has been proposed as a new area of interest for the next green revolution [139], and it has been shown that plants can adapt to changing environmental conditions not only by changing their own metabolism, but also by regulating the composition of the rhizosphere microbiota. This phenomenon was called "Cry for Help" [140–142].

Plant growth-promoting rhizobacteria (PGPR) [143] can activate a mechanism of plant resistance called "Induced Systemic Resistance" (ISR). Induced resistance is activated by various biotic and abiotic factors [144]. ISR activation by rhizosphere bacteria is similar to pathogen induced Systemic Acquired Resistance (SAR) in the sense that both of them lead to the development of resistance even in non-stressed plant parts. Strains with plant growth promoting activity have been identified from various genera, of which *Pseudomonas* and *Bacillus* have been studied most extensively and are capable of triggering ISR [145, 146].

The application of PGPR has been studied using a variety of agricultural plants such as canola, radicchio, soy, potato, maize, oat, peas, tomato, lentil, barley, wheat, and cucumber [147]. However the details of the direct and indirect mechanisms by which PGPRs promote plant growth and development are still widely discussed, but they are known to differ between bacteria [148]. Some PGPRs are able to produce plant hormones such as auxins, cytokinins, gibberellins, ABA, and ethylene [149, 150] and thus directly affect plant physiology, For example, cytokinins, which are produced by PGPR, stimulated cell division and led to an increase in the surface area of roots of agricultural plants through the enhanced formation of lateral and adventitious roots. There is evidence of the effect of PGPR-produced cytokinins on growth, development and productivity of wheat [151], soybeans [152], rape and lettuce [153]. Other strains increase mineral and nitrogen availability in the soil, thus improving plant growth. PGPRs have also been shown to inhibit the development of pathogenic soil microorganisms, thereby also promoting plant growth [154]. It has been shown that PGPR strains, individually or in a consortium, increase yields and growth of Chinese cabbage, even under conditions of infection with black rot caused by *Xanthomonas campestris pv. Campestris* [155] due to activation of ISR (see above). Other PGPR strains are capable of producing antibiotics, which leads to protection against phytopathogens [156]. The literature data show positive effects of plant PGPR colonization on various photosynthetic parameters, particularly chlorophyll content, transpiration rate, internal CO2 concentration, and stomatal conductance [157].

The positive effect of PGPR colonization on plant tolerance to drought [158], soil salinity [159], high temperatures, changes in atmospheric carbon dioxide content, *etc*. is well presented in the literature [for a review see 148]. However, an equally important factor affecting plant growth and development is light conditions. The influence of colonization of barley plants by soil nonpathogenic *Pseudomonas* bacteria on the structure of the photosynthetic apparatus of leaves and the effect of light intensity of the parameters studied was made in [61]. Rhizosphere bacteria *Pseudomonas putida* (*P. putida*) BS3701 which are a part of a consortium that effectively degrades petroleum products, were used for the colonization of barley plants in that work. It was shown that colonized plants at low light intensity were characterized by higher activity of antioxidant systems, reduced hydrogen peroxide content and larger PS II antenna size compared to control plants, allowing colonized plants to capture more light, resulting in higher values of photosynthetic efficiency parameters. Thus, it was shown that the change in the size of the light harvesting antenna of PS II, and hence the regulation of light energy absorption, occurs not only under the action of abiotic environmental factors (Sections 5 and 6), but also during colonization of plants by PGPR, i.e. under the action of biotic factors.

In the same study [61], the effect of plant PGPR colonization on the course of barley plant adaptation to increased light was investigated as well. To reveal differences in the course of adaptation responses in control plants and plants colonized with *P. putida* BS3701, the plants were grown at moderate light intensity (100 μmol quanta m^{-2} s^{-1}), and then transferred to conditions of high light intensity (1000 μmol quanta m^{-2} s^{-1}) without night period. It was found that the adaptive reduction in PS II antenna size occurred in both control and colonized plants and had the same molecular mechanisms. However, the barley plants colonized by *P. putida* BS3701 adapted more rapidly and were more resistant to increased illumination. Taking into account the fact that before the plants were transferred to the conditions of increased illumination, the colonized plants had a significantly larger PS II antenna size compared to control plants. The decrease in PS II antenna size was more pronounced in the colonized plants, which, apparently, was the reason for more effective and faster adaptation of the photosynthetic apparatus of *P. putida* BS3701 colonized plants to new light conditions. Thus, PGPR colonization of plants leads to the optimization of the photosynthetic apparatus structure and an increase in the efficiency of adaptation mechanisms of higher plants.

8. Conclusion and assumptions

The most significant function of chloroplasts is oxygenic photosynthesis. However, chloroplasts perform many other functions essential for proper plant growth and development, including synthesis of amino acids, nucleotides, fatty acids, phytohormones, some vitamins and secondary metabolites, as well as nitrogen and sulfur assimilation. Acclimation processes occurring in chloroplasts under both abiotic and biotic stresses are important for plant–environment interaction, which promotes plant adaptation to stress factors, including drought, salinity, increased light, colonization by microorganisms, and many others.

The chapter presents data showing that the change in the size of the PS II antenna pigment-protein complex occurs in plants not only under changes in light intensity, but also under other abiotic factors (drought, soil salinity), as well as under the action of a biotic factor – colonization by the rhizosphere bacteria *P. putida* BS3701. Thus, we hypothesized that the regulation of PS II antenna size is one of the universal mechanisms of regulation of the structural and functional organization of photosynthetic apparatus necessary for the adaptation of higher plants to stress conditions. It is suggested that the reaction of $O_2^{\bullet-}$ with PQH_2, leading to the formation of "membrane" H_2O_2 in thylakoids, plays a determining role in this process [132].

The Lhcb proteins of the PS II light-harvesting antenna complex are encoded in the nuclear genome [160], therefore, the regulation of the expression of these genes under stress conditions appears to occur through the retrograde chloroplast-nucleus signal transduction pathway. The question arises, how exactly the expression of genes of PS II antenna proteins is regulated with the involvement of hydrogen peroxide? It is assumed that transcription factors are one of the main participants in retrograde signaling [161]. For example, the transcription factor ABSCISCIC ACID INSENSITIVE 4 (ABI4) is a key factor in multiple retrograde signaling pathways generated by GUN1 (genome uncoupled; tetrapyrrole-dependent signal transduction from plastid to nucleus) [162]. GUN1 is known to transmit a signal that induces ABI4 binding to the promoter sequences of *lhcb* genes in the nucleus that blocks *lhcb* gene expression [161], leading then to a reduction in PS II antenna size. Another transcription factor, PTM, a chloroplast envelope-associated homeodomain (PHD) factor, also functions in multiple retrograde signaling pathways. PTM connects the GUN1 pathway in plastids to the ABI4 pathway in the nucleus [162, 163]; for this

DOI: http://dx.doi.org/10.5772/intechopen.97664

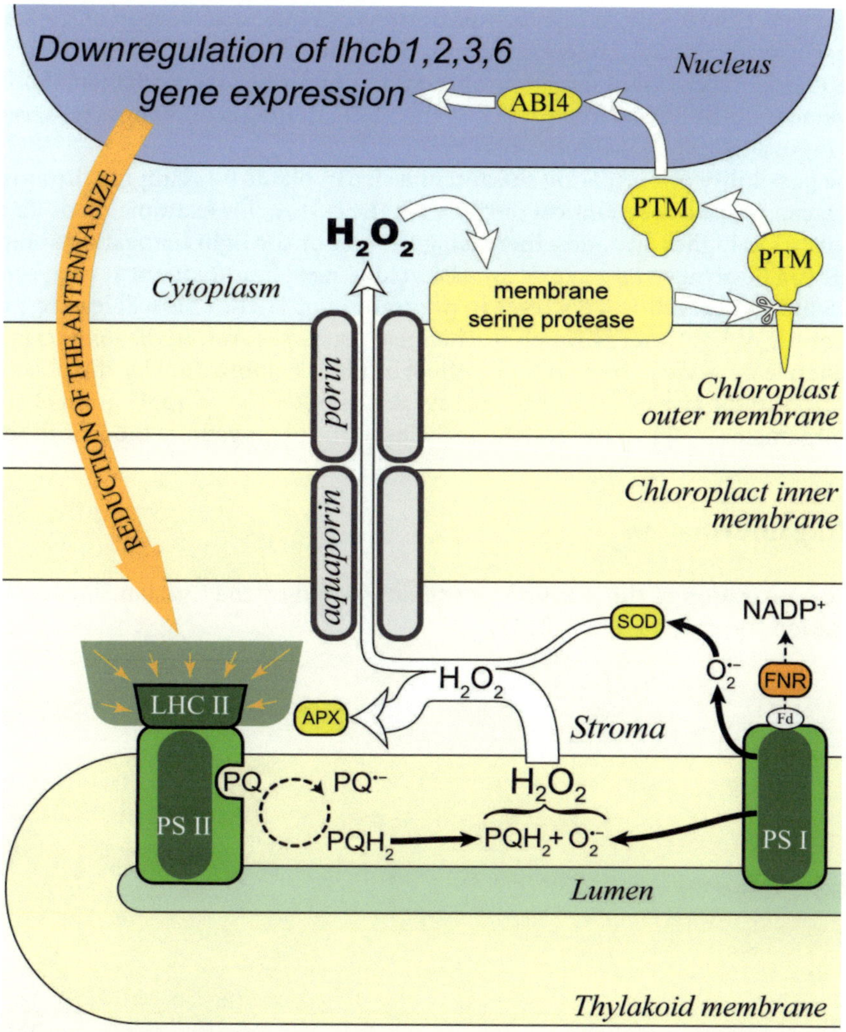

Figure 1.

Hypothetical mechanism of the involvement of "thylakoid membrane" H_2O_2 in the PS II (LHC II) antenna size regulation in the cells of higher plants. PS II, photosystem II; PS I, photosystem I; PQ, oxidized plastoquinone; $PQ^{\bullet-}$, plastosemiquinone; PQH_2, plastohydroquinone; PC, plastocyanine; Fd, ferredoxin; FNR, ferredoxin-$NADP^+$ reductase; SOD, superoxide dismutase; APX, ascorbate peroxidase; PTM, chloroplast envelope-associated homeodomain transcription factor; ABI4, nuclear transcription factor. According to our hypothesis, H_2O_2, which is formed in chloroplasts in the light as a result of the reaction of $O_2^{\bullet-}$ With PQH_2, diffuses through the chloroplast membrane, changes the activity of the serine protease, which, in turn, converts PTM into its soluble form and, thus, affects the expression of lhcb genes that encode LHC II proteins.

purpose, a soluble shortened form of PTM is released from the chloroplast envelope into the cytoplasm, moves into the nucleus, and activates ABI4 expression. The soluble form of PTM is formed as a result of proteolysis of this transcription factor by a serine protease, leading to its detachment from the transmembrane domains [164]. What exactly affects the activity of the serine protease is still unclear. SBcas3.3 serine protease activity in *Escherichia coli*, which belongs to the S8A subfamily of serine proteases, has been shown to increase when H_2O_2 is added at concentrations up to 10 g L^{-1}, but decreases when H_2O_2 is added at a higher concentration of 50 g L^{-1} [165]. It can be assumed that H_2O_2 formed in chloroplasts, when diffuses across the chloroplast membrane, changes the protease activity, affecting the transformation of PTM into the soluble form. At low concentrations, H_2O_2, enhances serine protease activity, leading to suppression of *lhcb* gene expression and to decreasing the PS II antenna size. Conversely, at high concentrations of H_2O_2, inactivation of the serine protease may occur and, as a consequence, no acclimatory change in the size of the PS II antenna complex should be observed. A hypothetical mechanism for the involvement of the "membrane" H_2O_2 in the PS II antenna size changes is presented in the **Figure 1**.

The possibility of regulating the antenna size in plants by changing the amount of H_2O_2 can be used for practical purposes in the future, for example, in order to grow plants in higher latitudes. Increasing the size of the light harvesting antenna of PS II can be achieved by hyperexpression of genes of antioxidant system proteins, which will result in a decrease in the amount of H_2O_2. This will lead to a more efficient use of light energy for photochemical processes and, in the long term, to a productive increase in biomass. The effect of plant colonization by rhizobacteria on changes in the size of PS II antenna reveals the potential of application of such microorganisms in agriculture without the need for plant genetic modifications.

Funding information

The preparation of the manuscript was supported by the Russian Science Foundation (project no. 17-14-01371).

Author details

Maria M. Borisova-Mubarakshina*, Ilya A. Naydov, Daria V. Vetoshkina, Marina A. Kozuleva, Daria V. Vilyanen, Natalia N. Rudenko and Boris N. Ivanov
Institute of Basic Biological Problems of the Russian Academy of Sciences, Federal Research Center, Pushchino Scientific Center for Biological Research of the Russian Academy of Sciences, Pushchino, Russia

*Address all correspondence to: mubarakshinamm@gmail.com

References

[1] Allen JF. Control of gene expression by redox potential and the requirement for chloroplast and mitochondrial genomes. J Theor Biol. 1993 Dec 21;165(4):609-631. doi: 10.1006/jtbi.1993.1210

[2] Karpinski S, Reynolds H, Karpinska B, Wingsle G, Creissen G, Mullineaux P. Systemic signaling and acclimation in response to excess excitation energy in Arabidopsis. Science. 1999 Apr 23; 284(5414):654-657. doi: 10.1126/science.284.5414.654

[3] Foyer CH, Noctor G. Redox regulation in photosynthetic organisms: signaling, acclimation, and practical implications. Antioxid Redox Signal. 2009 Apr;11(4):861-905. doi: 10.1089/ars.2008.2177

[4] Wientjes E, van Amerongen H, Croce R. LHCII is an antenna of both photosystems after long-term acclimation. Biochim Biophys Acta. 2013 Mar;1827(3):420-426. doi: 10.1016/j.bbabio.2012.12.009

[5] Jansson S. The light-harvesting chlorophyll a/b-binding proteins. Biochim Biophys Acta. 1994 Feb 8;1184(1):1-19. doi: 10.1016/0005-2728(94)90148-1

[6] Damkjaer JT, Kereïche S, Johnson MP, Kovacs L, Kiss AZ, Boekema EJ, et al. The photosystem II light-harvesting protein Lhcb3 affects the macrostructure of photosystem II and the rate of state transitions in Arabidopsis. Plant Cell. 2009 Oct;21(10):3245-3256. doi: 10.1105/tpc.108.064006

[7] Boekema EJ, van Roon H, Calkoen F, Bassi R, Dekker JP. Multiple types of association of photosystem II and its light-harvesting antenna in partially solubilized photosystem II membranes. Biochemistry. 1999 Feb 23;38(8):2233-2239. doi: 10.1021/bi9827161

[8] Minagawa J. Dynamic reorganization of photosynthetic supercomplexes during environmental acclimation of photosynthesis. Front Plant Sci. 2013 Dec 17;4:513. doi: 10.3389/fpls.2013.00513

[9] Kouřil R, Dekker JP, Boekema EJ. Supramolecular organization of photosystem II in green plants. Biochim Biophys Acta. 2012 Jan;1817(1):2-12. doi: 10.1016/j.bbabio.2011.05.024

[10] Lunde C, Jensen PE, Haldrup A, Knoetzel J, Scheller HV. The PSI-H subunit of photosystem I is essential for state transitions in plant photosynthesis. Nature. 2000 Nov 30;408(6812): 613-615. doi: 10.1038/35046121

[11] Zhang S, Scheller HV. Light-harvesting complex II binds to several small subunits of photosystem I. J Biol Chem. 2004 Jan 30;279(5):3180-3187. doi: 10.1074/jbc.M311640200

[12] Allen JF, Hall DO. Superoxide reduction as a mechanism of ascorbate-stimulated oxygen uptake by isolated chloroplasts. Biochemical and Biophysical Research Communications. 1973 Jun 8;52(3):856-862. doi: 10.1016/0006-291X(73)91016-4

[13] Badger MR. Photosynthetic Oxygen Exchange. Annual Review of Plant Physiology. 1985;36(1):27-53. doi: 10.1146/annurev.pp.36.060185.000331

[14] Allen JF. Oxygen reduction and optimum production of ATP in photosynthesis. Nature. 1975 Aug; 256(5518):599-600. doi: 10.1038/256599a0

[15] Ivanov BN, Red'ko TP, Shmeleva VL, Mukhin EN. Role of ferredoxin in pseudo-cyclic electron transport in isolated pea chloroplasts. Biochemistry Moscow. 1980 Aug 1;45(8):1425-1432.

[16] Asada K, Kiso K, Yoshikawa K. Univalent reduction of molecular

oxygen by spinach chloroplasts on illumination. Journal of Biological Chemistry. 1974;249(7):2175-2181.

[17] Golbeck J, Radmer R. Is the rate of oxygen uptake by reduced ferredoxin sufficient to account for photosystem I-mediated O2 reduction. Advances in photosynthesis research. 1984;1:561-561.

[18] Kozuleva M, Goss T, Twachtmann M, Rudi K, Trapka J, Selinski J, et al. Ferredoxin:NADP(H) Oxidoreductase Abundance and Location Influences Redox Poise and Stress Tolerance. Plant Physiology. 2016 Nov 1;172(3):1480-1493. doi: 10.1104/pp.16.01084

[19] Kozuleva MA, Ivanov BN. Evaluation of the participation of ferredoxin in oxygen reduction in the photosynthetic electron transport chain of isolated pea thylakoids. Photosynthesis research. 2010;105(1):51-61.

[20] Takahashi M, Asada K. Superoxide production in aprotic interior of chloroplast thylakoids. Archives of Biochemistry and Biophysics. 1988 Dec 1;267(2):714-722. doi: 10.1016/0003-9861(88)90080-X

[21] Hormann H, Neubauer C, Asada K, Schreiber U. Intact chloroplasts display pH 5 optimum of O2-reduction in the absence of methyl viologen: Indirect evidence for a regulatory role of superoxide protonation. Photosynth Res. 1993 Jul 1;37(1):69-80. doi: 10.1007/BF02185440

[22] Kozuleva M, Petrova A, Milrad Y, Semenov A, Ivanov B, Redding KE, et al. Phylloquinone is the principal Mehler reaction site within photosystem I in high light. Plant Physiology 2021, accepted

[23] Kruk J, Jemioła-Rzemińska M, Burda K, Schmid GH, Strzałka K. Scavenging of Superoxide Generated in Photosystem I by Plastoquinol and Other Prenyllipids in Thylakoid Membranes. Biochemistry. 2003 Jul 1;42(28):8501-8505. doi: 10.1021/bi034036q

[24] Kozuleva MA, Petrova AA, Mamedov MD, Semenov AY, Ivanov BN. O2 reduction by photosystem I involves phylloquinone under steady-state illumination. FEBS letters. 2014;588(23):4364-4368.

[25] McConnell MD, Cowgill JB, Baker PL, Rappaport F, Redding KE. Double Reduction of Plastoquinone to Plastoquinol in Photosystem 1. Biochemistry. 2011 Dec 27;50(51):11034-11046. doi: 10.1021/bi201131r

[26] Ptushenko VV, Cherepanov DA, Krishtalik LI, Semenov AYu. Semi-continuum electrostatic calculations of redox potentials in photosystem I. Photosynth Res. 2008 May 16;97(1):55. doi: 10.1007/s11120-008-9309-y

[27] Kozuleva M, Klenina I, Proskuryakov I, Kirilyuk I, Ivanov B. Production of superoxide in chloroplast thylakoid membranes: ESR study with cyclic hydroxylamines of different lipophilicity. FEBS Letters. 2011 Apr 6;585(7):1067-1071. doi: 10.1016/j.febslet.2011.03.004

[28] Kozuleva M, Klenina I, Mysin I, Kirilyuk I, Opanasenko V, Proskuryakov I, et al. Quantification of superoxide radical production in thylakoid membrane using cyclic hydroxylamines. Free Radical Biology and Medicine. 2015 Dec 1;89:1014-1023. doi: 10.1016/j.freeradbiomed.2015.08.016

[29] Ogawa K, Kanematsu S, Takabe K, Asada K. Attachment of CuZn-Superoxide Dismutase to Thylakoid Membranes at the Site of Superoxide Generation (PSI) in Spinach Chloroplasts: Detection by Immuno-Gold Labeling After Rapid Freezing and Substitution Method. Plant and Cell Physiology. 1995 Jun 1;36(4):565-573. doi: 10.1093/oxfordjournals.pcp.a078795

[30] Mubarakshina M, Khorobrykh S, Ivanov B. Oxygen reduction in chloroplast thylakoids results in production of hydrogen peroxide inside the membrane. Biochimica et Biophysica Acta (BBA) - Bioenergetics. 2006 Nov 1;1757(11):1496-1503. doi: 10.1016/j.bbabio.2006.09.004

[31] Mubarakshina MM, Khorobrykh SA, Kozuleva MA, Ivanov BN. Intramembrane formation of hydrogen peroxide during oxygen reduction in thylakoids of higher plants. Dokl Biochem Biophys. 2006 Jun;408(1):113-116. doi: 10.1134/S160767290603001X

[32] (Mubarakshina) Borisova MM, Kozuleva MA, Rudenko NN, Naydov IA, Klenina IB, Ivanov BN. Photosynthetic electron flow to oxygen and diffusion of hydrogen peroxide through the chloroplast envelope via aquaporins. Biochimica et Biophysica Acta (BBA) - Bioenergetics. 2012 Aug 1;1817(8):1314-1321. doi: 10.1016/j.bbabio.2012.02.036

[33] Ivanov B, Mubarakshina M, Khorobrykh S. Kinetics of the plastoquinone pool oxidation following illumination: Oxygen incorporation into photosynthetic electron transport chain. FEBS Letters. 2007 Apr 3;581(7):1342-1346. doi: 10.1016/j.febslet.2007.02.044

[34] Kruk J, Strzałka K. Dark reoxidation of the plastoquinone-pool is mediated by the low-potential form of cytochrome b-559 in spinach thylakoids. Photosynthesis Research. 1999 Dec 1;62(2):273-279. doi: 10.1023/A:1006374319191

[35] McCauley SW, Melis A. Quantitation of plastoquinone photoreduction in spinach chloroplasts. Photosynth Res. 1986 Jan;8(1):3-16. doi: 10.1007/BF00028472

[36] Borisova-Mubarakshina MM, Naydov IA, Ivanov BN. Oxidation of the plastoquinone pool in chloroplast thylakoid membranes by superoxide anion radicals. FEBS Letters. 2018 Oct 1;592(19):3221-3228. doi: 10.1002/1873-3468.13237

[37] Bailey S, Walters RG, Jansson S, Horton P. Acclimation of Arabidopsis thaliana to the light environment: the existence of separate low light and high light responses. Planta. 2001 Sep;213(5):794-801. doi: 10.1007/s004250100556

[38] Anderson JM. Photoregulation of the Composition, Function, and Structure of Thylakoid Membranes. Annual Review of Plant Physiology. 1986;37(1):93-136. doi: 10.1146/annurev.pp.37.060186.000521

[39] Poorter H, Niinemets Ü, Ntagkas N, Siebenkäs A, Mäenpää M, Matsubara S, et al. A meta-analysis of plant responses to light intensity for 70 traits ranging from molecules to whole plant performance. New Phytologist. 2019;223(3):1073-105. doi: https://doi.org/10.1111/nph.15754

[40] Ludlow MM, Wilson GL. Photosynthesis of Tropical Pasture Plants III. Leaf Age. Aust Jnl Of Bio Sci. 1971; 24(4):1077-1088. doi: 10.1071/bi9711077

[41] Wild A, Wolf G. The Effect of Different Light Intensities on the Frequency and Size of Stomata, the Size of Cells, the Number, Size and Chlorophyll Content of Chloroplasts in the Mesophyll and the Guard Cells during the Ontogeny of Primary Leaves of Sinapis alba. Zeitschrift für Pflanzenphysiologie. 1980 May 1;97(4):325-342. doi: 10.1016/S0044-328X(80)80006-7

[42] Boardman NK. Development of Chloroplast Structure and Function. In: Trebst A, Avron M, editors. Photosynthesis I: Photosynthetic Electron Transport and Photophosphorylation [Internet]. Berlin, Heidelberg: Springer; 1977 [cited 2021 Mar 26]. p. 583-600. (Encyclopedia of Plant Physiology). doi: 10.1007/978-3-642-66505-9_42

[43] Lichtenthaler HK, Buschmann C, Döll M, Fietz H-J, Bach T, Kozel U, et al. Photosynthetic activity, chloroplast ultrastructure, and leaf characteristics of high-light and low-light plants and of sun and shade leaves. Photosynth Res. 1981 Jun 1;2(2):115-141. doi: 10.1007/BF00028752

[44] Anderson JM, Goodchild DJ, Boardman NK. Composition of the photosystems and chloroplast structure in extreme shade plants. Biochimica et Biophysica Acta (BBA) - Bioenergetics. 1973 Dec 14;325(3):573-585. doi: 10.1016/0005-2728(73)90217-X

[45] Albanese P, Manfredi M, Meneghesso A, Marengo E, Saracco G, Barber J, et al. Dynamic reorganization of photosystem II supercomplexes in response to variations in light intensities. Biochimica et Biophysica Acta (BBA) - Bioenergetics. 2016 Oct 1;1857(10):1651-1660. doi: 10.1016/j.bbabio.2016.06.011

[46] Antal T, Mattila H, Hakala-Yatkin M, Tyystjärvi T, Tyystjärvi E. Acclimation of photosynthesis to nitrogen deficiency in Phaseolus vulgaris. Planta. 2010 Sep 1;232(4):887-898. doi: 10.1007/s00425-010-1227-5

[47] Brugnoli E, Cona A, Lauteri M. Xanthophyll cycle components and capacity for non-radiative energy dissipation in sun and shade leaves ofLigustrum ovalifolium exposed to conditions limiting photosynthesis. Photosynth Res. 1994 Sep 1;41(3):451-463. doi: 10.1007/BF02183047

[48] Demmig-Adams B, Adams WW. Xanthophyll cycle and light stress in nature: uniform response to excess direct sunlight among higher plant species. Planta. 1996 Sep 1;198(3):460-470. doi: 10.1007/BF00620064

[49] Park YI, Chow WS, Anderson JM. Chloroplast Movement in the Shade Plant Tradescantia albiflora Helps Protect Photosystem II against Light Stress. Plant Physiology. 1996 Jul 1;111(3):867-875. doi: 10.1104/pp.111.3.867

[50] Rudenko NN, Fedorchuk TP, Terentyev VV, Dymova OV, Naydov IA, Golovko TK, et al. The role of carbonic anhydrase α-CA4 in the adaptive reactions of photosynthetic apparatus: the study with α-CA4 knockout plants. Protoplasma. 2020;257(2):489-499.

[51] Dall'Osto L, Caffarri S, Bassi R. A Mechanism of Nonphotochemical Energy Dissipation, Independent from PsbS, Revealed by a Conformational Change in the Antenna Protein CP26. The Plant Cell. 2005 Apr 1;17(4):1217-1232. doi: 10.1105/tpc.104.030601

[52] Demmig-Adams B, Adams WW. Photoprotection in an ecological context: the remarkable complexity of thermal energy dissipation. New Phytologist. 2006;172(1):11-21. doi: https://doi.org/10.1111/j.1469-8137.2006.01835.x

[53] Galka P, Santabarbara S, Khuong TTH, Degand H, Morsomme P, Jennings RC, et al. Functional Analyses of the Plant Photosystem I–Light-Harvesting Complex II Supercomplex Reveal That Light-Harvesting Complex II Loosely Bound to Photosystem II Is a Very Efficient Antenna for Photosystem I in State II. The Plant Cell. 2012 Jul 1;24(7):2963-2978. doi: 10.1105/tpc.112.100339

[54] Rochaix J-D, Lemeille S, Shapiguzov A, Samol I, Fucile G, Willig A, et al. Protein kinases and phosphatases involved in the acclimation of the photosynthetic apparatus to a changing light environment. Philosophical Transactions of the Royal Society B: Biological Sciences. 2012 Dec 19;367(1608):3466-3474. doi: 10.1098/rstb.2012.0064

[55] Shapiguzov A, Ingelsson B, Samol I, Andres C, Kessler F, Rochaix J-D, et al. The PPH1 phosphatase is specifically involved in LHCII dephosphorylation

and state transitions in Arabidopsis. PNAS. 2010 Mar 9;107(10):4782-4787. doi: 10.1073/pnas.0913810107

[56] Depège N, Bellafiore S, Rochaix J-D. Role of Chloroplast Protein Kinase Stt7 in LHCII Phosphorylation and State Transition in Chlamydomonas. Science. 2003 Mar 7;299(5612):1572-1575. doi: 10.1126/science.1081397

[57] Lemeille S, Willig A, Depège-Fargeix N, Delessert C, Bassi R, Rochaix J-D. Analysis of the Chloroplast Protein Kinase Stt7 during State Transitions. PLOS Biology. 2009 Mar 3;7(3): e1000045. doi: 10.1371/journal.pbio.1000045

[58] Shapiguzov A, Chai X, Fucile G, Longoni P, Zhang L, Rochaix J-D. Activation of the Stt7/STN7 Kinase through Dynamic Interactions with the Cytochrome b6f Complex. Plant Physiology. 2016 May 1;171(1):82-92. doi: 10.1104/pp.15.01893

[59] Dumas L, Zito F, Blangy S, Auroy P, Johnson X, Peltier G, et al. A stromal region of cytochrome b6f subunit IV is involved in the activation of the Stt7 kinase in Chlamydomonas. PNAS. 2017 Nov 7;114(45):12063-12068. doi: 10.1073/pnas.1713343114

[60] Mekala NR, Suorsa M, Rantala M, Aro E-M, Tikkanen M. Plants Actively Avoid State Transitions upon Changes in Light Intensity: Role of Light-Harvesting Complex II Protein Dephosphorylation in High Light. Plant Physiology. 2015 Jun 1;168(2):721-734. doi: 10.1104/pp.15.00488

[61] Vetoshkina DV, Pozdnyakova-Filatova IY, Zhurikova EM, Frolova AA, Naydov IA, Ivanov BN, et al. The increase in adaptive capacity to high illumination of barley plants colonized by rhizobacteria P. putida BS3701. Applied Biochemistry and Microbiology. 2019; 55(2):173-181.

[62] Rintamäki E, Martinsuo P, Pursiheimo S, Aro E-M. Cooperative regulation of light-harvesting complex II phosphorylation via the plastoquinol and ferredoxin-thioredoxin system in chloroplasts. PNAS. 2000 Oct 10; 97(21):11644-11649. doi: 10.1073/pnas.180054297

[63] Trotta A, Suorsa M, Rantala M, Lundin B, Aro E-M. Serine and threonine residues of plant STN7 kinase are differentially phosphorylated upon changing light conditions and specifically influence the activity and stability of the kinase. The Plant Journal. 2016;87(5):484-94. doi: https://doi.org/10.1111/tpj.13213

[64] Lemeille S, Rochaix J-D. State transitions at the crossroad of thylakoid signalling pathways. Photosynth Res. 2010 Nov 1;106(1):33-46. doi: 10.1007/s11120-010-9538-8

[65] Vetoshkina DV, Kozuleva MA, Terentyev VV, Zhurikova EM, Borisova-Mubarakshina MM, Ivanov BN. Comparison of state transitions of the photosynthetic antennae in Arabidopsis and barley plants upon illumination with light of various intensity. Biochemistry (Moscow). 2019;84(9):1065-1073.

[66] Tikkanen M, Grieco M, Kangasjärvi S, Aro E-M. Thylakoid Protein Phosphorylation in Higher Plant Chloroplasts Optimizes Electron Transfer under Fluctuating Light. Plant Physiology. 2010 Feb 1;152(2):723-735. doi: 10.1104/pp.109.150250

[67] Bonardi V, Pesaresi P, Becker T, Schleiff E, Wagner R, Pfannschmidt T, et al. Photosystem II core phosphorylation and photosynthetic acclimation require two different protein kinases. Nature. 2005 Oct;437(7062):1179-1182. doi: 10.1038/nature04016

[68] Pursiheimo S, Rintamäki E, Aro E-M. Reversible phosphorylation of LHCII proteins in rye leaves — redox

control and physiological significance. In: Garab G, editor. Photosynthesis: Mechanisms and Effects: Volume I–V: Proceedings of the XIth International Congress on Photosynthesis, Budapest, Hungary, August 17-22, 1998 [Internet]. Dordrecht: Springer Netherlands; 1998 [cited 2021 Mar 26]. p. 1903-6. doi: 10.1007/978-94-011-3953-3_443

[69] Tikkanen M, Nurmi M, Kangasjärvi S, Aro E-M. Core protein phosphorylation facilitates the repair of photodamaged photosystem II at high light. Biochimica et Biophysica Acta (BBA) - Bioenergetics. 2008 Nov 1;1777(11):1432-1437. doi: 10.1016/j.bbabio.2008.08.004

[70] Kirchhoff H, Hall C, Wood M, Herbstová M, Tsabari O, Nevo R, et al. Dynamic control of protein diffusion within the granal thylakoid lumen. PNAS. 2011 Dec 13;108(50):20248-20253. doi: 10.1073/pnas.1104141109

[71] Herbstová M, Tietz S, Kinzel C, Turkina MV, Kirchhoff H. Architectural switch in plant photosynthetic membranes induced by light stress. PNAS. 2012 Dec 4;109(49):20130-20135. doi: 10.1073/pnas.1214265109

[72] Baena-González E, Barbato R, Aro E-M. Role of phosphorylation in the repair cycle and oligomeric structure of photosystem II. Planta. 1999 Apr 1;208(2):196-204. doi: 10.1007/s0042 50050550

[73] Barber J, Anderson JM, Baena–González E, Aro E. Biogenesis, assembly and turnover of photosystem II units. Philosophical Transactions of the Royal Society of London Series B: Biological Sciences. 2002 Oct 29;357(1426): 1451-1460. doi: 10.1098/rstb.2002.1141

[74] Allen JF. Why chloroplasts and mitochondria retain their own genomes and genetic systems: Colocation for redox regulation of gene expression. PNAS. 2015 Aug 18;112(33):10231-10238. doi: 10.1073/pnas.1500012112

[75] Allen JF, Pfannschmidt T. Balancing the two photosystems: photosynthetic electron transfer governs transcription of reaction centre genes in chloroplasts. Osmond CB, Foyer CH, Bock G, editors. Philosophical Transactions of the Royal Society of London Series B: Biological Sciences. 2000 Oct 29;355(1402):1351-1359. doi: 10.1098/rstb.2000.0697

[76] Puthiyaveetil S, Allen JF. Transients in chloroplast gene transcription. Biochemical and Biophysical Research Communications. 2008 Apr 18;368(4): 871-874. doi: 10.1016/j.bbrc.2008.01.167

[77] Ballottari M, Dall'Osto L, Morosinotto T, Bassi R. Contrasting Behavior of Higher Plant Photosystem I and II Antenna Systems during Acclimation*. Journal of Biological Chemistry. 2007 Mar 23;282(12):8947-8958. doi: 10.1074/jbc.M606417200

[78] Morosinotto T, Bassi R, Frigerio S, Finazzi G, Morris E, Barber J. Biochemical and structural analyses of a higher plant photosystem II supercomplex of a photosystem I-less mutant of barley. The FEBS Journal. 2006;273(20):4616-30. doi: https://doi.org/10.1111/j.1742-4658.2006.05465.x

[79] Żelisko A, García-Lorenzo M, Jackowski G, Jansson S, Funk C. AtFtsH6 is involved in the degradation of the light-harvesting complex II during high-light acclimation and senescence. PNAS. 2005 Sep 20;102(38):13699-13704. doi: 10.1073/pnas.0503472102

[80] Wagner R, Aigner H, Pružinská A, Jänkänpää HJ, Jansson S, Funk C. Fitness analyses of *Arabidopsis thaliana* mutants depleted of FtsH metalloproteases and characterization of three FtsH6 deletion mutants exposed to high light stress, senescence and chilling. New Phytologist. 2011;191(2):449-58. doi: https://doi.org/10.1111/j.1469-8137.2011.03684.x

[81] Lindahl M, Yang D-H, Andersson B. Regulatory Proteolysis of the Major Light-Harvesting Chlorophyll a/b Protein of Photosystem II by a Light-Induced Membrane-Associated Enzymic System. European Journal of Biochemistry. 1995;231(2):503-9. doi: https://doi.org/10.1111/j.1432-1033.1995.0503e.x

[82] Pfannschmidt T, Nilsson A, Allen JF. Photosynthetic control of chloroplast gene expression. Nature. 1999 Feb;397(6720):625-628. doi: 10.1038/17624

[83] Escoubas JM, Lomas M, LaRoche J, Falkowski PG. Light intensity regulation of cab gene transcription is signaled by the redox state of the plastoquinone pool. PNAS. 1995 Oct 24;92(22):10237-10241. doi: 10.1073/pnas.92.22.10237

[84] Frigerio S, Campoli C, Zorzan S, Fantoni LI, Crosatti C, Drepper F, et al. Photosynthetic Antenna Size in Higher Plants Is Controlled by the Plastoquinone Redox State at the Post-transcriptional Rather than Transcriptional Level*. Journal of Biological Chemistry. 2007 Oct 5;282(40):29457-29469. doi: 10.1074/jbc.M705132200

[85] Dietzel L, Gläßer C, Liebers M, Hiekel S, Courtois F, Czarnecki O, et al. Identification of Early Nuclear Target Genes of Plastidial Redox Signals that Trigger the Long-Term Response of Arabidopsis to Light Quality Shifts. Molecular Plant. 2015 Aug 3;8(8):1237-1252. doi: 10.1016/j.molp.2015.03.004

[86] Borisova-Mubarakshina MM, Ivanov BN, Vetoshkina DV, Lubimov VY, Fedorchuk TP, Naydov IA, et al. Long-term acclimatory response to excess excitation energy: evidence for a role of hydrogen peroxide in the regulation of photosystem II antenna size. J Exp Bot. 2015 Dec 1;66(22):7151-7164. doi: 10.1093/jxb/erv410

[87] Zandalinas SI, Mittler R. ROS-induced ROS release in plant and animal cells. Free Radical Biology and Medicine. 2018 Jul 1;122:21-27. doi: 10.1016/j.freeradbiomed.2017.11.028

[88] Lawlor DW. Limitation to Photosynthesis in Water-stressed Leaves: Stomata vs. Metabolism and the Role of ATP. Ann Bot. 2002 Jun 15;89(7):871-885. doi: 10.1093/aob/mcf110

[89] Chaves MM, Flexas J, Pinheiro C. Photosynthesis under drought and salt stress: regulation mechanisms from whole plant to cell. Annals of Botany. 2009 Feb 1;103(4):551-560. doi: 10.1093/aob/mcn125

[90] Basu S, Ramegowda V, Kumar A, Pereira A. Plant adaptation to drought stress. F1000Res. 2016 Jun 30;5:1554. doi: 10.12688/f1000research.7678.1

[91] De Micco V, Aronne G. Morpho-Anatomical Traits for Plant Adaptation to Drought. In: Aroca R, editor. Plant Responses to Drought Stress: From Morphological to Molecular Features [Internet]. Berlin, Heidelberg: Springer; 2012 [cited 2021 Apr 11]. p. 37-61. doi: 10.1007/978-3-642-32653-0_2

[92] Medrano H, Parry MAJ, Socias X, Lawlor DW. Long term water stress inactivates Rubisco in subterranean clover. Annals of Applied Biology. 1997;131(3):491-501.

[93] Edwards G, Walker D. C Three C Four: Mechanisms, Cellular and Environmental Regulation of Photosynthesis. University of California Press; 1983. 558 p.

[94] Dalal VK, Tripathy BC. Water-stress induced downsizing of light-harvesting antenna complex protects developing rice seedlings from photo-oxidative damage. Scientific Reports. 2018 Apr 13;8(1):5955. doi: 10.1038/s41598-017-14419-4

[95] Zhou Y, Lam HM, Zhang J. Inhibition of photosynthesis and energy dissipation induced by water and high light stresses

in rice. Journal of Experimental Botany. 2007 Mar 1;58(5):1207-1217. doi: 10.1093/jxb/erl291

[96] Alberte RS, Fiscus EL, Naylor AW. The Effects of Water Stress on the Development of the Photosynthetic Apparatus in Greening Leaves. Plant Physiology. 1975 Feb 1;55(2):317-321. doi: 10.1104/pp.55.2.317

[97] Giardi MT, Cona A, Geiken B, Kučera T, Masojídek J, Mattoo AK. Long-term drought stress induces structural and functional reorganization of photosystem II. Planta. 1996 May 1;199(1):118-125. doi: 10.1007/BF0019 6888

[98] He JX, Wang J, Liang HG. Effects of water stress on photochemical function and protein metabolism of photosystem II in wheat leaves. Physiologia Plantarum. 1995;93(4):771-7. doi: https://doi.org/10.1111/j.1399-3054.1995.tb05130.x

[99] Chakraborty N, Tripathy BC. Involvement of Singlet Oxygen in 5-Aminolevulinic Acid-Induced Photodynamic Damage of Cucumber (Cucumis sativus L.) Chloroplasts. Plant Physiology. 1992 Jan 1;98(1):7-11. doi: 10.1104/pp.98.1.7

[100] Murata N, Allakhverdiev SI, Nishiyama Y. The mechanism of photoinhibition in vivo: Re-evaluation of the roles of catalase, α-tocopherol, non-photochemical quenching, and electron transport. Biochimica et Biophysica Acta (BBA) - Bioenergetics. 2012 Aug 1;1817(8):1127-1133. doi: 10.1016/j.bbabio.2012.02.020

[101] Schmitt F-J, Renger G, Friedrich T, Kreslavski VD, Zharmukhamedov SK, Los DA, et al. Reactive oxygen species: Re-evaluation of generation, monitoring and role in stress-signaling in phototrophic organisms. Biochimica et Biophysica Acta (BBA) - Bioenergetics. 2014 Jun 1;1837(6):835-848. doi: 10.1016/j.bbabio.2014.02.005

[102] Wilkinson S, Kudoyarova GR, Veselov DS, Arkhipova TN, Davies WJ. Plant hormone interactions: innovative targets for crop breeding and management. J Exp Bot. 2012 May;63(9): 3499-3509. doi: 10.1093/jxb/ers148

[103] Wilkinson S, Davies WJ. Drought, ozone, ABA and ethylene: new insights from cell to plant to community. Plant Cell Environ. 2010 Apr;33(4):510-525. doi: 10.1111/j.1365-3040.2009.02052.x

[104] Spollen WG, Sharp RE. Spatial distribution of turgor and root growth at low water potentials. Plant Physiology. 1991;96(2):438-443.

[105] Deak KI, Malamy J. Osmotic regulation of root system architecture. The Plant Journal. 2005;43(1):17-28.

[106] Chen H, Li Z, Xiong L. A plant microRNA regulates the adaptation of roots to drought stress. FEBS Letters. 2012;586(12):1742-7. doi: https://doi.org/10.1016/j.febslet.2012.05.013

[107] Jaffe MJ, Takahashi H, Biro RL. A Pea Mutant for the Study of Hydrotropism in Roots. Science. 1985 Oct 25;230(4724):445-447. doi: 10.1126/science.230.4724.445

[108] Blilou I, Xu J, Wildwater M, Willemsen V, Paponov I, Friml J, et al. The PIN auxin efflux facilitator network controls growth and patterning in Arabidopsis roots. Nature. 2005; 433(7021):39-44.

[109] Chaves MM, Oliveira MM. Mechanisms underlying plant resilience to water deficits: prospects for water-saving agriculture. Journal of Experimental Botany. 2004 Nov 1;55(407): 2365-2384. doi: 10.1093/jxb/erh269

[110] Oosterhuis DM, Wullschleger SD. Osmotic Adjustment in Cotton (Gossypium hirsutum L.) Leaves and Roots in Response to Water Stress. Plant Physiol. 1987 Aug;84(4):1154-1157. doi: 10.1104/pp.84.4.1154

[111] Sauter A, Davies WJ, Hartung W. The long-distance abscisic acid signal in the droughted plant: the fate of the hormone on its way from root to shoot. J Exp Bot. 2001 Oct;52(363):1991-1997. doi: 10.1093/jexbot/52.363.1991

[112] He X, Chen Z, Wang J, Li W, Zhao J, Wu J, et al. A sucrose:fructan-6-fructosyltransferase (6-SFT) gene from Psathyrostachys huashanica confers abiotic stress tolerance in tobacco. Gene. 2015 Oct 10;570(2):239-247. doi: 10.1016/j.gene.2015.06.023

[113] Ge L-F, Chao D-Y, Shi M, Zhu M-Z, Gao J-P, Lin H-X. Overexpression of the trehalose-6-phosphate phosphatase gene OsTPP1 confers stress tolerance in rice and results in the activation of stress responsive genes. Planta. 2008 Jun;228(1):191-201. doi: 10.1007/s00425-008-0729-x

[114] Ashraf M, Foolad MR. Roles of glycine betaine and proline in improving plant abiotic stress resistance. Environmental and Experimental Botany. 2007 Mar 1;59(2):206-216. doi: 10.1016/j.envexpbot.2005.12.006

[115] Isayenkov SV, Maathuis FJM. Plant Salinity Stress: Many Unanswered Questions Remain. Front Plant Sci [Internet]. 2019 [cited 2021 Mar 26];10. doi: 10.3389/fpls.2019.00080

[116] Isayenkov SV. Physiological and molecular aspects of salt stress in plants. Cytol Genet. 2012 Sep 1;46(5):302-318. doi: 10.3103/S0095452712050040

[117] Tsugane K, Kobayashi K, Niwa Y, Ohba Y, Wada K, Kobayashi H. A Recessive Arabidopsis Mutant That Grows Photoautotrophically under Salt Stress Shows Enhanced Active Oxygen Detoxification. The Plant Cell. 1999 Jul 1;11(7):1195-1206. doi: 10.1105/tpc.11.7.1195

[118] Kim H-J, Fonseca JM, Choi J-H, Kubota C, Kwon DY. Salt in irrigation water affects the nutritional and visual properties of romaine lettuce (Lactuca sativa L.). J Agric Food Chem. 2008 May 28;56(10):3772-3776. doi: 10.1021/jf0733719

[119] Gilroy S, Swanson SJ. Gravitropic Signaling in Plants. In: eLS [Internet]. American Cancer Society; 2014 [cited 2021 Mar 26]. doi: 10.1002/9780470015902.a0025267

[120] Roy SJ, Negrão S, Tester M. Salt resistant crop plants. Current Opinion in Biotechnology. 2014 Apr 1;26:115-124. doi: 10.1016/j.copbio.2013.12.004

[121] Munns R, Passioura JB. Hydraulic Resistance of Plants. III. Effects of NaCl in Barley and Lupin. Functional Plant Biol. 1984;11(5):351-359. doi: 10.1071/pp9840351

[122] Munns R, Termaat A. Whole-Plant Responses to Salinity. Functional Plant Biol. 1986;13(1):143-160. doi: 10.1071/pp9860143

[123] Rajendran K, Tester M, Roy SJ. Quantifying the three main components of salinity tolerance in cereals. Plant, Cell & Environment. 2009;32(3):237-49. doi: https://doi.org/10.1111/j.1365-3040.2008.01916.x

[124] Munns R, Tester M. Mechanisms of salinity tolerance. Annu Rev Plant Biol. 2008;59:651-681. doi: 10.1146/annurev.arplant.59.032607.092911

[125] Allakhverdiev SI, Sakamoto A, Nishiyama Y, Inaba M, Murata N. Ionic and Osmotic Effects of NaCl-Induced Inactivation of Photosystems I and II in Synechococcus sp. Plant Physiology. 2000 Jul 1;123(3):1047-1056. doi: 10.1104/pp.123.3.1047

[126] Yamane K, Kawasaki M, Taniguchi M, Miyake H. Differential effect of NaCl and polyethylene glycol on the ultrastructure of chloroplasts in rice seedlings. Journal of Plant Physiology. 2003;160(5):573-575.

[127] Yamane K, Oi T, Enomoto S, Nakao T, Arai S, Miyake H, et al. Three-dimensional ultrastructure of chloroplast pockets formed under salinity stress. Plant Cell Environ. 2018 Mar;41(3):563-575. doi: 10.1111/pce.13115

[128] Locy RD, Chang CC, Nielsen BL, Singh NK. Photosynthesis in Salt-Adapted Heterotrophic Tobacco Cells and Regenerated Plants. Plant Physiology. 1996 Jan 1;110(1):321-328. doi: 10.1104/pp.110.1.321

[129] Foyer CH, Allen JF. Lessons from redox signaling in plants. Antioxidants and Redox Signaling. 2003;5(1):3-5.

[130] Monroe JD, Storm AR, Badley EM, Lehman MD, Platt SM, Saunders LK, et al. β-Amylase1 and β-Amylase3 Are Plastidic Starch Hydrolases in Arabidopsis That Seem to Be Adapted for Different Thermal, pH, and Stress Conditions. Plant Physiology. 2014 Dec 1;166(4):1748-1763. doi: 10.1104/pp.114.246421

[131] Thalmann M, Santelia D. Starch as a determinant of plant fitness under abiotic stress. New Phytol. 2017 May;214(3):943-951. doi: 10.1111/nph.14491

[132] Borisova-Mubarakshina MM, Vetoshkina DV, Ivanov BN. Antioxidant and signaling functions of the plastoquinone pool in higher plants. Physiologia Plantarum. 2019;166(1):181-198. doi: 10.1111/ppl.12936

[133] Borisova-Mubarakshina MM, Vetoshkina DV, Naydov IA, Rudenko NN, Zhurikova EM, Balashov NV, et al. Regulation of the size of photosystem II light harvesting antenna represents a universal mechanism of higher plant acclimation to stress conditions. Functional Plant Biology. 2020;47(11):959-969.

[134] Balashov NV, Zhurikova EM, Naydov IA, Rudenko NN, Vetoshkina DV, Ivanov BN, et al. Regulation of PSII antenna size under drought and salt stress. In: Proceedings of the 2nd Russia-Japan Joint Forum for Education and Research. Moscow, Russia: Moscow State University; 2018.

[135] Hirth M, Dietzel L, Steiner S, Ludwig R, Weidenbach H, and JP, et al. Photosynthetic acclimation responses of maize seedlings grown under artificial laboratory light gradients mimicking natural canopy conditions. Front Plant Sci [Internet]. 2013 Sep 12 [cited 2021 Apr 11];4. doi: 10.3389/fpls.2013.00334

[136] Meacham K, Sirault X, Quick WP, von Caemmerer S, Furbank R. Diurnal Solar Energy Conversion and Photo-protection in Rice Canopies. Plant Physiol. 2017 Jan;173(1):495-508. doi: 10.1104/pp.16.01585

[137] Cammarisano L, Donnison IS, Robson PRH. Producing Enhanced Yield and Nutritional Pigmentation in Lollo Rosso Through Manipulating the Irradiance, Duration, and Periodicity of LEDs in the Visible Region of Light. Front Plant Sci. 2020;11:598082. doi: 10.3389/fpls.2020.598082

[138] Horton P, Murchie EH, Ruban AV, Walters RG. Increasing Rice Photosynthesis by Manipulation of the Acclimation and Adaptation to Light. In: Novartis Foundation Symposium 236 - Rice Biotechnology: Improving Yield, Stress Tolerance and Grain Quality [Internet]. John Wiley & Sons, Ltd; [cited 2021 Apr 11]. p. 117-34. doi: 10.1002/9780470515778.ch9

[139] Rodriguez R, Durán P. Natural holobiome engineering by using native extreme microbiome to counteract the climate change effects. Frontiers in Bioengineering and Biotechnology. 2020;8.

[140] Carrión VJ, Perez-Jaramillo J, Cordovez V, Tracanna V, De Hollander M, Ruiz-Buck D, et al. Pathogen-induced activation of disease-suppressive functions in the endophytic

root microbiome. Science. 2019; 366(6465):606-612.

[141] Li X, Jousset A, de Boer W, Carrión VJ, Zhang T, Wang X, et al. Legacy of land use history determines reprogramming of plant physiology by soil microbiome. The ISME journal. 2019;13(3):738-751.

[142] Rolfe SA, Griffiths J, Ton J. Crying out for help with root exudates: adaptive mechanisms by which stressed plants assemble health-promoting soil microbiomes. Current opinion in microbiology. 2019;49:73-82.

[143] Schroth MN, Kloepper JW. Plant growth promoting rhizobacteria on radish. In: Proceedings of the fourth conference plant pathogenic bacteria Ed Station de Pathogenic Vegetable ET (Phytobacteriologic) INRA Angers. 1978. p. 876-82.

[144] Van Loon LC, Bakker P, Pieterse CMJ. Systemic resistance induced by rhizosphere bacteria. Annual review of phytopathology. 1998;36(1):453-483.

[145] Kumar A, Prakash A, Johri BN. Bacillus as PGPR in crop ecosystem. In: Bacteria in agrobiology: crop ecosystems. Springer; 2011. p. 37-59.

[146] Van Wees SC, Van der Ent S, Pieterse CM. Plant immune responses triggered by beneficial microbes. Current opinion in plant biology. 2008;11(4): 443-448.

[147] Gray EJ, Smith DL. Intracellular and extracellular PGPR: commonalities and distinctions in the plant–bacterium signaling processes. Soil biology and biochemistry. 2005;37(3):395-412.

[148] Dimkpa C, Weinand T, Asch F. Plant–rhizobacteria interactions alleviate abiotic stress conditions. Plant, cell & environment. 2009;32(12):1682-1694.

[149] Patten CL, Glick BR. Bacterial biosynthesis of indole-3-acetic acid. Can J Microbiol. 1996 Mar;42(3):207-220. doi: 10.1139/m96-032

[150] Arkhipova TN, Prinsen E, Veselov SU, Martinenko EV, Melentiev AI, Kudoyarova GR. Cytokinin producing bacteria enhance plant growth in drying soil. Plant Soil. 2007 Mar 1;292(1):305-315. doi: 10.1007/s11104-007-9233-5

[151] Timmusk S, Nicander B, Granhall U, Tillberg E. Cytokinin production by Paenibacillus polymyxa. Soil Biology and Biochemistry. 1999;31(13):1847-1852.

[152] García de Salamone IE, Hynes RK, Nelson LM. Cytokinin production by plant growth promoting rhizobacteria and selected mutants. Canadian Journal of microbiology. 2001;47(5):404-411.

[153] Noel TC, Sheng C, Yost CK, Pharis RP, Hynes MF. Rhizobium leguminosarum as a plant growth-promoting rhizobacterium: direct growth promotion of canola and lettuce. Canadian Journal of Microbiology. 1996;42(3):279-283.

[154] Glick RE, Schlagnhaufer CD, Arteca RN, Pell EJ. Ozone-Induced Ethylene Emission Accelerates the Loss of Ribulose-1,5-Bisphosphate Carboxylase/Oxygenase and Nuclear-Encoded mRNAs in Senescing Potato Leaves. Plant Physiology. 1995 Nov 1;109(3):891-898. doi: 10.1104/pp.109.3.891

[155] Liu K, Garrett C, Fadamiro H, Kloepper JW. Induction of systemic resistance in Chinese cabbage against black rot by plant growth-promoting rhizobacteria. Biological Control. 2016;99:8-13.

[156] Haas D, Défago G. Biological control of soil-borne pathogens by fluorescent pseudomonads. Nature reviews microbiology. 2005;3(4):307-319.

[157] Ansari WA, Atri N, Ahmad J, Qureshi MI, Singh B, Kumar R, et al. Drought mediated physiological and

molecular changes in muskmelon (*Cucumis melo* L.). PLOS ONE. 2019 Sep 24;14(9):e0222647. doi: 10.1371/journal.pone.0222647

[158] Naveed M, Mitter B, Reichenauer TG, Wieczorek K, Sessitsch A. Increased drought stress resilience of maize through endophytic colonization by Burkholderia phytofirmans PsJN and Enterobacter sp. FD17. Environmental and Experimental Botany. 2014;97:30-39.

[159] Jha Y, Subramanian RB. PGPR regulate caspase-like activity, programmed cell death, and antioxidant enzyme activity in paddy under salinity. Physiology and Molecular Biology of Plants. 2014;20(2):201-207.

[160] Jansson S. A guide to the Lhc genes and their relatives in Arabidopsis. Trends in Plant Science. 1999 Jun 1;4(6):236-240. doi: 10.1016/S1360-1385(99)01419-3

[161] Yurina NP, Odintsova MS. Chloroplast Retrograde Signaling System. Russ J Plant Physiol. 2019 Jul 1;66(4):509-520. doi: 10.1134/S1021443719040149

[162] Sun X, Feng P, Xu X, Guo H, Ma J, Chi W, et al. A chloroplast envelope-bound PHD transcription factor mediates chloroplast signals to the nucleus. Nature communications. 2011;2(1):1-10.

[163] Zhang Z-W, Zhang G-C, Zhu F, Zhang D-W, Yuan S. The roles of tetrapyrroles in plastid retrograde signaling and tolerance to environmental stresses. Planta. 2015;242(6):1263-1276.

[164] Adam Z. Plastid intramembrane proteolysis. Biochimica et Biophysica Acta (BBA)-Bioenergetics. 2015;1847(9):910-4.

[165] Biver S, Portetelle D, Vandenbol M. Characterization of a new oxidant-stable serine protease isolated by functional metagenomics. Springerplus. 2013;2(1):1-10.

Section 3

Bioindicators

Chapter 15

Rockbee Repellent Endemic Plant Species of Andaman-Nicobar Archipelago in the Bay of Bengal

Sam Paul Mathew and Raveendranpillai Prakashkumar

Abstract

The article concisely illustrates the vegetation dynamics and interrelationships among Man, Animal and Plant species of the insular tropical rain forests found to occur in Andaman-Nicobar Islands in the Bay of Bengal. There are two aboriginal groups known as 'Onges' and 'Shompens' living in different islands such as the Little Andaman Island and the Great Nicobar Island are mostly depends on the forest vegetation for their livelihood. The delicacy of these two ancient aboriginal groups towards wild honey in their food habits is quite time immemorial. Interestingly, it is found that the insular vegetation dynamics has a key role to gratify their needs in this regard. The two endemic plant species of the insular vegetation *viz*. *Orophea katschallica* Kurz and *Etlingera fenzlii* (Kurz) Škorničk. & M. Sabu is traditionally used by these primitive aboriginal communities for honey collection from the large wild hives of the ferocious rockbee (*Apis dorsata* Fabricius). The article also included details on chemical characterisation of *Etlingera fenzlii*.

Keywords: Andaman-Nicobar Islands, *Apis dorsata*, *Orophea katschallica*, *Etlingera fenzlii*, Onges, Shompens, Honeybee repellent endemics

1. Introduction

The Andaman-Nicobar Islands, located approximately 650 nautical miles far off from the Coromandel Coast of the Peninsular India between the latitudes 6° 45″ to 13° 41" N and the longitudes 92° 12″ to 94° 16″ E, are characterised with enchanting seascapes bordering rocky or sandy beaches with lush green tropical rainforests. This archipelago comprises around 325 major islands and islets which offer a total insular landmass of roughly 8249sq km in the Bay of Bengal. According to periodical enumerations on flora of the Andaman-Nicobar Islands by various botanists from the Botanical Survey of India could ascertain the occurrence of 2649 plant taxa comprising 2508 species, 32 subspecies, 103 varieties and 6 forma under 1109 genera within 238 families belonging to 4 different plant groups, namely Bryophytes (mosses), Pteridophytes, Gymnosperms and Angiosperms [1–5]. The plant genetic resources (PGR) of the Andaman-Nicobar Islands have a wide range of economically significant gene pools of lesser known plant taxa and wild prototypes of several popular cultivars. Wild occurrence of popular cultivars like coconut palms, betel nut palms, betel vines were recorded since centuries ago from these islands [6–11]. The apparent wild occurrence of these popular cultivars among several uninhabited islands over a century ago substantiate to suggest an interesting argument in phytogeographical studies on Andaman-Nicobar Islands as these islands might be a centre of origin of these species.

The tropical rainforests occurring in Andaman-Nicobar Islands is the last stronghold of pristine rainforests within the Indian territory, perhaps only exception being the slopes of the Western Ghats where it has been remarkably disturbed and degraded by human interventions (**Figures 1–17**). Andaman-Nicobar Islands obviously represents one of the richest repositories of insular biodiversity in the Bay of Bengal within a limited geographical area. The unique geographical location of this archipelago between the two major biodiversity

Figure 1.
Etlingera fenzlii – *Habit (Great Nicobar Island, 2014).*

Figure 2.
Etlingera fenzlii – *Inflorescence (Great Nicobar Island, 2014).*

Figure 3.
Etlingera fenzlii – *Inflorescence (Kamorta Island, 2014).*

Figure 4.
Etlingera fenzlii – *Fruiting specimen (Great Nicobar Island, 2014).*

areas (Indian Subcontinent and Malesian Islands) bestowed with an incomparable distribution of plant species with representatives of the Indian, Myanmarese, Thai, Malaysian and Indonesian floras. The flora of the Andaman group of islands shows closer affinity towards Indo-Myanmarese-Thai floras, while the Nicobar groups of islands demonstrates similarity towards the flora of Indonesia and Malaysia [2, 9].

Figure 5.
Etlingera fenzlii- *single leaf (Great Nicobar Island, 2014).*

Figure 6.
Orophea katschallica – *Flowering specimen (Little Andaman Island, source, BSI, Port Blair 2014).*

2. Materials and methods

The interrelationship between the Plant Kingdom and the Animal Kingdom is one of the intriguing subjects in vegetation dynamics, especially among isolated insular regions. The insular human population of the Andaman-Nicobar Islands

Figure 7.
Orophea katschallica – *Fruiting specimen (Little Andaman Island, source, BSI, Port Blair 2014).*

Figure 8.
Onge couple at settlement in Little Andaman Island (1994).

could be classified into ethnic tribes or original inhabitants, old settlers who came before Indian independence and the migrants after independence. The Andaman-Nicobar tribal groups (Great Andamanense, Jarawas, Sentinelese, Onges and Shompens) except Nicobarese (Nicobar Islands) could precisely be considered as the stakeholders of insular genetic diversity, since they are mostly depend on insular biodiversity for their livelihood. The native Negritude tribes of the Andaman Islands principally Jarawas, Sentinelese and Onges are rather hunter-gatherers, sustaining on wild or marine food resources and have practically no cultivation practice; Unlike the tribes of the Andaman Islands, the Nicobaries, the indigenous people of Car Nicobar, Katchal, Kamorta, Nancowry, Chowra etc. are maintaining some genetic diversity of cultivars in their native islands. The *Shompens* of Great Nicobar Island are semi-nomadic and mostly depend on wild resources for their livelihood; however, they have some crude forms in cultivation practice of wild

Figure 9.
An old photograph of an Onge hut (source Kloss C. Boden 1903).

Figure 10.
An Onge climbing on a tree for honey collection (source – PRD, A & N Administration 1980).

species like *Tacca leontopetaloides* (L.) Kuntze popularly known as '*Nicobari aalu*'. There are several interesting lesser known endemics used by these primitive insular aboriginals for their sustainable living. Honey bee repellent plants are one such group of insular endemics traditionally used by the insular aboriginals.

3. Insular wild honey bees: taxonomy and distribution

Apis creana Indica Fabricius, *Apis florae* Fabricius and *Apis dorsata* Fabricius are the three indigenous wild insular taxa of honey bees found to occur among the islands of Andaman-Nicobar Archipelago. The former two species usually nesting in cavities and have several combs while latter with open air single comb nesting habit generally on tall tree limbs and rock cliffs of the interior forests and rarely colonising on undisturbed building corners. *Apis creana Indica* is a subspecies of

Figure 11.
Shompen man along with a woman (Great Nicobar) source: Anthropological survey of India, Port Blair.

Apis creana spread across the Peninsular India, Sri Lanka, Thailand, Cambodia, Vietnam, Bangladesh, Myanmar, Andaman-Nicobar Islands, Indonesia, Malaysia and Philippines. This taxon is well demarcated with the smallest body size among the three. Geographical occurrence of *Apis florae* is reported from South Asia towards Southeast Asiatic regions and some warmer regions of the Middle East. This arthropod is rather frequent among the tropical forests, woods and sometimes in farming yards of Indian Subcontinent, Sri Lanka, Thailand, Andaman-Nicobar Islands, Indonesia, Malaysia and Philippines and has also been reported from Persian Gulf (warmer regions of Oman). *Apis dorsata* is a wide spread wild honey-bee species reported from South to Southeast Asian countries.

Apis dorsata is a major pollinating agent for several plant species of the Andaman-Nicobar rainforests as well as a remarkable source of natural forest honey. It is found that each colony of this species may yield 50 to 70 Kgs of honey in favourable season [12]. Forest honey has remarkable influence with the food habits of the various aboriginals of the Andaman-Nicobar Islands. Domestication of *Apis dorsata* is rather impossible owing to its aggressive nature, frequent migratory habit,

Figure 12.
Shompen lady along with her children (source – PRD, A & N Administration, 1980).

Figure 13.
Shompen hut (source Kloss C. Boden 1903).

quick temperedness etc. [13–15]. According to Koeniger [16] there are only two species *viz*. *A. mellifera* L. and *A. cerana* Fabricius under domesticated groups.

4. The 'Giants' or 'Rockbees': taxonomy and distribution

Apis dorsata Fabricius and *A. laboriosa* Smith are ferocious wild bees generally referred to as the 'Giant' or 'Rockbee' among the honeybees. *Apis dorsata* is with light orange brown or tawny body colour more frequent among interior insular forest habitats or infrequently establish on undisturbed shade walls and corners of old buildings in Andaman-Nicobar Islands. The nesting pattern is open air single-comb habit on tall tree limbs or rock cliffs, usually from 03 to 30 m above ground level. In Andaman-Nicobar Islands, generally the hives are nesting beneath horizontal limps of insular tall trees such as *Dipterocarpus* spp., *Terminalia* spp., *Parishia insignis* Hook. f. etc. and sometimes more than 05 hives harbouring in a single tree. Large

Figure 14.
Apis dorsata – *A Rockbee colony on building corner of BSI at Port Blair (2014).*

Figure 15.
Apis dorsata – *A close-up view of a Rockbee colony (2014).*

hives are also being located in some islands on solitary tall trees along the open spaces (deforested areas) of the interior forests as well as along the forest edges and even at corners of buildings. One of such hives was noted by the senior author (SPM) during his tenure in Andaman's at one corner of the Botanical Survey of India building for about three years. The hanging single-comb nest may be as much as one meter in width. *Apis dorsata* is a migratory species with large hives comprising with population ranging from 25,000 to 80,000 bees [17]. The event of migration is solely correlated with seasons from rainy to winter and summer and said to be migrated towards far inter islands of the Andaman-Nicobar Archipelago up to 50 to 70 Km every year. The 'Rockbees' are extremely ferocious and well-known for its viciousness even to a minor disturbance. The huge mass of defending 'workers' can able to intruders for long distance of several kilometres through mark by their stings releasing the alarm pheromones *viz*. Isopentyl (3-methylbutyl) acetate [18]

Figure 16.
Apis dorsata – *A view of a Rockbee colony formation at Mount Harriet (South Andaman, 2019).*

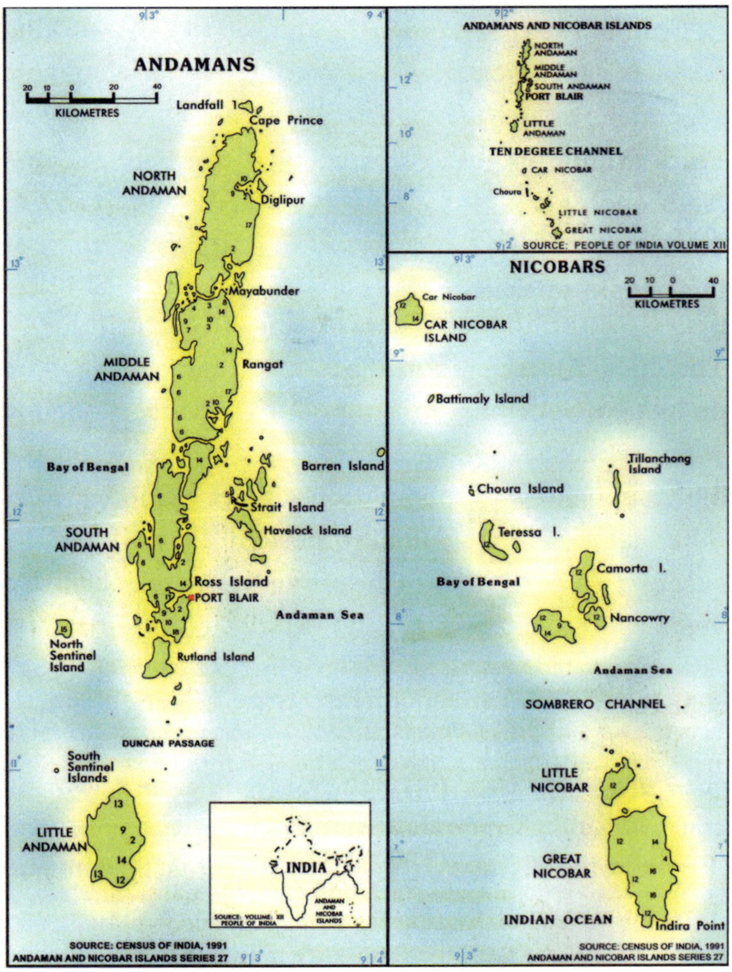

Figure 17.
Andaman–Nicobar Islands (source: Census of India, 1991).

and their massive assault may turn to be fatal for animals and even to human beings. Veith *et al.* [19] isolated a new alarm pheromone, *viz*. 2-decen-1-yl-acetate, from the stings of *Apis dorsata*. According to Veith *et al.* [19] a single sting may inject 06 μg of this compound into the body of the prey. The worker group of *Apis dorsata* can fly even in night hours, if adequate moon light is available.

Apis dorsata is the well-known 'Giant Honeybee Proper' located along lower altitudes across South to Southeast Asia. According to one school of taxonomists, the taxonomical status of *Apis dorsata* is recognised with different subspecies, *viz. A. dorsata breviligula,* the Indonesian subspecies with short tongue, medium forewing length. *Apis dorsata binghami* found to occur from South to Southeast Asian regions with long tongue and long forewing [20]. *Apis laboriosa* Smith, the 'Himalayan Honeybee' found to distribute along higher altitudes from 1200 to 4000 m was originally described as a distinct species and currently treated as a subspecies of *Apis dorsata* as *Apis dorsata laboriosa* [21] based upon biological species concept. According to Arias and Sheppard [22], it is a distinct species based upon the genetic species concept with remarkable behavioural adaptations for nesting along high altitudes regardless of low ambient temperature. The organisation of Rockbee comb is similar to other species. Honey storage at the top portion of the hives, followed by pollen storage beneath and then worker brood and drone brood. The lower part of the hives generally referred to as 'mouth' of the colony where workers take-off and landing and communications on food sources by the scouts, takes place.

5. Onges and Shompens: the stakeholders of honey bee repellent plant species

The aboriginals of Andaman group of islands are Negritude race of 13 primitive tribes, mostly being extinct, and the remaining are collectively termed as 'Great Andamanese'. Currently, the indigenous people of Andaman group of islands are being classified into four groups, namely, the Great Andamanese, the Jarawa, the Onge and the Sentinelese. Interestingly, the indigenous people of Nicobar group, Nicobarese and Shompens belong to the Mongoloid race.

The Onges, presently around 100 in number on the road of extinction, are confined to within two pockets of the Little Andaman Island *viz*. Dugong Creek and South Bay. They were the semi nomadic people exclusively dependent on nature for their livelihood prior to amenities provided by the Andaman-Nicobar Administration. Even now, Rockbee honey is a delicious item in their food habits other than turtle, fish, tubers, jack fruits and screw pine fruits.

The Shompens are another primitive tribe living in Great Nicobar with two distinct divisions, the smaller division being designated as Coastal Shompens or Mawa Shompens inhabited East Coast of the island. They are very shy people; while the other group, designated as Forest Shompens, are rather hostile and are living in interior regions of Alexendra River and Galathia River areas and both groups are totally isolated with each other. According to Patnaik *et al.* [23], the segregation within the Shompen tribe as Forest Shompen and Coastal Shompen has been shown to exist on the basis of ecological analysis. The Forest Shompens are still exclusively depends on nature for their livelihood. Screw pine fruits and Rockbee honey are their staple food; while, the Mawa Shompens have a settlement provided by the Andaman-Nicobar Administration at Campbell Bay in Great Nicobar. The Forest Shompens occasionally barter Rockbee honey with East Coast Nicobarese for machetes and iron scraps useful for spear tips and knives.

6. Bee repellent insular endemic plant species

Two endemic plant taxa in Andaman-Nicobar Islands are reported as honey bee repellent species. *Orophea katschallica* Kurz, an endemic species belonging to the family Annonaceae is widely distributed from the Little Andaman Island towards the Nicobar group of Islands *viz*. Car Nicobar Island, Katchal Island, Kamorta Island, Great Nicobar Island. However, the natural distribution of the taxon is rather infrequent towards northern islands of the Andaman group beyond the Little Andaman. This taxon is originally described by Wilhelm Sulpiz Kurz in 1875 from Kactchal Island of the Nicobar group [24]. The type specimen is deposited at K! (Acc. No. K 000574819). Thoththari [25] was the pioneer who reported the ethno botanical use of *Orophea katschallica* by the Onges for honey collection from the wild Rockbee combs. Later Bhargava [26] stated the tranquillising property of *Orophea katschallica* on wild Rockbee. This species is known under different accents among Onges as *Toyoge* or *Tanjoge* or *Tonyoge* [27]. Interestingly, the insular honey-bee species was confirmed as *Apis dorsata* recently only in 1983 [28]. The traditional practice of Onges on gathering honey from the large Rockbee hives is very simple by chewing the leaves of *Orophea katschallica* and smears the juice mixed with their own saliva on their bodies just prior to climbing tree to get the hives. On reaching the tree limb where the hives are hanging, a coarse spray of juice mixed with saliva spits over the honey comb makes the wild bees get tranquillised and able to gather honey. Interestingly, they used to eat honey during gathering while on the trees has been observed by the authors.

As regards to the taxonomy of *Etlingera fenzlii* (Kurz) Škorničk. & M. Sabu, the species is originally described by Wilhelm Sulpiz Kurz in 1876 from Kamota Island of the Nicobar group as *Amomum fenzlii* Kurz [7]. Later the taxon treated as *Hornstedtia fenzlii* (Kurz) K. Schum. and currently as *Etlingera fenzlii* (Kurz) Škorničk. and M. Sabu. Shompans are widely used this species for honey collection from the hanging combs of Rockbees by smearing the leaf juice over their body and chewing leaves and coarse spray of juice mixed with saliva by spits over the honey comb, as if the Onges are performing for honey collection. This species widely distributed in Great Nicobar Island and also found growing in Kamorta and Katchal Island of the Nicobar group. Recently, the extended distribution of this taxon is being recorded from the Mount Harriet Hill ranges of South Andaman's during an exploration in 2006 by the author [29].

Orophea katschallica Kurz is a small tree or a tree let with black coloured branches and oblong-lanceolate coriaceous leaves. Trimerous creamy yellowish flowers with triangular sepals and clawed petals cohering at apex. Stamens are 6 in numbers with broad connectives. Tricarpellary ovary, apocarpous. Fruits are linear beaded follicles. *Etlingera fenzlii* is around 03 m tall herbs with long linear oblong subcoriaceous to coriaceous leaflets. Inflorescence is arising from the rhizome. Flowers are light rose to deep Red in colour on maturity. The Great Nicobar specimens are with deep Red flowers while specimens from Kamorta Island are with light rose flowers. Living accessions of *Etlingera fenzlii* from Great Nicobar Island and Kamorta Island are conserving at the field gene bank of JNTBGRI as part of the studies on insect repellent species.

7. Result and discussion

The Onges and Shompens are two distinct ethnic indigenous dwindling communities of the Andaman-Nicobar Islands. The Shompens were constitutes one-sixth of the total indigenous insular population during remote past [30]. Curiously, the present population of both communities are lower than 100 numbers of individuals and being endangered owing to various genetic and environmental reasons. Interestingly, the Shompen vernacular name of the species, *Etlingera fenzlii* has not been found in literature except in one article by Elancezhian *et al.* [31] where they referred to as '*hami*'; however, the senior author during an exploration in Great Nicobar Island met with Mawa Shompens and gathered the vernacular name of the species with the help of a translator. The Shompen accent on this species name sounds like '*uijau-koaun*'. In this context, it would also be relevant to mention that Shompens are also using this species for post-delivery care. This information was also collected from the Mawa Shompens during the Nicobar exploration. Rhizome and flowers are boiled with water and used to wash the uterus after child birth.

The terms 'honeybee repellent or insect repellent' is rather confusing with reality. In fact the active principles or phytochemical molecules from the plant extracts block the receptors of the insects or honeybees to detect the intruder's presence on their combs. Precisely, the phytochemical molecules act as 'bite preventing element' rather than 'insect repellent ingredient' and obliviously, the efficacy varies with the quantity and quality of the active ingredients. The studies on *Etlingera fenzlii* carried out in JNTBGRI based on the traditional know-how of 'Shompen' tribe of the Great Nicobar Island unveiled that this endemic species having active components with insect 'repellent' properties towards worker bees and mosquitoes. On analysis, it is found that essential oil yield is remarkably high in leaves rather than rhizome. The chemical characterisation on essential oils of the species carried out proved that *Etlingera fenzlii* has effective 'tranquillizing' property towards insects. The essential oil extracted both from rhizome and leaves has a pungent odour obviously with quantitative and qualitative variations. The rhizome oil (0.02%) is yellowish while the leaf oil (0.4%) is colourless. Qualitative analysis by gas chromatography-out of the 98% components in the rhizome/leaf, 4 components has been identified *viz.* eugenol-14.864%, geraniol-29.41%, inalool-18.673% and methyl chavicol-41.094%. Twenty four essential oils including p-cymene, linalool and eugenol have proved to be effective repellents against worker bees and mosquitoes [32].

8. Conclusion

Insular habitats are of remarkable significance in conservation of global plant diversity, although they comprise only 5% of the total landmass of the world, approximately one quarter of all known extant vascular plants are endemic to insular habitats [33]. Apart from this, insular landmasses also have a remarkable function to the livelihood, economy and cultural diversity of 600 million islanders, approximately one-tenth of the present human population of the world [34]. The plant genetic resource (PGR) of the Andaman-Nicobar Islands ranges from sea weeds to several economically important higher plant species. It includes several endemics with promising economic values such as medicinal species used by primitive insular aborigines, wild occurrence of popular cultivars (coconut, betel nut, betel vine, etc.), wild relatives of popular cultivars (spice plants, rice, pluses, yams, aroids, fruit plants etc) landraces of cultivars (rice, coconut, betel nut, betel vine etc), timber yielding species (about 60 classified tree species), lesser known endemic insect repellent species (used by aborigines for honey collections), economically promising minor forest produce (canes, bamboos etc), endemic wild relatives of plantains (*Musa* spp.), economically important seaweeds etc. However, detailed floristic survey, evaluation and screening of economically valuable insular botanical entities of Andaman-Nicobar Islands are still remaining incomplete. Several insular medicinal species used by the aborigines of the Andaman-Nicobar

Islands for their health care have promising medicinal values and would certainly be rewarded in the field of modern medicines during the forthcoming decades. From taxonomic point of view, the geographical isolation encourages insular plants to have as much of variations as possible and exhibits remarkable degree of genetic variations within specific level. There are several insular species with minor variations on the way of become independent taxa.

Acknowledgements

The authors are thankful to all Forest officials of the Andaman Nicobar Administration at Port Blair and Great Nicobar Island for all logistic and manpower supports for exploration at Great Nicobar Island and also Dr. Rajan, Zoological Survey of India, Port Blair for all technical support during preparation of the manuscript. The authors also wish to record their sincere thanks to the Director, Anthropological Survey of India, Port Blair and Andaman-Nicobar Administration for providing rare old photographs of Onges and Shompens. The authors also would like to acknowledge Kerala State Council for Science Technology and Environment (KSCSTE) for financial assistance for the studies.

Author details

Sam Paul Mathew* and Raveendranpillai Prakashkumar
Jawaharlal Nehru Tropical Botanic Garden and Research Institute (JNTBGRI), Thiruvananthapuram, Kerala, India

*Address all correspondence to: sampmatthew@gmail.com

References

[1] Lekshminarasimhan, P. and P. S. N. Rao. A supplementary list of Angiospem recorded (1983-1993) from Andaman and Nicobar Islands, India, Journal of Economic and Taxonomic Boany 20 (1996): 175– 185.

[2] Mathew, S. P. A supplementary report on the flora and vegetation of the Bay Islands, India, 22 (1998): Journal of Economic and Taxonomic Botany 249 – 272.

[3] Murugan C., S. Prabhu, R. Sathiyaseelan & R.P. Pandey. "A check list of plants of Andaman – Nicobar Islands. http://www.bsienvis.nic.in/Database/Checklist-of-Andaman-Nicobar-Islands_24427.aspx. (2016).

[4] Pandey, R. P. and P. G. Diwakar. An integrated check-list flora of Andaman-Nicobar Isalands, India. Journal of Economic and Taxonomic Botany 32. (2008): 403 – 500.

[5] Vasudeva Rao, M. K. A preliminary report on the Angiosperms of Andaman-Nicobar Isalands, Journal of Economic and Taxonomic Botany 8 (1986): 107 – 184.

[6] Balakrishnan, N. P. & R. B. Nair. Wild populations of Areca and Cocos in Andaman Islands, Indian Journal of Forestry 2 (1979): 350-363.

[7] Kurz, W. S. "A sketch of the vegetation of the Nicobar Islands". Journal of Asiatic Society of Bengal. Pt 2 Nat. Hist. (1876): 45: 154. Pl. XII.

[8] Mathew, S. P and S. Abraham: The vanishing palms of Andaman-Nicobar Islands, India. Palms 38 (1994): 100-104.

[9] Mathew, S. P. On floristic diversity of Andaman-Nicobar Islands with special reference to insular germplasm conservation outside the islands. Abrahamia 3 (2017): 104-130.

[10] Prain, D. The vegetation of the Coco group. Journal Asiatic Society Bengal 60 (1891) : 283-306.

[11] Sreekumar, P. V. & J. L. Ellis. Six wild relatives of Betel vine from Great Nicobar, Journal Andaman Science Association 6 (1991):105-107.

[12] Duta, T. R, R. Ahamed, S. R. Abbas and M. K. V. Rao. "Plants Used by Andaman Aborigines in Gathering Rock-Bee Honey Source". Economic Botany. 39 (1985): 130-138.

[13] Mishra, R. S, G. S. Dogra, and P. R. Gupta. An attempt to domesticate rock-bee *Apis dorsata*. Indian Bee Journal 39 (1977): 21.

[14] Muttoo, R. N. "Tree apiaries." Indian Bee Journal 14 (1952): 87-89.

[15] Thakar, C. V. "A preliminary note on hiving *Apis dorsata* colonies." Bee World. 54 (1973): 24-27.

[16] Koeniger N. "Interspecific competition between *Apis florea* and *Apis mellifera* in the tropics". Bee World. 57 (1976): 110-112.

[17] Reddy. Studies on the nest site selection by *A. dorsata* F. Ph.D Thesis. Bangalore University, pp: 205. (1988).

[18] Morse, R. A, D. A. Shearer, R. Boch & A. W. Benton. "Observations on the alarm substance of the genus *Apis*." Journal of Apicultural Research 6 (1967): 113-118.

[19] Veith, H. J, Weiss, J and N. Koeniger. "A new alarm pheromone (2-decen-1-yl-acetate) isolated from the stings of *Apis dorsata* and *Apis florea* (Hymenoptera: Apidae)." Experientia 34 (1978): 423-424.

[20] Gupta, R. K. Taxonomy and Distribution of Different Honeybee

Species (in Rakesh Kumar Gupta Wim Reybroeck, Johan W. van Veen , Anuradha Gupta (eds.), DOI 10.1007/978-94-017-9199-1_2). 63-103. 2014.

[21] Engel, M. S. The taxonomy of recent and fossil honeybees (Hymenoptera: Apidae; *Apis*). Journal of Hymenoptera Research 8 (1999):165-196.

[22] Arias, M. C., and W. S. Sheppard. Phylogenetic relationships of Honeybees (Hymenoptera: Apinae: Apini) inferred from nuclear and mitochondrial DNA sequence data. Molecular Phylogenetics and Evolution 37(2005):25-35.

[23] Patnaik R, G.P. Reddy and V. Sudersan. “An Ecological dimension to Ethnic identity: Forest and Coastal Shom Pen of the Great Nicobar Island.” Proceedings of National Seminar on Man and Environment, Andhra University, Visakhapatnam. 1986.

[24] Kurz, W. S. Description of new plants from the Nicobar Islands, Journal of Botany, London 13 (1875) :321-333.

[25] Thothathri, K. “The 'tonyoge' plant of Little Andaman.” Indian Forester. 92 (1966): 530-532.

[26] Bhargava, N. “Ethnobotanical studies of the tribes of Andaman and Nicobar Islands, India. I.”. Onge. Economic Botany 37 (1983): 110-119.

[27] Sharief, U. M. Plants folk medicine of Negrito tribes of Bay Islands. Indian Journal of Treditional Knowledge. 6 (2007): 468-476.

[28] Zoological Survey of India (ZSI), Port Blair, personal communication, 2014.

[29] Mathew, S. P, R. K Radha, P. N. Krishnan and S. Seeni. Phytogeography of Andaman-Nicobar Islands with special reference to *Hornstedtia fenzilii*. Current Science. 98 (2010): 905-907.

[30] Das, H. H., and R. Rath. The land of coral beds: Andaman and Nicobar Islands. Aruna Printing Works, Orissa. 43-45. 1989.

[31] Elancezhian, R, R. Rajamanickam, R. S. Kumar and M. A. Suryanarayna. “Ethnobotany of Shom Pens- A primive tribe of Great Nicobar Island”. Indian journal of Traditional Knowledge. 6 (2007): 342-345.

[32] Garge, S, and Banerjee, A. K. Indian Perfumer. 41 (1997): 73-84.

[33] Kreft, H., W. Jetz, J. Mutke, G. Kier, W. Barthlott. Global diversity of island floras from a macroecological perspective. Ecology Letters 11 (2008): 116-127.

[34] Lutchman, I., B. Aalbersberg, D. Hinchley, G. Miles, A. Tiraa, S. Wells (2005): Marine Protected Areas: Benefits and Costs for Islands. WWF the Netherlands. http://www.panda.org/coral S.

Chapter 16

Evaluating Insects as Bioindicators of the Wetland Environment Quality (Arid Region of Algeria)

*Brahimi Djamel, Rahmouni Abdelkader,
Brahimi Abdelghani and Mesli Lotfi*

Abstract

The wetland of Naâma situated in the arid region of Alegria offers an important fauna and flora diversity due to its geographical location it constitutes the main resting place in North Africa for migratory birds. Insects are used as bioindicators, due to their sensitivity to environmental conditions which, because of their ecological peculiarities, gives information on the characteristics of terrestrial and aquatic environments. The aim of this study is to know and specify the entomofauna bioindicator of the quality of the aquatic environment of the wetland Naâma (SW Algeria). The study carried out in the wetland from September 2017 to September 2020. Benthic insects were sampled according to the IBGN protocol (Standard Global Biological Index). Study and statistical analysis of insects communities was based by the use of the structural and statistical index, Correspondence factor analysis (CFA), and The ascending hierarchical classification (C.H.A). The results show that the collected insect 51 species, belong to 9 orders, The Coleoptera order is the most represented with 11 species, followed by the Odonata with six species, Lepidoptera ranks third with five species followed by Diptera with 03 species. The various indicators used, namely the specific richness (51 species), the Shannon index (1.01 bits), and fairness (0.56) show that this environment is characterized by significant fauna biodiversity. The study of the hydro-biological quality of the water courses of this site, assessed by the IBGN method showed a good hydro-biological quality with moderate pollution (IBGN = 14). This pollution is precisely marked by the requirement of Ephemeroptera and the disappearance of Plecoptera. These results lay the foundation for any biomonitoring action of the ecological quality of the waters of this wetland.

Keywords: arid, entomofauna, bioindicator, wetland, Naama

1. Introduction

The arid regions of Algeria undergo total degradation and desertification of environment, due to its geographical location, the Naâma region is also threatened by human actions and locust invasions.

The wetland of Naâma classified as Ramsar, situated in the arid region of Alegria, offers an important fauna and flora diversity due to its geographical location it constitutes the main resting place in north Africa for migratory birds, animal and plant species, including many endemic species.

Bioindicators are species used to appraise the health conditions of the environment or ecosystem and they are capable of determining the environmental integrity using their functions and populations. Wetlands are fragile ecosystems that perform major functions, such as storage and the restitution of water as well as the natural filtering of mineral and organic matter, they also host a rich biodiversity and particularly adapted to this environment [1].

The insects are responsible for many processes in the ecosystem and its loss can have negative effects on entire ecosystem, Insects are used as bioindicators, because of their sensitivity to environmental conditions which, due to its ecological peculiarities, gives information on the characteristics of the environment in which it is present or on the evolution of this environment under the influence of certain practices are used to detect changes in the environment and the presence of pollution.

Insects are the most abundant animals in almost all ecosystems and can be used to evaluate the impact of environmental change.

Entomofauna studies to furnish information about ecosystems conservation status their productivity and levels of water contamination and pollution. Therefore, bioindicator species identification is essential, due to the important role that these organisms have as transformers and regulators of ecosystems [2].

The aim of this study is:

- study the entomofauna structure in the wetland of Naâma (SW Algeria).
- identify bioindicator species and assess the quality of aquatic environments in the wetland.

2. Materials and methods

2.1 Description of the study area

The resort is a wetland listed by Ramsar, localized at (longitude 0°west and latitude 33°north). The water of wetland concerned two hundred hectares surrounded by several units or peripheral areas; immediate area of water is characterized by *Tamarix* and *Alfa* formation. The gausses diagram and Ombrothermic bagnouls shows the dry period in the Naâma region is longer from April until October during the period (2012–2019). The rainfall climagramme emberger quotient (Q2) show that three stations located in the fresh winter upper arid area (Brahimi et al. 2020) (**Figure 1**).

2.2 Sampling sites

We studied the insects in 5 sampling sites: 3 sites on the water bodies of the hawdh edaira wetland and 2 sites along the lake. The plots covered several types of environmental gradients (water body and wild natural areas (wooded, grassy, rocky and forest areas) Land cover of the sampling sites was classified into 5 types based on Google Earth images and field observation.

Insect field collections were carried out continuously for 36 months from September 2017 to September 2020 to obtain baseline data on insect composition. Inventories were carried out in the middle of each month. The transect method (300 m × 200 m) was used to study insects in open areas.

Figure 1.
Satellite images and photos of the wetland of Naâma region.

2.3 Sampling technique

Extrapolating sample data from these methods to the square metre could lead to the overestimation of species abundance and diversity at the site due to the generally patchy distribution of invertebrates in streams [3].

Two qualitative sampling tools (D-frame net and square net) and a quantitative sampling tool (Surber sampler) were used to collect the aquatic insects the wetland of Naâma. A D-frame net with a 1.2-m-long handle and a 60 cm long cone-shaped net with 0.3 mm mesh and a diameter of 0.38 m was used in this study. A square frame net with a 0.5 m × 0.5 m opening, a 90 cm cone shaped net with 0.3 mm mesh, and a 1.2 m long handle were used to collect samples. A sampling tool's efficiency is also determined by the time required to process the samples [4].

Other techniques have been used namely; – Butterfly net - Catching insects by hand - Pheromone insect trap.

2.4 Study of plants

The species inventory is essential for the structural analysis of a station.

The harvest of the plant in the field is the prospecting which aims to know the totality of the flora of the region of Naâma. In this case, it is essential to visit the sites during all seasons to collect the maximum number of plants of different species. In the course of surveys, plastic bags are used. The collected samples are then dried and placed in paper folders with a label, mentioning the date, place, and other interesting observations. All the species have been kept in a herbarium.

The determination of the plant species was carried out using the two guides; Nouvelle Flore d'Algérie of QUÉZEL and SANTA S and the Flore du Sahara of Paul Ozenda.

2.5 Ecological indices

Total and average Richness: Is the number of species that make up a population. It is one of the fundamental characteristics of a settlement. In this study, two types of wealth are calculated, namely total wealth and average wealth [5].

Shannon-Weaver Diversity Index H′: According to Ramade [5], (H′) translated by a determined abundance distribution, closely related that of specific diversity:

$$H' = -\Sigma\left((ni/n) * \log 2\,(ni/n)\right)$$

ni: number of individuals of a given species, i ranging from 1 to S (total number of species).

n: total number of individuals.

Equitability (Fairness E): The knowledge of H′ and H′ max makes it possible to determine E. E varies between 0 and 1, E tends to 0 when the quasi-totality of the populations corresponds to a single species of the stand, E tends to 1 when each species is represented by the same number of individuals [5].

Dispersion index: The knowledge of the distribution mode is useful during a density evaluation population by sampling [6].

$$S2 = \Sigma\ (x - m)/n - 1$$

n: collection set; m: the average number of individuals in each sample; x: number of individuals from each sample. If: S2 = 0: the distribution is uniform or regular; S2 < m: the distribution is contagious .

The relative frequency: According to Dajoz [7], The principle consists of noting the presence or absence of species in the records, it is expressed as follows:

$$F\ (i) = ni/N \times 100$$

F (i): Relative frequency of the species contained in the statement as a percentage.

ni: The number of times the insect (i) are present.

N: Total number of individuals.

2.6 Statistic study

The Shapiro–Wilks test: The Shapiro–Wilks test for normality is one of three general normality tests designed to detect all departures from normality. It is comparable in power to the other two tests.

$$\boldsymbol{b} = \sum_{i=1}^{m} \boldsymbol{a_i}\ (\boldsymbol{x_{n+1-i}} - \boldsymbol{x_i})$$

2.7 Descriptive statistics of a numeric variable

Descriptive statistics are the first pieces of information used to understand and represent a dataset. Their goal, in essence, is to describe the main features of numerical and categorical information with simple summaries. These summaries can be presented with a single numeric measure, using summary tables, or via graphical representation. Here, I illustrate the most common forms of descriptive statistics for numerical data but keep in mind there are numerous ways to describe and illustrate key features of data.

2.8 Student's t test for single sample

The t test tells you how significant the differences between groups are; In other words it lets you know if those differences (measured in means) could have happened by chance.

2.9 Factorial Correspondence Analysis (CFA)

Factorial correspondence analysis is a descriptive method. It aims at the representation with the minimum loss of information in a space with n dimension [2].

The purpose of this analysis is to realize several graphs from data table. The observation of the graph can give an idea of the interpretation of the factors and show which variables are responsible for the proximity between this or that observation. **CFA** is a method that consists of summarizing the information contained in a table with n rows (the stations in this case) and p columns or variables (Orthoptera species). In addition, a technique has for describing in particular in a graphical form the maximum of the information contained in a rectangular array of data.

2.10 Logiciels Statistique

The key to determining insects order Plecoptera, Ephemera and Odonata, is performed by the software Xper3, This key is based on a list of taxa and associated descriptors, such as morphological characters.

Factorial correspondence analysis (CFA) was studied by minitab version 19 software.

Student's t test for single sample, Descriptive statistics of a numeric variable and The Shapiro–Wilks test have been studied by R++ statistics software.

2.11 Study of the quality of aquatic environment

2.11.1 Global Normalized Biological Index IBGN

Benthic insects are one of the most famous organism groups used for rivers biomonitoring surveys.

The use of IBGN is especially indicated for disturbances that induce a modification of the nature of the substrate and the organic quality of the water: rejection urban predominantly organic, pollution by suspended matter, side effects of certain types of rejection (organic, metallic) and eutrophication. In addition, the IBGN reflecting the structure of a biocenosis made up of organisms long-term integrators are especially sensitive to chronic disturbances or well to disturbances of the intermittent type but sufficiently intense to cause a immediate mortality.

- Class 1A blue in color which indicates excellent water quality
- Class 1B green in color which indicates good quality water (with pollution moderate)
- Class 2 yellow in color which indicates average water quality (with a clear Pollution)
- Class 3 orange color which indicates poor quality water (with pollution important)
- Excluding class 4 red which indicates poor quality (with pollution excessive)

2.11.2 Faunistic analysis and interpretation (IBGN)

After identifying the macroinvertebrates, a list faunistic is established, listing all the taxa found by faunistic groups, and indicating the total number of taxa. The index is calculated from the table "IBGN values according to type and the taxonomic variety of macrofauna. We first determine the taxonomic variety (Σt), i.e. the total number of taxa identified (the number of individuals per taxon is not taken into account). Then look for the faunistic indicator group (GI) in the list provided

IBGN	Class 1A > ou = à17	Class 1B 16–13	Class 2 12–19	Class 3 8–5	Class 4 < ou = à 4
color	blue	green	yellow	orange	Red

Table 1.
Reference values of the IBGN.

and select the taxon that has the highest degree of pollutant sensitivity of the full sample of the station studied.

The index can then be read in the table of values of the IBGN: it is at the intersection of the column for the taxonomic variety and the row for the indicator faunistic group (**Table 1**).

3. Results

3.1 Data description

The insect species recorded are divided into 9 orders; Ephemeroptera, Coleoptera, Orthoptera, Hymenoptera, Lepidoptera, Hemiptera, Odonata, Diptera and Plecoptera. From this present work, 27 insect families were found. The most represented order is Coleoptera with six families and 11 species Orthoptera represent 5 families and 20 species. The Hemiptera, Plecoptera and Ephemeroptera represent only one family and one species for each order (**Table 2**).

3.2 Shannon-Weaver diversity, maximum diversity and Equitability

The Shannon-Weaver diversity index obtained in the study area is 2.24bit, maximum diversity is 1.89. Equitability is of the order of 0.56. These values indicate that the wetland is characterized by a very important entomofauna diversity, these results show a great diversity of insects in the naama wetland (**Figure 2**).

3.3 Dispersion index and The relative frequency

The study of the Dispersion index and The relative frequency of each species identified in the hawdh ed. daira wetland of Naâma allowed to know the frequency and the type of distribution of each species;distribution uniform, regular or contagious (**Table 2**). This information is necessary for the monitoring and control of these species over time (**Figure 3**).

3.4 Shapiro: Wilk normality test

In this study the calculation of the Shapiro-Wilk normality test is equal:

W = 0.6428, p-value = 1.842e-07.

this value means, the p-value is greater than 0.05, the data of this stady are normally distributed.

3.5 Descriptive statistics of a numeric variable

the calculation of the Descriptive statistics of a numeric variable of insect species in the Naâma wetland show that the population insects are normally distributed with a Standard deviation equal 3.4345 (**Table 3**).

Class	Order	Family	Species	Relative frequencies	Dispersion index
Insect	Coleoptera	Coccinellidae	*Coccinella septempunctata (Linnaeus, 1758)*	11.98	8.51
		Tenebrionidae	*Blaps lethifera (Marsham, 1802)*	7.85	3.73
			Gnaptor spinimanus (Pallas, 1781)	9.22	9.22
			Pimelia bipunctata (Fabricius, 1781)	12.16	9.14
		Carabidae	*lophyra flexuosa (Fabricius, 1787)*	13.22	8.13
			Poecilus sp (Schaller 1783)	14.22	9.65
		Cantharidae	*Cantaris fuxa (Linnaeus, 1758)*	6.34	3.33
		Geotrupidae	*Geotrupes sp (Latreille, 1796)*	19.34	9.65
		Meloidae	*Mylabris variabilis (Pallas, 1781)*	2.77	1.22
			Mylabris quadripunctata (Linnaeus, 1767)	2.87	1.39
			Mylabris sp (Pallas, 1781)	1.92	0.43
	orthoptera	Acrididae	*Chorthippus sp (Fieber, 1852)*	1.34	0.87
			Acrotylus fischeri (Fieber, 1853)	6.03	0.75
			Oedipoda fuscocincta (Lucas 1849)	15.47	12.3
			Oedipoda miniata (Lucas 1849)	6.41	9.63
			Sphingonnotus rebescens (Walker, 1870)	14.13	3.25
			Sphingonotus octofasciatus (Serville, 1838)	4.15	1.2
			Sphingoderus carinatus (Saussure, 1888)	4.52	0.23
			Sphingonotus lucasii (Saussure, 1888)	3.01	2.05
			Calliptamus barbarus (Costa 1836)	4.15	5.36
			Calliptamus wattenwylianus (Pantel, 1896)	1.88	1.11
			Pezotettix giornai (Rossi, 1794)	0.67	0.13
			Anacridium aegyptium (Linnaeus, 1774)	6.79	4.7
			Omocestus lepineyi (Chopard, 1937)	3.39	0.27
			Omocestus lecerfi (Chopard, 1937)	4.15	1.03
		Pyrgomorphidae	*Pyrgomorpha conica (Olivier 1791)*	1.88	0.17

Class	Order	Family	Species	Relative frequencies	Dispersion index
		Pamphagidae	*Tmethis marocanus (Bolívar, 1908)*	15.84	7.4
			Tmethis cisti (Bolívar, 1908)	1.13	1.54
			Ocneridia volxemii (Bolívar, 1878)	5.28	1.9
		Gryllidae	*Melanogryllus desertus (Pallas, 1774)*	0.75	0.16
		Tettigonidae	*Tettigonia albifrons (Fabricius, 1775)*	0.75	0.16
	Hymenoptera	Vespidae	*Polistes dominula (Christ, 1791)*	3.87	2.26
		Pompilidae	*Arachnospila sp (Kohl, 1898)*	4.74	2.37
		Apidae	*Apis mellifera (Linnaeus, 1758)*	9.76	8.14
		Formicidae	*Dinoponera sp (Guérin-Méneville, 1838)*	3.98	1.92
	Lepidoptera	Nymphalidae	*Issoria sp (Doherty, 1886)*	3.65	1.72
			Vanessa atalanta (Linnaeus, 1758)	0.23	1.23
		Pieridae	*Pieris sp (Schrank, 1801)*	0.83	0.11
		Sphingidae	*Agrius convolvuli (Linnaeus, 1758*	1.98	1.10
		Noctuidae	*Acronicta (Linnaeus, 1758)*	3.99	0.11
	Diptera	Muscidae	*Musca domestica (Linnaeus, 1758)*	6.87	4.84
		Calliphoridae	*Calliphora vicina (Robineau-Desvoidy, 1830)*	2.65	1.56
		Sarcophagidae	*Sarophage carnaria (Linnaeus, 1758)*	1.88	0.82
	Odonata	Libellulidae	*Orthetrum brunneum (Fonscolombe, 1837)*	1 .09	0.17
			Orthetrum coerulescens (Fabricius, 1798)	3.87	0.88
			Portecoupe holartique (Charpentier, 1840)	1.73	0.27
		Coenagrionidae	*Enallagma civile (Hagen, 1861)*	2.54	1.88
			Ceriagrion tenellum (Villers, 1789)	0.54	1.36
	Hemiptera	Pyrrhocoridae	*Pyrrhocoris apterus (Linnaeus, 1758)*	2.77	1.02
	Plecoptera	Leuctridae	*Leuctra sp (Klapálek, 1905)*	6.58	3.71
	Ephemeroptera	Baetidae	*Baetis rhodani (Pictet, 1843)*	3.84	1.77

Table 2.
Summary of insect species identified in the wetland of Naâma.

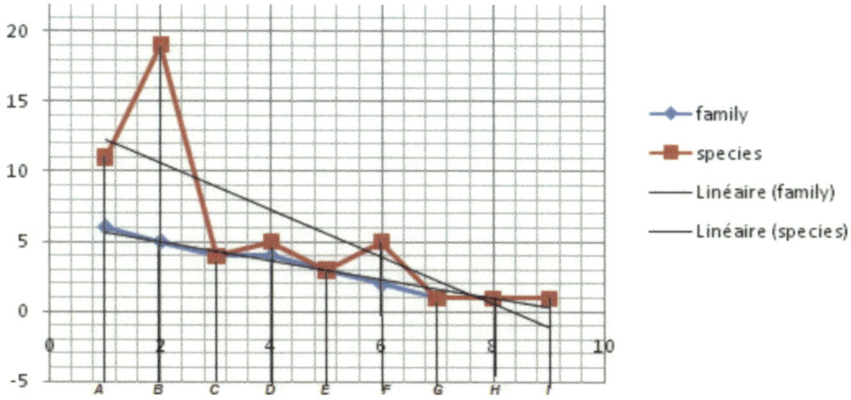

Figure 2
Distribution of insect orders by families and species in the study area.

3.6 Student's t test for single sample

The calculation of the student test for the identified insect population in the Naama wetland shows that the test statistic would follow a normal distribution. The sample mean is equal to the population mean with a Standard deviation equal 3.4345 and Student's t test equal 4.26 (**Table 4**).

3.7 Correspondence factor analysis (CFA); CFA of insect order species

The initial table (1) corresponding to 20 surveys show the presence of species in the stations according to the type of environment; plants environment, rocky environment and aquatic environment An AFC conducted on this matrix allowed to build a hierarchical classification calculated from the coordinates of species. Dendrogram clearly differentiates three groups of species of unequal size:

Group A: It is mainly represented in the plants environment.

Group B: It includes species specific to degraded and rocky environment.

Group C: species specific to aquatic environment*(Portecoupe sp, Orthetrum sp, Orthetrum sp, Pieris sp,Vanessa sp, Issoria sp)* (**Figure 4**).

3.8 The ascending hierarchical classification (C.H.A)

From the Euclidean distances based on the scores of the three factors A.F.C (**Figure 2**), it is possible to recognize three groups: **Group A**: It is mainly represented in the plants environment.

Group B: It includes species specific to degraded and rocky environment.

Group C: species specific to aquatic environment.

The ascending hierarchical classification (C.H.A) confirm our results of Correspondence factor analysis (**Figure 5**).

3.9 CFA of orthoptera order species

In the plants and rocky environment., very high faunal diversity of orthoptera (locust)species was recorded; this is supported by the fact that high environmental quality usually correlates with the greatest species richness and diversity.

Group A: It is mainly represented in the plants environment.

Group B: It includes species specific to degraded and rocky environment.

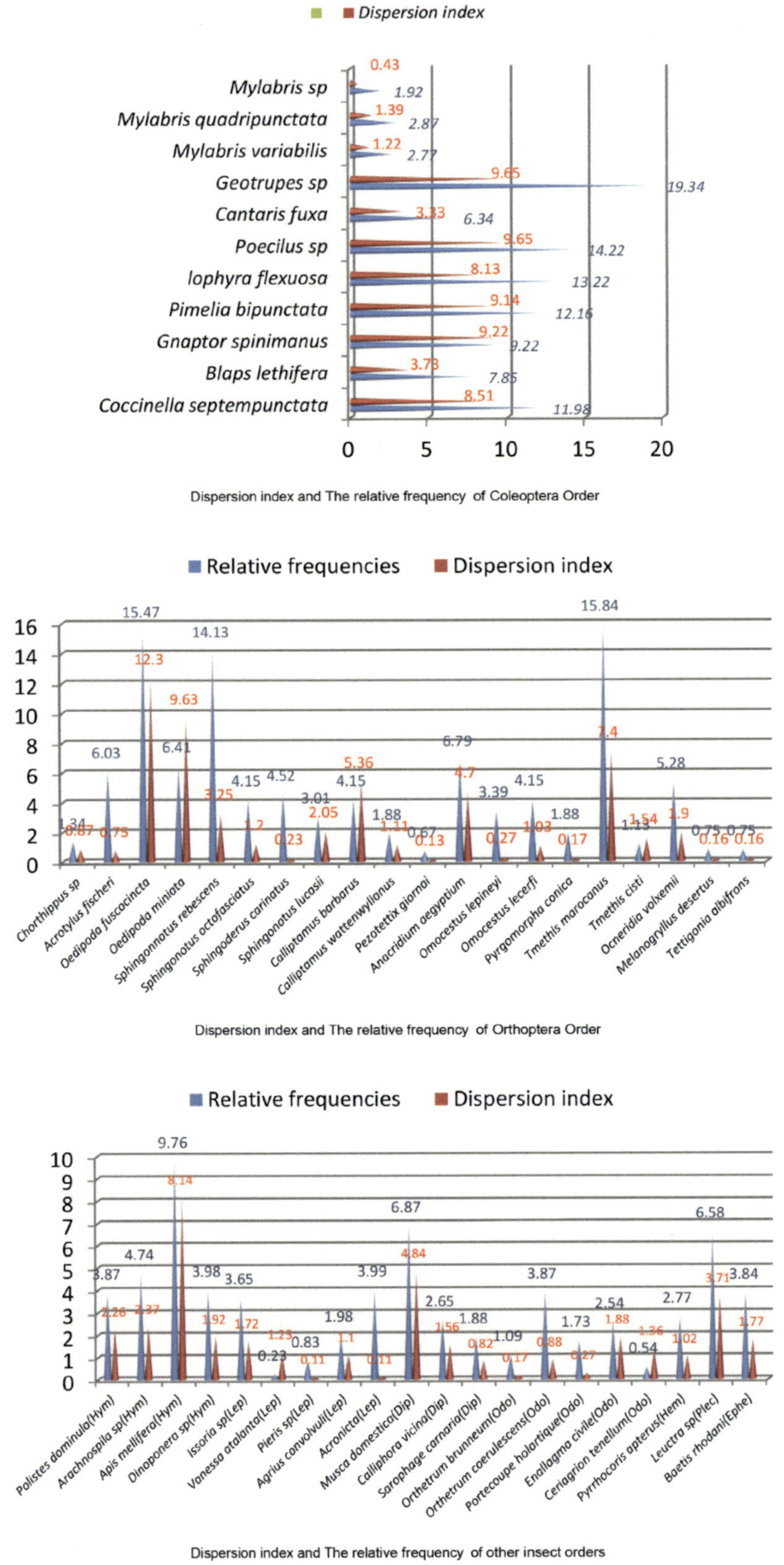

Figure 3.
Dispersion index and The relative frequency of insect species order in the wetland of Naâma.

Minimum	Quartile 1	Median	Mean	Quartile 3	Maximum	Standard deviation
1	1	1	3.065	3	12	3.4345

Table 3.
Descriptive statistics of a numeric variable of insect species in the Naâma wetland.

t	Df	p-value	Confidence interval, 95%
4.268	51	2.55e-05	[1.8047, 4.3243]
	Size	**Average**	**Standard deviation**
Variable	51	3.0645	3.4345

Table 4.
Student's t test for single sample of insect species in the Naâma wetland.

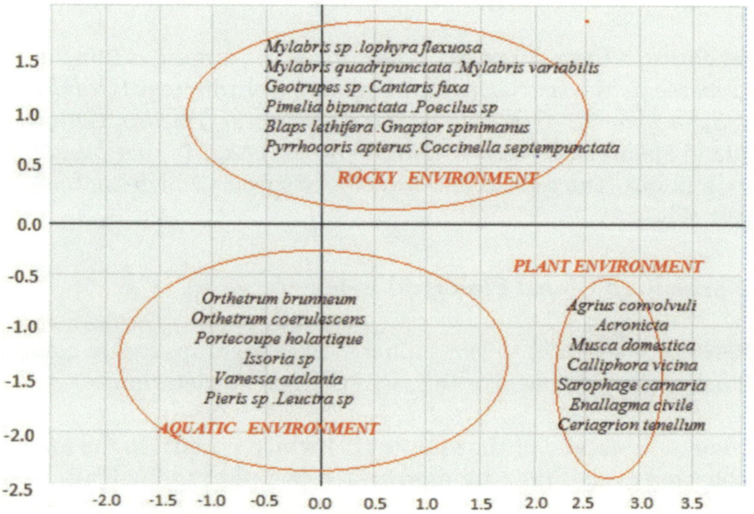

Figure 4.
Factorial analysis of the correspondence of insect species in the wetland of Naâma.

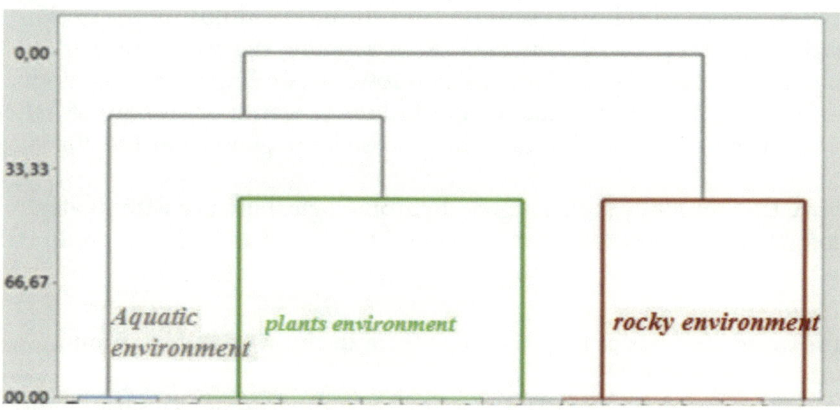

Figure 5.
Hierarchical ascending classification of insects species in the wetland of Naâma.

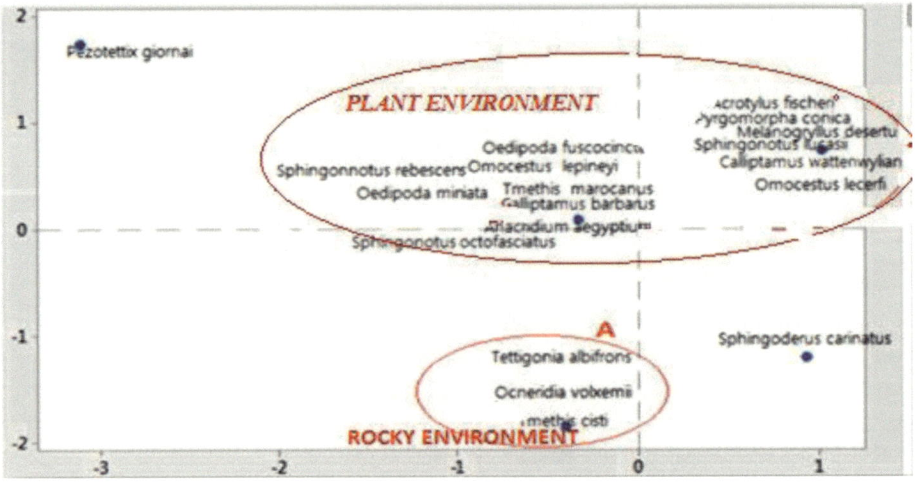

Figure 6.
Factorial analysis of the correspondence of Orthoptera species of Naâma (Algeria).

The first entity in the right of the projection is the largest as it includes 75% of species. It represents the species caught in plants environment (*Oedipoda fuscocincta, Sphingonnotus, Oedipoda miniata, Omocestus, Tmethis, Calliptamus, Anacridium*and *Sphingonotus octofasciatus. Pyrgomorpha, Melanogryllus, sp, Sphingonotus*and *sp.).* The second entity includes species which are found in rocky environment (**Figure 6**).

3.10 The Normalized Global Biological Index

The IBGN is organized in rows 9 faunal indicator groups and in columns 14 classes of taxonomic varieties. For this, we successively determine:

- The taxonomic variety of the sample (Σt) which is equal to the total number of taxa collected even if they are represented by only 1 individual.

- The faunistic indicator group (GI) by taking into account only the indicator taxa represented in the sample by 3 individuals or 10 individuals depending on the taxa. The determination of GI is carried out by prospecting the columns of the table from top to bottom and by selecting the taxon which represents the highest degree of pollutant sensitivity of the entire sample of the station studied. Then the IBGN value be read by crossing the taxonomic variety column and the indicator faunal group row. Depending on the taxonomic diversity of the OglatedDaira station and the presence or absence of indicator taxa, a variant hydrobiological quality score is assigned from 1 to 20.

We see that the study area has good hydrobiological quality with moderate pollution (IBGN = 14: the taxonomic variety ST = 51 and an indicator group (GI = 9) (**Table 1**).

3.11 Study and analysis of the structure flora in the wetland of naama

The diversity analysis by Shannon index (H'), shows that the most important diversity is marked in the wetland of Ain Ben Khelil with (H' = 2.43). the maximum diversity is equal 3.13.

Equitability is equal 0.77, which is usually taken as an indicator of a balanced population and which reflects the stability of the environment.

3.12 Overall recoveries of plant species recorded in the wetland of Naâma

In the wetland of naama,*Stipa tenacissima* is the species most representative, it covers more than 50%. Against in the wetland of Ain Ben Khelil, *Stipa tenacissima, Tamarix gallica, Lygeum spartum, Ziziphus lotus* are the representative species of this station.

3.13 The relative abundance of plant species recorded in the wetland of Naâma

15 families are present in the wetland of Naama, the most representative families are Asteraceae with a relative abundance of 17.39%, followed by the Amaranthaceae and Poaceae with 13.04%.

3.14 Biological type of plant species recorded in the wetland of Naâma

The biological type leads to the natural form of the plant that is one of the basic criteria for classifying species in biological types, composed of perennial species, woody or herbaceous and annual species. The specific aspect of the form obtained is dependent on environmental variations.

In our study area, therophytes dominate other biological types with 40.62% followed by chamaephytes with31.25%, hemicryptophytes remain in third with 15.62%.

4. Discussion

This study was carried out in the arid region of Naâma, this region characterized by steppe formations dominated by *Stipa tenacissima*, According to Tormo [8], Woody species play a key role in *Stipa tenacissima* steppes, they affect ecosystem functioning, facilitate establishment of other plants and increase plant richness.

The insect species recorded are divided into 9 orders; Ephemeroptera, Coleoptera, Orthoptera, Hymenoptera, Lepidoptera, Hemiptera, Odonata, Diptera and Plecoptera. From this present work, 27 insect families were found. The most represented order is Coleoptera with six families and 11 species Orthoptera represent 5 families and 20 species. The Hemiptera, Plecoptera and Ephemeroptera represent only one family and one species for each order. Invertebrates are more severely and quickly affected than other taxa by changes in the landscape. The insects are responsible for many processes in the ecosystem and its loss can have negative effects on entire communities. Thus, a strong understanding of insect responses to human activity is necessary both to support policy decisions for conservation and to evaluate functional consequences of human disturbance on ecosystems [6].

many diversity indices have been developed to describe responses of a community to environment variation, combining the three components of community structure, namely richness (number of species present), Shannon-Wiener Index [9].

The Shannon-Weaver diversity index obtained in the study area is 2.24bit, maximum diversity is 1.89. Equitability is of the order of 0.56. These values indicate that the wetland is characterized by a very important fauna diversity.

Orthoptera (locust) are able to threaten arid ecosystems, human health and resistant to pesticides, a study was carried by (Brahimi et al., 2020) on the chemical mechanism of locust resistance in the arid region of Naama. Hardersen [10] reported the potential of aquatic insects as indicators of water quality. Several other species of the families Gyrinidae, Dytiscidae, Hydrophilidae (Coleoptera), Notonectidae, Veliidae (Heteroptera) and Plecoptera and Ephemeroptera Orders have high adaptive capacity, colonizing most of the environments and occurring throughout the year, reflecting ecological and geographical changes, and hence their conservation status.

Davis [11] confirm beetles species (Coleoptera: Scarabaeidae) have a high potential as environmental indicators in forest area .

In this study an important diversity of beetles (coleoptera) was recorded in this area, beetles play an essential role as decomposers of organic matter in the balance of ecosystems. Beetles from Order Coleoptera and Family Carabidae are important predators. They participate of biological control, biological monitoring of pollution from oil, sulfur, herbicides, CO2, insecticides and radioactive phosphorus. The moths and butterflies (Lepidoptera), besides having basic requirements, have ecological faithfulness in temperate and tropical regions and are very sensitive to changes in the environment [12]. the trapping methods used have a great importance to better collect the species in quantity and quality. This explains the differentiation from other orders in number of species.

According to Rizo-Patrón et al. [13], group of macroinvertebrates (*Baetis sp., Fallceon sp., Leptohyphes sp., Tricorythodes sp., Farrodes sp., Phyllogomphoides sp., Hydroptila sp., Mayatrichia sp., Neotrichia sp., Oxyethira sp., Nectopsyche sp.1, Nectopsyche sp.2, Oecetis* sp.) can be used as a bioindicators of water quality in management practices agroecosystems.

In this study we noticed a scarcity of pollinating species, this scarcity is due to the uncontrolled and anarchic use of phytosanitary products and insecticides in the action of the Locust control. Pollinators, especially honeybees (*Apis mellifera*), are considered reliable biological indicators because they show environment chemical impairment due to high mortality rate and intercept particles suspended in air or flowers. These substances can then be detected using methods of analysis [14]. Among these organisms, the insects may contribute to a practical assessment of the sustainability degree [15].

According to Eggleton et al. [16], termites are important decomposers in land ecosystems. Its activity increases soil infiltration capacity, leading to water retention and soil productivity. Read and Anderson [17] showed The value of ants as bio-warning indicators in the Australian arid rangelands.

According to Nummelin et al. [18], A study of the heavy metal concentrations of different predatory insects (Gerridae), dragon fly larvae (Odonata), anteater larvae (Myrmeleontidae) and ants (Formicidae) showed higher metal concentrations and that these groups of insects can be used as indicators of heavy metals.

Goncharov et al. [19] indicates that using the density index of Ephemeroptera Plecoptera-Trichoptera (EPT); showed a capacity for regeneration of these indicator groups under unfavorable conditions. Cortelezzia [20] indicates that In some taxa of chironomids, their potential as a bioindicator increased as the taxonomic level decreased (eg, Chironominae). However, in other taxa, this potential as a bioindicator of water quality remains at the subfamily level. The Ephemeroptera and Plecoptera larvae are recognized as good bioindicators of eutrophication in running water due to their sensitivity to oxygen depletion. The pollution indicators can be attributed by the disappearance of certain more or less sensitive species or, on the contrary, by the appearance of other so-called resistant species. The specific river environment sampled may influence the proliferation of certain taxa and the specific behaviours of those taxa in certain habitats [21].

5. Conclusion

The insect species recorded are divided into 9 orders; Ephemeroptera, Coleoptera, Orthoptera, Hymenoptera, Lepidoptera, Hemiptera, Odonata, Diptera and Plecoptera. From this present work, 27 insect families were found. The most represented order is Coleoptera with six families and 11 species Orthoptera represent 5 families and 20 species. The Hemiptera, Plecoptera and Ephemeroptera represent only one family and one species for each order. by this study we noticed a scarcity of pollinating species, this scarcity is due to the uncontrolled and anarchic use of phytosanitary products and insecticides in the action of the Locust control.

The Shannon-Weaver diversity index obtained in the study area is 2.24bit, maximum diversity is 1.89. Equitability is of the order of 0.56. These values indicate that the wetland is characterized by a very important fauna diversity. Statistical study of the entomofauna identified in the Naama wetland show that the population insects are normally distributed in this environment.

The study of the hydrobiological quality in the wetland of Naama, assessed by the IBGN method, showed good hydrobiological quality with moderate pollution (IBGN = 14). This pollution is precisely marked by the requirement of Ephemeroptera and Plecoptera. Ephemeroptera are polluo-resistant species in polluted aquatic environments, unlike stoneflies which are polluo-sensitive species. These two orders are used as good biological indicators of polluted aquatic environments.

This study concluded that the Class Insecta has many potential representatives that can be used as environmental bioindicators, among which are some species from the Coleoptera, Diptera, Lepidoptera, Hymenoptera, Hemiptera, Isoptera Orders and others.

These data constitute a first database on the structure of the entomofauna in the wetland classified by Ramsar Hawdh ed. daira of arid region ofAlgeria, the results obtained also make it possible to control and monitor pollution in this wetland, in order to protect these fragile and arid ecosystems. Threatened by desertification and human actions .

Acknowledgements

This work was supported by the General Directorate of Scientific Research and Technological Development DGRSDT of Algeria republic.

Declaration of competing interest

The authors declare that they have no conflict of interest.

Author details

Brahimi Djamel[1], Rahmouni Abdelkader[2*], Brahimi Abdelghani[3] and Mesli Lotfi[4]

1 Department of Sciences of Nature and Life, Laboratory "Sustainable Management of Natural Resources in Arid and Semi-Arid Zones", Team, Valorization and Conservation of the Biodiversity in Arid and Semi-Arid Zones, Faculty of Sciences and Technology, University of Salhi Ahmed, Naâma, Algeria

2 Department of Chemistry, Laboratory of Polymer Chemistry, University of Oran1-Ahmed Benbella, Oran-Algeria

3 Department of Civil Engineering, University of Sidi Bel Abbess-Algeria

4 Faculty of Sciences Natural and Life Sciences, University Abou Bakr Belkaid-Tlemcen-Algeria

*Address all correspondence to: ramaek23@yahoo.fr

References

[1] Delaunay M.2016. Clé d'identification numérique des insectes aquatiques d'Ile-de-France: un outil pour les acteurs de terrain. Conference: 7èmes Recontres Naturalistes .

[2] Brown, K. 1991.Conservation of neotropical environments: Insects as indicators. The conservation of insects and their habitats. Collins N.J. Thomas. Chap. 14, p. 350- 420.

[3] Che Salmah MR, Al-Shami SA, Madziatul Rosemahanie M, Abu Hassan A. 2013. Biological and ecological diversity of aquatic macroinvertebrates in response to hydrological and physicochemical parameters in tropical forest streams of Gunung Tebu, Malaysia: Implications for ecohydrological assessment. Ecohydrology.;7(2):496–507. [Google Scholar].

[4] Taylor BW, McIntosh AR, Peckarsky BL. Sampling stream invertebrates using electroshocking techniques: Implications for basic and applied research. Canadian Journal of Fisheries and Aquatic Sciences. 2001;58 (3):437–445. [Google Scholar].

[5] Ramade F.2003. Eléments d'écologie – Ecologie fondamentale. № 03, Ed. DUNOD, Paris, 690p.

[6] Nicholsa, E. Larsenb, T. Spectora, S. Davise, A. L. Escobarc, F. Favilad, M. Vulinece, K. 2007.Global dung beetle response to tropical forest modification and fragmentation: A quantitative literature review and meta-analysis. Biologic. Conserv. v. 137, p. 1– 19.

[7] Dajoz R .1970. Précis d'écologie. Ed. Dunod, Paris,). 357 pp.

[8] Tormo J a, Amat B b, Cortina J b. 2020.Litter as a filter for germination in semi-arid *Stipa tenacissima* steppes Journal of Arid Environments. https://doi.org/10.1016/j.jaridenv.2020.104258.

[9] Li Li, Binghui Z, Lusan L .2010. Biomonitoring and Bioindicators Used for River Ecosystems: Definitions, Approaches and Trends .Procedia Environmental Sciences 2 1510–1524International Society for Environmental Information Sciences. Annual Conference (ISEIS) .Elsevier . Ltd.Open access under. CC BY-NC-ND license.

[10] Hardersen, S. 2000. The role of behavioural ecology of damselflies in the use of fluctuating asymmetry as a bioindicator of water popպution. Ecolog. Entomol., v. 25, p. 45 – 53.

[11] Davis, A 2000. Does reduced-impact logging help preserve biodiversity in tropical rainforest? A case study from Borneo using dung beetles (Coleoptera: Scarabaeoidea) as indicators. Environ. Entomol., v. 29, n.3, p. 469-473.

[12] Da Rocha J R M, Almeida J R, Lins G A, Durval A.2010. Insects as indicators of environmental changing and pollution: a review of appropriate species and their monitoring. HOLOS Environment, v.10 n.2, 2010 - P. 250.

[13] Rizo-Patrón F V. Anjali Kumar, Michael B. Colton M .2013. Macroinvertebrate communities as bioindicators of water quality in conventional and organic irrigated rice fields in Guanacaste, Costa Rica Ecological Indicators. Volume 29, June 2013, Pages 68-78. ELSEVIER.

[14] Ghini, S. Fernández, M. Picó, Y . marín, R. Fini, F. Mañes, J. Girotti, S. 2004.Occurrence and distribution of pesticides in the province of Bologna, Italy, using honeybees as bioindicators. Arch. Environ. Contam.Toxicol., v. 47, p. 479–488.

[15] Lopes, B. G. C. 2008.Levantamento da entomofauna bioindicadora da

qualidade ambiental em diferentes áreas do Alto Jequitinhonha – Minas Gerais. 2008. 47f. Trabalho de Conclusão de Curso - Escola Agrotécnica Federal de Inconfidentes.Inconfidentes, Minas Gerais.

[16] Eggleton, P, Williams, P. H, Gaston, K. J. 1994. Explaining global termite diversity: productivity or history? Biodivers. Conserv., v. 3, p.318–330,

[17] Read J L. Andersen N..2000.The value of ants as early warning bioindicators: responses to pulsed cattle grazing at an Australian arid zone locality Journaof Arid Environments Volume 45, Issue 3, July, Pages 231 251 https://doi.org/10.1006/jare.2000.0634

[18] Nummelina M, Lodeniusa M, Tulisaloa E, Hirvonena H, Alankob T. 2007. Predatory insects as bioindicators of heavy metal pollution Environmental Pollution Volume 145, Issue 1, January 2007, Pages 339-347 Elsevier. https://doi.org/10.1016/j.envpol.2006.03.002 .

[19] Goncharov A.V., Baturinab N.S, Maryinskyc V.V., Kausd A.K., Chalova S.R.2020.Ecological assessment of the Selenga River basin, the main tributary of Lake Baikal, using aquatic macroinvertebrate communities as bioindicators. Journal of Great Lakes Research. Volume 46, Issue 1, February 2020, Pages 53-61. https://doi.org/10.1016/j.jglr.2019.11.005

[20] Cortelezzia A, Simoya M, Sirib A, Donatob M, Cepedaa R, Marinellia C, Berkunskya I .2020 .New insights on bioindicator value of Chironomids by using occupancy Ecological Indicators modelling.Volume 117, October 2020, 106619 Elsevier. https://doi.org/10.1016/j.ecolind.2020.106619.

[21] Merritt RW, Cummins KW. 1996. An introduction to the aquatic insects of North America. 3rd ed. Iowa, USA: Kendall/Hunt Publishing Company; [Google Scholar]